D0858158

Adaptive Digital Filters
and Signal Analysis

ELECTRICAL ENGINEERING AND ELECTRONICS

A Series of Reference Books and Textbooks

Editors

Marlin O. Thurston
Department of Electrical
Engineering
The Ohio State University
Columbus, Ohio

William Middendorf
Department of Electrical
and Computer Engineering
University of Cincinnati
Cincinnati, Ohio

1. Rational Fault Analysis, *edited by Richard Saeks and S. R. Liberty*
2. Nonparametric Methods in Communications, *edited by P. Papantoni-Kazakos and Dimitri Kazakos*
3. Interactive Pattern Recognition, *Yi-tzuu Chien*
4. Solid-State Electronics, *Lawrence E. Murr*
5. Electronic, Magnetic, and Thermal Properties of Solid Materials, *Klaus Schröder*
6. Magnetic-Bubble Memory Technology, *Hsu Chang*
7. Transformer and Inductor Design Handbook, *Colonel Wm. T. McLyman*
8. Electromagnetics: Classical and Modern Theory and Applications, *Samuel Seely and Alexander D. Poularikas*
9. One-Dimensional Digital Signal Processing, *Chi-Tsong Chen*
10. Interconnected Dynamical Systems, *Raymond A. DeCarlo and Richard Saeks*
11. Modern Digital Control Systems, *Raymond G. Jacquot*
12. Hybrid Circuit Design and Manufacture, *Roydn D. Jones*
13. Magnetic Core Selection for Transformers and Inductors: A User's Guide to Practice and Specification, *Colonel Wm. T. McLyman*
14. Static and Rotating Electromagnetic Devices, *Richard H. Engelmann*
15. Energy-Efficient Electric Motors: Selection and Application, *John C. Andreas*
16. Electromagnetic Compossibility, *Heinz M. Schlicke*
17. Electronics: Models, Analysis, and Systems, *James G. Gottling*

18. Digital Filter Design Handbook, *Fred J. Taylor*
19. Multivariable Control: An Introduction, *P. K. Sinha*
20. Flexible Circuits: Design and Applications, *Steve Gurley, with contributions by Carl A. Edstrom, Jr., Ray D. Greenway, and William P. Kelly*
21. Circuit Interruption: Theory and Techniques, *Thomas E. Browne, Jr.*
22. Switch Mode Power Conversion: Basic Theory and Design, *K. Kit Sum*
23. Pattern Recognition: Applications to Large Data-Set Problems, *Sing-Tze Bow*
24. Custom-Specific Integrated Circuits: Design and Fabrication, *Stanley L. Hurst*
25. Digital Circuits: Logic and Design, *Ronald C. Emery*
26. Large-Scale Control Systems: Theories and Techniques, *Magdi S. Mahmoud, Mohamed F. Hassan, and Mohamed G. Darwish*
27. Microprocessor Software Project Management, *Eli T. Fathi and Cedric V. W. Armstrong (Sponsored by Ontario Centre for Microelectronics)*
28. Low Frequency Electromagnetic Design, *Michael P. Perry*
29. Multidimensional Systems: Techniques and Applications, *edited by Spyros G. Tzafestas*
30. AC Motors for High-Performance Applications: Analysis and Control, *Sakae Yamamura*
31. Ceramic Materials for Electronics: Processing, Properties, and Applications, *edited by Relva C. Buchanan*
32. Microcomputer Bus Structures and Bus Interface Design, *Arthur L. Dexter*
33. End User's Guide to Innovative Flexible Circuit Packaging, *Jay J. Miniet*
34. Reliability Engineering for Electronic Design, *Norman B. Fuqua*
35. Design Fundamentals for Low-Voltage Distribution and Control, *Frank W. Kussy and Jack L. Warren*
36. Encapsulation of Electronic Devices and Components, *Edward R. Salmon*
37. Protective Relaying: Principles and Applications, *J. Lewis Blackburn*
38. Testing Active and Passive Electronic Components, *Richard F. Powell*
39. Adaptive Control Systems: Techniques and Applications, *V. V. Chalam*
40. Computer-Aided Analysis of Power Electronic Systems, *Venkatachari Rajagopalan*
41. Integrated Circuit Quality and Reliability, *Eugene R. Hnatek*
42. Systolic Signal Processing Systems, *edited by Earl Swartzlander*
43. Adaptive Digital Filters and Signal Analysis, *Maurice G. Bellanger*
44. Electronic Ceramics: Properties, Configuration, and Applications, *edited by Lionel M. Levinson*
45. Computer Systems Engineering Management, *Robert S. Alford*

Additional Volumes in Preparation

Systems Modeling and Computer Simulation, *edited by Naim A. Kheir*
Transformer and Inductor Design Handbook, Second Edition, Revised and
    Expanded, *Colonel Wm. T. McLyman*
Signal Processing Handbook, *edited by C. H.Chen*

Electrical Engineering-Electronics Software

1. Transformer and Inductor Design Software for the IBM PC,
   Colonel Wm. T. McLyman
2. Transformer and Inductor Design Software for the Macintosh,
   Colonel Wm. T. McLyman
3. Digital Filter Design Software for the IBM PC,
   Fred J. Taylor and Thanos Stouraitis

Adaptive Digital Filters and Signal Analysis

Maurice G. Bellanger

Laboratoire des Signaux et Systèmes (LSS/ESE)
University of Paris
Orsay, France

and

Télécommunications Radioélectriques
et Téléphoniques (TRT)
Le Plessis-Robinson, France

WITHDRAWN

Marcel Dekker, Inc.   **New York and Basel**

374705   Tennessee Tech. Library
Cookeville. Tenn.

Library of Congress Cataloging-in-Publication Data

Bellanger, Maurice.
    Adaptive digital filters and signal analysis / Maurice Bellanger.
        p.   cm. – (Electrical engineering and electronics : 43)
    Includes bibliographies and index.
    ISBN 0-8247-7784-0
    1. Adaptive filters.   2. Adaptive signal processing.   I. Title.
II. Series.
TK7872.F5B45   1987
621.3815'324 – dc 19   87-18920
                                                       CIP

Copyright © 1987 by MARCEL DEKKER, INC. All Rights Reserved.

Neither this book nor any part may be reproduced or transmitted in any form or
by any means, electronic or mechanical, including photocopying, microfilming,
and recording, or by any information storage and retrieval system, without per-
mission in writing from the publisher.

MARCEL DEKKER, INC.
270 Madison Avenue, New York, New York 10016

Current printing (last digit):
10 9 8 7 6 5 4 3 2 1

PRINTED IN THE UNITED STATES OF AMERICA

*To the many friends in industry and research
who made it possible to write this book.*

Preface

The main idea behind this book, and the incentive for writing it, is that strong connections exist between adaptive filtering and signal analysis, to the extent that it is not realistic—at least from an engineering point of view—to separate them. In order to understand adaptive filters well enough to design them properly and apply them successfully, a certain amount of knowledge of the analysis of the signals involved is indispensable. Conversely, several major analysis techniques can be made really efficient and useful in products only when designed and implemented in an adaptive fashion. It seemed worthwhile to dedicate a book to the intricate relationships between these two areas. Moreover, the approach can lead to new ideas and new techniques in either field.

The areas of adaptive filters and signal analysis use concepts from several different theories, among which are estimation, information, and circuit theories, in connection with sophisticated mathematical tools. As a consequence, they present a problem to the application-oriented reader. However, if these concepts and tools are introduced with adequate justification and illustration, and if their physical and practical meaning is emphasized, they become easier to understand, retain, and exploit. The work has therefore been made as complete and self-contained as possible, presuming a background in discrete time signal processing and stochastic processes.

The book is organized to provide a smooth evolution from a basic knowledge of signal representations and properties, to simple gradient algorithms, to more elaborate adaptive techniques, to spectral analysis methods, and finally to implementation aspects and applications. The characteristics of determinist, random, and natural signals are given in Chapter 2, and fundamental results for analysis are derived. Chapter 3 concentrates on the correlation matrix and spectrum and their relationships; it is intended to familiarize the reader with concepts and properties which have to be fully understood for an in-depth knowledge of necessary adaptive techniques in engineering. The gradient or least mean squares (LMS) adaptive filters are treated in Chapter 4. The theoretical aspects, engineering design options, finite word-length effects, and implementation structures are covered in turn. Chapter 5 is entirely devoted to linear prediction theory and techniques, which are crucial in deriving and understanding fast algorithms operations. Fast least squares (FLS) algorithms of the transversal type are derived and studied in Chapter 6, with emphasis on design aspects and performance. Several complementary algorithms of the same family are presented in Chapter 7 to cope with various practical situations and signal types. Time and order recursions which lead to FLS lattice algorithms are presented in Chapter 8, which ends with an introduction to the unified geometric approach for deriving all sorts of FLS algorithms. The major spectral analysis and estimation techniques are described in Chapter 9, and the connections with adaptive methods are emphasized. Chapter 10 discusses circuits and architecture issues, and a wide range of illustrative applications, taken from different technical fields, are briefly presented, to show the significance and versatility of adaptive techniques.

At the end of several chapters, FORTRAN listings of computer subroutines are given to help the reader start practicing and evaluating the major techniques.

The book has been written with engineering in mind, so that it should be most useful to practicing engineers and professional readers. However, it can also be used as a textbook and is suitable for use in a graduate course. It is worth pointing out that researchers should also be interested, as a number of new results and ideas have been included which may deserve further work.

I am indebted to many friends and colleagues from industry and research for contributions in various forms and wish to thank them all for their help. For their direct contributions, special thanks are due to J. M. Travassos-Romano from the Laboratoire des Signaux et Systèmes (LSS/ESE), to R. Lamberti from the Institut National des Télécommunications (INT) and to S. Hethuin and C. Evci from the Télécommunications Radioélectriques et Téléphoniques Company (TRT). Stimulating interaction with O. Macchi, M. Benidir and B. Picinbono (LSS/ESE) is also gratefully acknowledged.

Maurice G. Bellanger

Contents

Preface   *v*

1. Adaptive Filtering and Signal Analysis   **1**

   1.1   Signal Analysis   2
   1.2   Characterization and Modeling   4
   1.3   Adaptive Filtering   6
   1.4   Normal Equations   7
   1.5   Recursive Algorithms   9
   1.6   Implementation and Applications   11
   1.7   Further Reading   12
       References   13

2. Signals and Noise   **15**

   2.1   The Damped Sinusoid   15
   2.2   Periodic Signals   18
   2.3   Random Signals   22
   2.4   Gaussian Signals   24
   2.5   Synthetic Moving Average and Autoregressive Signals   26
   2.6   ARMA Signals   30
   2.7   Markov Signals   35

vii

2.8   Linear Prediction and Interpolation   37
2.9   Predictable Signals   41
2.10   The Fundamental (Wold) Decomposition   42
2.11   Harmonic Decomposition   44
2.12   Multidimensional Signals   46
2.13   Nonstationary Signals   48
2.14   Natural Signals   49
2.15   Summary   51
        Exercises   52
        References   53

3. **Correlation Function and Matrix**   **55**

3.1   Cross-Correlation and Autocorrelation   55
3.2   Estimation of Correlation Functions   58
3.3   Recursive Estimation   63
3.4   The Autocorrelation Matrix   65
3.5   Solving Linear Equation Systems   67
3.6   Eigenvalue Decomposition   69
3.7   Eigenfilters   73
3.8   Properties of Extremal Eigenvalues   77
3.9   Signal Spectrum and Eigenvalues   79
3.10   Iterative Determination of Extremal Eigenparameters   81
3.11   Estimation of the AC Matrix   83
3.12   Eigen (KL) Transform and Approximations   86
3.13   Summary   88
        Exercises   88
        Annex 3.1   FORTRAN Subroutine to Solve a Linear
        System with Symmetrical Matrix   90
        Annex 3.2   FORTRAN Subroutine to Compute the Eigen
        Vector Corresponding to the Minimum Eigenvalue by the
        Conjugate Gradient Method   92
        References   95

4. **Gradient Adaptive Filters**   **97**

4.1   The Gradient—LMS Algorithm   97
4.2   Stability Condition and Specifications   99
4.3   Residual Error   100
4.4   Learning Curve and Time Constant   104
4.5   Word-Length Limitations   106
4.6   Leakage Factor   112
4.7   The LMAV and Sign Algorithms   114

4.8	Normalized Algorithms for Nonstationary Signals	117
4.9	Delayed LMS Algorithms	121
4.10	FIR Filters in Cascade Form	123
4.11	IIR Gradient Adaptive Filters	125
4.12	Nonlinear Filtering	128
4.13	Strengths and Weaknesses of Gradient Filters	130
	Exercises	130
	References	132

5. Linear Prediction Error Filters   **135**

5.1	Definition and Properties	135
5.2	First- and Second-Order FIR Predictors	138
5.3	Forward and Backward Prediction Equations	140
5.4	Order Iterative Relations	142
5.5	The Lattice Linear Prediction Filter	146
5.6	The Inverse AC Matrix	148
5.7	The Notch Filter and Its Approximations	150
5.8	Zeros of FIR Prediction Error Filters	151
5.9	Poles of IIR Prediction Error Filters	155
5.10	Gradient Adaptive Predictors	156
5.11	Linear Prediction and Harmonic Decomposition	161
5.12	Conclusion	163
	Exercises	164
	Annex 5.1 Levinson Algorithm	166
	Annex 5.2 Leroux-Gueguen Algorithm	166
	References	167

6. Fast Least Squares Transversal Adaptive Filters   **169**

6.1	The First-Order LS Adaptive Filter	169
6.2	Recursive Equations for the Order N Filter	174
6.3	Relationships Between LS Variables	176
6.4	Fast Algorithm Based on A Priori Errors	179
6.5	Algorithm Based on All Prediction Errors	183
6.6	Stability Conditions for LS Recursive Algorithms	187
6.7	Initial Values of the Prediction Error Energies	189
6.8	A Stabilization Constant	191
6.9	Roundoff Error Accumulation and Its Control	193
6.10	A Simplified Algorithm	194
6.11	Performance of LS Adaptive Filters	195
6.12	Selecting FLS Parameter Values	200

6.13   Word-Length Limitations and Implementation   203
6.14   Comparison of FLS and LMS Approaches—Summary   205
        Exercises   207
        Annex 6.1.   FLS Algorithm Based on A Priori Errors 208
        Annex 6.2.   FLS Algorithm Based on All the Prediction
        Errors and with Roundoff Error Control   209
        References   210

7. **Other Adaptive Filter Algorithms**   **213**

7.1   Covariance Algorithms   213
7.2   A Sliding Window Algorithm   216
7.3   The Case of Complex Signals   220
7.4   Multidimensional Input Signals   222
7.5   M-D Algorithm Based on All Prediction Errors   227
7.6   Filters of Nonuniform Length   230
7.7   FLS Pole-Zero Modeling   230
7.8   Multirate Adaptive Filters   234
7.9   Frequency Domain Adaptive Filters   235
7.10   Unified General View and Conclusion   237
        Exercises   240
        Annex 7.1 . FLS Algorithm with Multidimensional
        Input Signal   241
        References   243

8. **Lattice Algorithms and Geometrical Approach**   **215**

8.1   Order Recurrence Relations for Prediction Coefficients 245
8.2   Order Recurrence Relations for the Filter Coefficients 248
8.3   Time Recurrence Relations   251
8.4   FLS Algorithms for Lattice Structures   252
8.5   Normalized Lattice Algorithms   255
8.6   Calculation of Transversal Filter Coefficients   260
8.7   Multidimensional Lattice Algorithms   263
8.8   Block Processing   266
8.9   Geometrical Description   267
8.10   Order and Time Recursions   270
8.11   Unified Derivation of FLS Algorithms   274
8.12   Summary and Conclusion   276
        Exercises   277
        Annex 8.1   FLS Algorithm for a Predictor in Lattice
        Structure   278
        References   279

9. Spectral Analysis   **281**

9.1   Definition and Objectives   281
9.2   The Periodogram Method   282
9.3   The Correlogram Method   284
9.4   The minimum Variance (MV) Method   286
9.5   Harmonic Retrieval Techniques   289
9.6   Autoregressive Modeling   292
9.7   ARMA Modeling   296
9.8   Signal and Noise Space Methods   298
9.9   Estimation Bounds   300
9.10 Conclusion   303
      Exercises   303
      References   304

10. Circuits and Applications   **307**

10.1   Division and Square Root   307
10.2   A Multibus Architecture   310
10.3   Line Canceling and Enhancement   311
10.4   Adaptive Differential Coding   312
10.5   Echo Cancellation   313
10.6   Channel Equalization and Measurement   313
10.7   Adaptive Deconvolution   314
10.8   Adaptive Processing in Radar   316
10.9   Adaptive Antennas   317
10.10 Image Signal Prediction   319
10.11 Artificial Intelligence   320
10.12 Conclusion   322
       References   323

Index   *325*

1

Adaptive Filtering and Signal Analysis

Digital techniques are characterized by flexibility and accuracy, two properties which are best exploited in the rapidly growing technical field of adaptive signal processing.

Among the processing operations, linear filtering is probably the most common and important. It is made adaptive if its parameters, the coefficients, are varied according to a specified criterion as new information becomes available. That updating has to follow the evolution of the system environment as fast and accurately as possible, and, in general, it is associated with real-time operation. Applications can be found in any technical field as soon as data series and particularly time series are available; they are remarkably well developed in communications and control.

Adaptive filtering techniques have been successfully used for many years. As users gain more experience from applications and as signal processing theory matures, these techniques become more and more refined and sophisticated. But to make the best use of the improved potential of these techniques, users must reach an in-depth understanding of how they really work, rather than simply applying algorithms. Moreover, the number of algorithms suitable for adaptive filtering has grown enormously. It is not unusual to find more than a dozen algorithms to complete a given task. Finding the best algorithm is a crucial engineering problem. The key to properly using adaptive techniques is an intimate knowledge of signal makeup. That is why signal analysis is so tightly connected to adaptive processing. In reality, the class of the most performant algorithms rests on a real-time analysis of the signals to be processed.

Conversely, adaptive techniques can be efficient instruments for performing signal analysis: For example, an adaptive filter can be designed as an intelligent spectrum analyzer.

So, for all these reasons, it appears that learning adaptive filtering goes with learning signal analysis, and both topics are jointly treated in this book.

First, the signal analysis problem is stated in very general terms.

1.1 SIGNAL ANALYSIS

By definition a signal carries information from a source to a receiver. In the real world, several signals, wanted or not, are transmitted and processed together, and the signal analysis problem may be stated as follows.

Let us consider a set of N sources which produce N variables $x_0, x_1, \ldots, x_{N-1}$ and a set of N corresponding receivers which give N variables $y_0, y_1, \ldots, y_{N-1}$, as shown in Figure 1.1. The transmission medium is assumed to be linear, and every receiver variable is a linear combination of

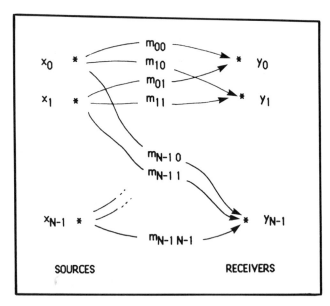

Figure 1.1   A transmission system of order N.

the source variables:

$$y_i = \sum_{j=0}^{N-1} m_{ij} x_j, \qquad 0 \leqslant i \leqslant N - 1 \tag{1.1}$$

The parameters m_{ij} are the transmission coefficients of the medium.

Now the problem is how to retrieve the source variables, assumed to carry the useful information looked for, from the receiver variables. It might also be necessary to find the transmission coefficients. Stated as such, the problem might look overly ambitious. It can be solved, at least in part, with some additional assumptions.

For clarity, conciseness, and thus simplicity, let us write equation (1.1) in matrix form:

$$Y = MX \tag{1.2}$$

with

$$X = \begin{bmatrix} x_0 \\ x_1 \\ \vdots \\ x_{N-1} \end{bmatrix}, \qquad Y = \begin{bmatrix} y_0 \\ y_1 \\ \vdots \\ y_{N-1} \end{bmatrix}$$

$$M = \begin{bmatrix} m_{00} & m_{01} & \cdots & m_{0\,N-1} \\ m_{10} & m_{11} & \cdots & m_{1\,N-1} \\ \vdots & & & \vdots \\ m_{N-1\,0} & \cdots & & m_{N-1\,N-1} \end{bmatrix}$$

Now assume that the x_i are random centered uncorrelated variables and consider the $N \times N$ matrix

$$YY^t = MXX^tM^t \tag{1.3}$$

where M^t denotes the transpose of the matrix M. Taking its mathematical expectation and noting that the transmission coefficients are deterministic variables, we get

$$E[YY^t] = ME[XX^t]M^t \tag{1.4}$$

Since the variables $x_i (0 \leqslant i \leqslant N - 1)$ are assumed to be uncorrelated, the $N \times N$ source matrix is diagonal:

$$E[XX^t] = \begin{bmatrix} P_{x_0} & 0 & \cdots & 0 \\ 0 & P_{x_1} & \cdots & 0 \\ \vdots & & \ddots & \vdots \\ 0 & 0 & \cdots & P_{x_{N-1}} \end{bmatrix} = \text{diag}[P_{x_0}, P_{x_1}, \ldots, P_{x_{N-1}}]$$

where

$$P_{x_i} = E[x_i^2]$$

is the power of the source with index i. Thus, a decomposition of the receiver covariance matrix has been achieved:

$$E[YY^t] = M \operatorname{diag}[P_{x_0}, P_{x_1}, \ldots, P_{x_{N-1}}]M^t \tag{1.5}$$

Finally, it appears possible to get the source powers and the transmission matrix from the diagonalization of the covariance matrix of the receiver variables. In practice, the mathematical expectation can be reached, under suitable assumptions, by repeated measurements, for example. It is worth noticing that if the transmission medium has no losses, the power of the sources is transferred to the receiver variables in totality, which corresponds to the relation $MM^t = 1$; the transmission matrix is unitary in that case.

In practice, useful signals are always corrupted by unwanted externally generated signals, which are classified as noise. So, besides useful signal sources, noise sources have to be included in any real transmission system. Consequently, the number of sources can always be adjusted to equal the number of receivers. Indeed, for the analysis to be meaningful, the number of receivers must exceed the number of useful sources.

The technique presented above is used in various fields for source detection and location (for example, radio communications or acoustics); the set of receivers is an array of antennas. However, the same approach can be applied as well to analyze a signal sequence when the data $y(n)$ are linear combinations of a set of basic components. The problem is then to retrieve these components. It is particularly simple when $y(n)$ is periodic with period N, because then the signal is just a sum of sinusoids with frequencies that are multiples of $1/N$, and the matrix M in decomposition (1.5) is the discrete Fourier transform (DFT) matrix, the diagonal terms being the power spectrum. For an arbitrary set of data, the decomposition corresponds to the representation of the signal as sinusoids with arbitrary frequencies in noise; it is a harmonic retrieval operation or a principal component analysis procedure.

Rather than directly searching for the principal components of a signal to analyze it, extract its information, condense it, or clear it from spurious noise, we can approximate it by the output of a model, which is made as simple as possible and whose parameters are attributed to the signal. But to apply that approach, we need some characterization of the signal.

1.2 CHARACTERIZATION AND MODELING

A straightforward way to characterize a signal is by waveform parameters. A concise representation is obtained when the data are simple functions of the index n. For example, a sinusoid is expressed by

$$x(n) = S \sin(n\omega + \varphi) \tag{1.6}$$

where S is the sinusoid amplitude, ω is the angular frequency, and φ is the phase. The same signal can also be represented and generated by the recurrence relation

$$x(n) = (2 \cos \omega)x(n - 1) - x(n - 2) \tag{1.7}$$

for $n \geqslant 0$, and the initial conditions

$$\begin{aligned}
x(-1) &= S \sin(-\omega + \varphi) \\
x(-2) &= S \sin(-2\omega + \varphi) \\
x(n) &= 0 \qquad \text{for } n < -2
\end{aligned}$$

Recurrence relations play a key role in signal modeling as well as in adaptive filtering. The correspondence between time domain sequences and recurrence relations is established by the z-transform, defined by

$$X(z) = \sum_{n=-\infty}^{\infty} x(n)z^{-n} \tag{1.8}$$

Waveform parameters are appropriate for synthetic signals, but for practical signal analysis the correlation function $r(p)$, in general, contains the relevant characteristics, as pointed out in the previous section:

$$r(p) = E[x(n)x(n - p)] \tag{1.9}$$

In the analysis process, the correlation function is first estimated and then used to derive the signal parameters of interest, the spectrum, or the recurrence coefficients.

The recurrence relation is a convenient representation or modeling of a wide class of signals, which are those obtained through linear digital filtering of a random sequence. For example, the expression

$$x(n) = e(n) - \sum_{i=1}^{N} a_i x(n - i) \tag{1.10}$$

where $e(n)$ is a random sequence or noise input, defines a model called autoregressive (AR). The corresponding filter is of the infinite impulse response (IIR) type. If the filter is of the finite impulse response (FIR) type, the model is called moving average (MA), and a general filter FIR/IIR is associated to an ARMA model.

The coefficients a_i in (1.10) are the FIR, or transversal, linear prediction coefficients of the signal $x(n)$; they are actually the coefficients of the inverse FIR filter defined by

$$e(n) = \sum_{i=0}^{N} a_i x(n - i), \qquad a_0 = 1 \tag{1.11}$$

The sequence $e(n)$ is called the prediction error signal. The coefficients are designed to minimize the prediction error power, which, expressed as a matrix form equation is

$$E[e^2(n)] = A^t E[XX^t]A \qquad (1.12)$$

So, for a given signal whose correlation function is known or can be estimated, the linear prediction (or AR modeling) problem can be stated as follows: find the coefficient vector A which minimizes the quantity $A^t E[XX^t]A$ subject to the constraint $a_0 = 1$. In that process, the power of a white noise added to the useful input signal is magnified by the factor $A^t A$.

To provide a link between the direct analysis of the previous section and AR modeling, and to point out their major differences and similarities, we note that the harmonic retrieval, or principal component analysis, corresponds to the following problem: find the vector A which minimizes the value $A^t E[XX^t]A$ subject to the constraint $A^t A = 1$. The frequencies of the sinusoids in the signal are then derived from the zeros of the filter with coefficient vector A. For determinist signals without noise, direct analysis and AR modeling lead to the same solution; they stay close to each other for high signal-to-noise ratios.

The linear prediction filter plays a key role in adaptive filtering because it is directly involved in the derivation and implementation of least squares (LS) algorithms, which in fact are based on real-time signal analysis by AR modeling.

1.3 ADAPTIVE FILTERING

The principle of an adaptive filter is shown in Figure 1.2. The output of a programmable, variable-coefficient digital filter is subtracted from a reference signal $y(n)$ to produce an error sequence $e(n)$, which is used in combination with elements of the input sequence $x(n)$, to update the filter coefficients, following a criterion which is to be minimized. The adaptive filters can be classified according to the options taken in the following areas:

The optimization criterion
The algorithm for coefficient updating
The programmable filter structure
The type of signals processed—mono- or multidimensional.

The optimization criterion is in general taken in the LS family in order to work with linear operations. However, in some cases, where simplicity of implementation and robustness are of major concern, the least absolute value (LAV) criterion can also be attractive; moreover, it is not restricted to minimum phase optimization.

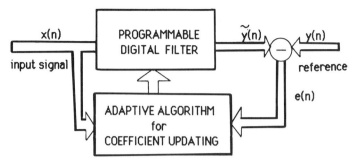

Figure 1.2 Principle of an adaptive filter.

The algorithms are highly dependent on the optimization criterion, and it is often the algorithm that governs the choice of the optimization criterion, rather than the other way round. In broad terms, the least mean squares (LMS) criterion is associated with the gradient algorithm, the LAV criterion corresponds to a sign algorithm, and the exact LS criterion is associated with a family of recursive algorithms, the most efficient of which are the fast least squares (FLS) algorithms.

The programmable filter can be a FIR or IIR type, and, in principle, it can have any structure: direct form, cascade form, lattice, ladder, or wave filter. Finite word-length effects and computational complexity vary with the structure, as with fixed coefficient filters. But the peculiar point with adaptive filters is that the structure reacts on the algorithm complexity. It turns out that the direct-form FIR, or transversal, structure is the simplest to study and implement, and therefore it is the most popular.

Multidimensional signals can use the same algorithms and structures as their monodimensional counterparts. However, computational complexity constraints and hardware limitations generally reduce the options to the simplest approaches.

The study of adaptive filtering begins with the derivation of the normal equations, which correspond to the LS criterion combined with the FIR direct form for the programmable filter.

1.4 NORMAL EQUATIONS

In the following, we assume that real-time series, resulting, for example, from the sampling with period $T = 1$ of a continuous-time real signal, are processed.

Let $H(n)$ be the vector of the N coefficients $h_i(n)$ of the programmable filter at time n, and let $X(n)$ be the vector of the N most recent input signal samples:

$$H(n) = \begin{bmatrix} h_0(n) \\ h_1(n) \\ \vdots \\ h_{N-1}(n) \end{bmatrix}, \qquad X(n) = \begin{bmatrix} x(n) \\ x(n-1) \\ \vdots \\ x(n+1-N) \end{bmatrix} \tag{1.13}$$

The error signal $\varepsilon(n)$ is

$$\varepsilon(n) = y(n) - H^t(n)X(n) \tag{1.14}$$

The optimization procedure consists of minimizing, at each time index, a cost function $J(n)$, which, for the sake of generality, is taken as a weighted sum of squared error signal values, beginning after time zero:

$$J(n) = \sum_{p=1}^{n} W^{n-p}[y(p) - H^t(n)X(p)]^2 \tag{1.15}$$

The weighting factor, W, is generally taken close to 1 ($0 \ll W \leqslant 1$).

Now, the problem is to find the coefficient vector $H(n)$ which minimizes $J(n)$. The solution is obtained by setting to zero the derivatives of $J(n)$ with respect to the entries $h_i(n)$ of the coefficient vector $H(n)$, which leads to

$$\sum_{p=1}^{n} W^{n-p}[y(p) - H^t(n)X(p)]X(p) = 0 \tag{1.16}$$

In concise form, (1.16) is

$$H(n) = R_N^{-1}(n)r_{yx}(n) \tag{1.17}$$

with

$$R_N(n) = \sum_{p=1}^{n} W^{n-p}X(p)X^t(p) \tag{1.18}$$

$$r_{yx}(n) = \sum_{p=1}^{n} W^{n-p}X(p)y(p) \tag{1.19}$$

If the signals are stationary, let R_{xx} be the $N \times N$ input signal autocorrelation matrix and let r_{yx} be the vector of cross-correlations between input and reference signals:

$$R_{xx} = E[X(p)X^t(p)], \qquad r_{yx} = E[X(p)y(p)] \tag{1.20}$$

Now

$$E[R_N(n)] = \frac{1-W^n}{1-W}R_{xx}, \qquad E[r_{yx}(n)] = \frac{1-W^n}{1-W}r_{yx} \tag{1.21}$$

So $R_N(n)$ is an estimate of the input signal autocorrelation matrix, and $r_{yx}(n)$ is an estimate of the cross-correlation between input and reference signals.

The optimal coefficient vector H_{opt} is reached when n goes to infinity:

$$H_{opt} = R_{xx}^{-1} r_{yx} \tag{1.22}$$

Equations (1.22) and (1.17) are the normal (or Yule–Walker) equations for stationary and evolutive signals, respectively. In adaptive filters, they can be implemented recursively.

1.5 RECURSIVE ALGORITHMS

The basic goal of recursive algorithms is to derive the coefficient vector $H(n + 1)$ from $H(n)$. Both coefficient vectors satisfy (1.17). In these equations, autocorrelation matrices and cross-correlation vectors satisfy the recursive relations

$$R_N(n + 1) = W R_N(n) + X(n + 1)X^t(n + 1) \tag{1.23}$$

$$r_{yx}(n + 1) = W r_{yx}(n) + X(n + 1)y(n + 1) \tag{1.24}$$

Now,

$$H(n + 1) = R_N^{-1}(n + 1)[W r_{yx}(n) + X(n + 1)y(n + 1)]$$

But

$$W r_{yx}(n) = [R_N(n + 1) - X(n + 1)X^t(n + 1)]H(n)$$

and

$$H(n + 1) = H(n) + R_N^{-1}(n + 1)X(n + 1)[y(n + 1) - X^t(n + 1)H(n)] \tag{1.25}$$

which is the recursive relation for the coefficient updating. In that expression, the sequence

$$e(n + 1) = y(n + 1) - X^t(n + 1)H(n) \tag{1.26}$$

is called the a priori error signal because it is computed by using the coefficient vector of the previous time index. In contrast, (1.14) defines the a posteriori error signal $\varepsilon(n)$.

For large values of the filter order N, the matrix manipulations in (1.25) lead to an often unacceptable hardware complexity. We obtain a drastic simplification by setting

$$R_N^{-1}(n + 1) \approx \delta I_N \tag{1.27}$$

where I_N is the $(N \times N)$ unity matrix and δ is a positive constant called the adaptation step size. The coefficients are then updated by

$$H(n + 1) = H(n) + \delta X(n + 1)e(n + 1) \tag{1.28}$$

which leads to just doubling the computations with respect to the fixed-coefficient filter. The optimization process no longer follows the exact LS criterion, but LMS criterion. The product $X(n + 1)e(n + 1)$ is proportional to the gradient of the square of the error signal with opposite sign, because differentiating equation (1.26) leads to

$$-\frac{\partial e^2(n + 1)}{\partial h_i(n)} = 2x(n + 1 - i)e(n + 1), \qquad 0 \leqslant i \leqslant N - 1 \tag{1.29}$$

hence the name gradient algorithm.

The value of the step size δ has to be chosen small enough to ensure convergence; it controls the algorithm speed of adaptation and the residual error power after convergence. It is a trade-off based on the system engineering specifications.

The gradient algorithm is useful and efficient in many applications; it is flexible, can be adjusted to all filter structures, and is robust against implementation imperfections. However, it has some limitations in performance and weaknesses which might not be tolerated in various applications. For example, its initial convergence is slow, its performance depends on the input signal statistics, and its residual error power may be large. If one is prepared to accept an increase in computational complexity by a factor usually smaller than an order of magnitude (typically 4 or 5), then the exact recursive LS algorithm can be implemented. The matrix manipulations can be avoided in the coefficient updating recursion by introducing the vector

$$G(n) = R_N^{-1}(n)X(n) \tag{1.30}$$

called the adaptation gain, which can be updated with the help of linear prediction filters. The corresponding algorithms are called FLS algorithms.

Up to now, time recursions have been considered, based on the cost function $\hat{J}(n)$ defined by equation (1.15) for a set of N coefficients. It is also possible to work out order recursions which lead to the derivation of the coefficients of a filter of order $N + 1$ from the set of coefficients of a filter of order N. These order recursions rely on the introduction of a different set of filter parameters, called the partial correlation (PARCOR) coefficients, which correspond to the lattice structure for the programmable filter. Now, time and order recursions can be combined in various ways to produce a family of LS lattice adaptive filters. That approach has attractive advantages from the theoretical point of view—for example, signal orthogonalization, spectral whitening, and easy control of the minimum phase property—and also from the implementation point of view, because it is robust to word-length limitations and leads to flexible and modular realizations.

The recursive techniques can easily be extended to complex and multidimensional signals. Overall, the adaptive filtering techniques provide a wide range of means for fast and accurate processing and analysis of signals.

1.6 IMPLEMENTATION AND APPLICATIONS

The circuitry designed for general digital signal processing can also be used for adaptive filtering and signal analysis implementation. However, a few specificities are worth pointing out. First, several arithmetic operations, such as divisions and square roots, become more frequent. Second, the processing speed, expressed in millions of instructions per second (MIPS) or in millions of arithmetic operations per second (MOPS), depending on whether the emphasis is on programming or number crunching, is often higher than average in the field of signal processing. Therefore specific efficient architectures for real-time operation can be worth developing. They can be special multibus arrangements to facilitate pipelining in an integrated processor or powerful, modular, locally interconnected systolic arrays.

Most applications of adaptive techniques fall into one of two broad classes: system identification and system correction.

The block diagram of the configuration for system identification is shown in Figure 1.3. The input signal $x(n)$ is fed to the system under analysis, which produces the reference signal $y(n)$. The adaptive filter parameters and specifications have to be chosen to lead to a sufficiently good model for the system under analysis. That kind of application occurs frequently in automatic control.

System correction is shown in Figure 1.4. The system output is the adaptive filter input. An external reference signal is needed. If the reference

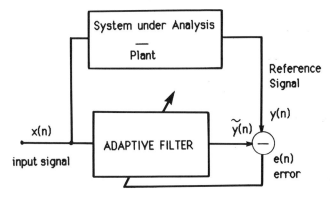

Figure 1.3 Adaptive filter for system identification.

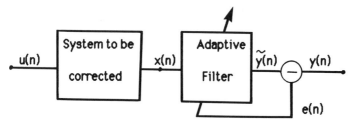

Figure 1.4 Adaptive filter for system correction.

signal $y(n)$ is also the system input signal $u(n)$, then the adaptive filter is an inverse filter; a typical example of such a situation can be found in communications, with channel equalization for data transmission. In both application classes, the signals involved can be real or complex valued, mono- or multidimensional. Although the important case of linear prediction for signal analysis can fit into either of the aforementioned categories, it is often considered as an inverse filtering problem, with the following choice of signals: $y(n) = 0$, $u(n) = e(n)$.

Another field of applications corresponds to the restoration of signals which have been degraded by addition of noise and convolution by a known or estimated filter. Adaptive procedures can achieve restoration by deconvolution.

The processing parameters vary with the class of applications as well as with the technical fields. The computational complexity and the cost efficiency often have a major impact on final decisions, and they can lead to different options in control, communications, radar, underwater acoustics, biomedical systems, broadcasting, or the different areas of applied physics.

1.7 FURTHER READING

The basic results, which are most necessary to read this book, in signal processing, mathematics, and statistics are recalled in the text as close as possible to the place where they are used for the first time, so the book is, to a large extent, self-sufficient. However, the background assumed is a working knowledge of discrete-time signals and systems and, more specifically, random processes, discrete Fourier transform (DFT), and digital filter principles and structures. Some of these topics are treated in [1]. Textbooks which provide thorough treatment of the above-mentioned topics are [2–4]. A theoretical view of signal analysis is given in [5], and spectral estimation techniques are described in [6], with emphasis on radar applications. Books on adaptive algorithms include [7] and [8]. A study of the architectures and the aspects of integration in silicon chips can be found in [9]. Various applications of adaptive digital filters in the field of communications are presented in [10].

REFERENCES

1. M. Bellanger, *Digital Processing of Signals—Theory and Practice*, John Wiley, New York, 1984.
2. A. V. Oppenheim, A. S. Willsky, and I. T. Young, *Signals and Systems*, Prentice-Hall, Englewood Cliffs, N.J., 1983.
3. F. J. Taylor, *Digital Filter Design Handbook*, Marcel Dekker, New York, 1983.
4. L. Ljung and T. Soderstrom, *Theory and Practice of Recursive Identification*, MIT Press, Cambridge, Mass., 1983.
5. A. Papoulis, *Signal Analysis*, McGraw-Hill, New York, 1977.
6. S. Haykin, *Non Linear Methods of Spectral Analysis*, Springer-Verlag, Berlin, 1983.
7. M. L. Honig and D. G. Messerschmitt, *Adaptive Filters—Structures, Algorithms and Applications*, Kluwer Academic, Boston, 1984.
8. B. Widrow and S. D. Stearns, *Adaptive Signal Processing*, Prentice-Hall, Englewood Cliffs, N.J., 1985.
9. B. A. Bowen and W. R. Brown, *VLSI Systems Design for Digital Signal Processing*, Prentice-Hall, Englewood Cliffs, N.J., 1982.
10. C. F. N. Cowan and P. M. Grant, *Adaptive Filters*, Prentice-Hall, Englewood Cliffs, N.J., 1985.

2

Signals and Noise

Signals carry information from sources to receivers, and they take many different forms. In this chapter a classification is presented for the signals most commonly used in many technical fields.

A first distinction is between useful, or wanted, signals and spurious, or unwanted, signals, which are often called noise. In practice, noise sources are always present, so any actual signal contains noise, and a significant part of the processing operations is intended to remove it. However, useful signals and noise have many features in common and can, to some extent, follow the same classification.

Only data sequences or time series are considered here, and the leading thread for the classification proposed is the set of recurrence relations, which can be established between consecutive data and which are the basis of several major analysis methods [1–3]. In the various categories, signals can be characterized by waveform functions, autocorrelation, and spectrum.

An elementary, but fundamental, signal is introduced first—the damped sinusoid.

2.1 THE DAMPED SINUSOID

Let us consider the following complex sequence, which is called the damped complex sinusoid, or damped cisoid:

$$y(n) = \begin{cases} e^{(\alpha + j\omega_0)n}, & n \geqslant 0 \\ 0, & n < 0 \end{cases} \tag{2.1}$$

where α and ω_0 are real scalars.

The z-transform of that sequence is, by definition

$$Y(z) = \sum_{n=0}^{\infty} y(n)z^{-n} \tag{2.2}$$

Hence

$$Y(z) = \frac{1}{1 - e^{(\alpha + j\omega_0)}z^{-1}} \tag{2.3}$$

The two real corresponding sequences are shown in Fig. 2.1(a). They are

$$y(n) = y_R(n) + jy_I(n) \tag{2.4}$$

with

$$y_R(n) = e^{\alpha n} \cos n\omega_0, \quad y_I(n) = e^{\alpha n} \sin n\omega_0, \quad n \geqslant 0 \tag{2.5}$$

The z-transforms are

$$Y_R(z) = \frac{1 - (e^\alpha \cos \omega_0)z^{-1}}{1 - (2e^\alpha \cos \omega_0)z^{-1} + e^{2\alpha}z^{-2}} \tag{2.6}$$

$$Y_I(z) = \frac{(e^\alpha \sin \omega_0)z^{-1}}{1 - (2e^\alpha \cos \omega_0)z^{-1} + e^{2\alpha}z^{-2}} \tag{2.7}$$

In the complex plane, these functions have a pair of conjugate poles, which are shown in Figure 2.1(b) for $\alpha < 0$ and $|\alpha|$ small. From (2.6) and (2.7) and also by direct inspection, it appears that the corresponding signals satisfy the recursion

$$y_R(n) - 2e^\alpha \cos \omega_0 \, y_R(n-1) + e^{2\alpha}y_R(n-2) = 0 \tag{2.8}$$

with initial values

$$y_R(-1) = e^{-\alpha} \cos(-\omega_0), \qquad y_R(-2) = e^{-2\alpha} \cos(-2\omega_0) \tag{2.9}$$

and

$$y_I(-1) = e^{-\alpha} \sin(-\omega_0), \qquad y_I(-2) = e^{-2\alpha} \sin(-2\omega_0) \tag{2.10}$$

More generally, the one-sided z-transform, as defined by (2.2), of equation (2.8) is

$$Y_R(z) = -\frac{b_1 y_R(-1) + b_2[y_R(-2) + y_R(-1)z^{-1}]}{1 + b_1 z^{-1} + b_2 z^{-2}} \tag{2.11}$$

with $b_1 = -2e^\alpha \cos \omega$ and $b_2 = e^{2\alpha}$.

The above-mentioned initial values are then obtained by identifying (2.11) and (2.6), and (2.11) and (2.7), respectively.

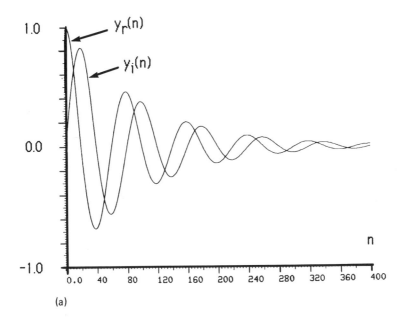

(a)

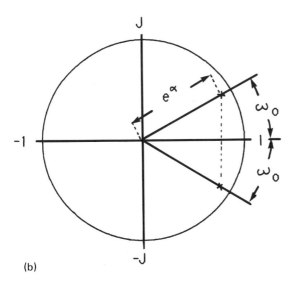

(b)

Figure 2.1(a) Waveform of a damped sinusoid. **(b)** Poles of the z-transform of the damped sinusoid.

The energy spectra of the sequences $y_R(n)$ and $y_I(n)$ are obtained from the z-transforms by replacing z by $e^{j\omega}$ [4]. For example, the function $|Y_I(\omega)|$ is shown in Figure 2.2; it is the frequency response of a purely recursive second-order filter section.

As n grows to infinity the signal $y(n)$ vanishes; it is nonstationary. Damped sinusoids can be used in signal analysis to approximate the spectrum of a finite data sequence.

2.2 PERIODIC SIGNALS

Periodic signals form an important category, and the simplest of them is the single sinusoid, defined by

$$x(n) = S \sin(n\omega_0 + \varphi) \tag{2.12}$$

where S is the amplitude, ω_0 is the radial frequency, and φ is the phase.

For $n \geqslant 0$, the results of the previous section can be applied with $\alpha = 0$. So the recursion

$$x(n) - 2 \cos \omega_0 \, x(n-1) + x(n-2) = 0 \tag{2.13}$$

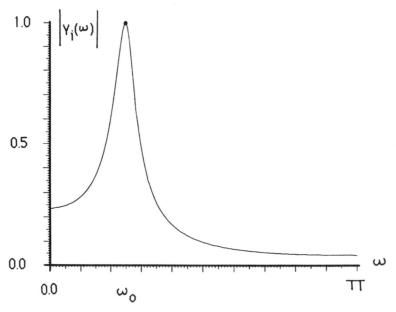

Figure 2.2 Spectrum of the damped sinusoid.

with initial conditions

$$x(-1) = S \sin(-\omega_0 + \varphi), \qquad x(-2) = S \sin(-2\omega_0 + \varphi) \qquad (2.14)$$

is satisfied. The z-transform is

$$X(z) = S \frac{\sin \varphi - \sin(-\omega_0 + \varphi)z^{-1}}{1 - (2 \cos \omega_0)z^{-1} + z^{-2}} \qquad (2.15)$$

Now the poles are exactly on the unit circle, and we must consider the power spectrum. It cannot be directly derived from the z-transform. The sinusoid is generated for $n > 0$ by the purely recursive second-order filter section in Figure 2.3 with the above-mentioned initial conditions, the circuit input being zero. For a filter to cancel a sinusoid, it is necessary and sufficient to implement the inverse filter—that is, a filter which has a pair of zeros on the unit circle at the frequency of the sinusoid; such filters appear in linear prediction.

The autocorrelation function (ACF) of the sinusoid, which is a real signal, is defined by

$$r(p) = \lim_{N \to \infty} \frac{1}{N} \sum_{n=0}^{N-1} x(n)x(n-p) \qquad (2.16)$$

Hence,

$$r(p) = \frac{S^2}{2} \cos p\omega_0 - \lim_{N \to \infty} \frac{1}{N} \frac{S^2}{2} \sum_{n=0}^{N-1} \cos 2\left(\frac{2n-p}{2}\omega_0 + \varphi\right) \qquad (2.17)$$

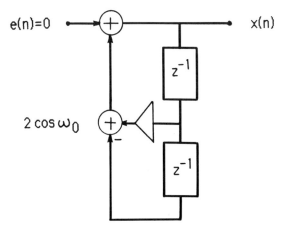

Figure 2.3 Second-order filter section to generate a sinusoid.

and for any ω_0,

$$r(p) = \frac{S^2}{2} \cos p\omega_0 \tag{2.18}$$

The power spectrum of the signal is the Fourier transform of the ACF; for the sinusoid it is a line with magnitude $S^2/2$ at frequency ω_0.

Now, let us proceed to periodic signals. A periodic signal with period N consists of a sum of complex sinusoids, or cisoids, whose frequencies are integer multiples of $1/N$ and whose complex amplitudes S_k are given by the discrete Fourier transform (DFT) of the signal data:

$$\begin{bmatrix} S_0 \\ S_1 \\ \vdots \\ S_{N-1} \end{bmatrix} = \frac{1}{N} \begin{bmatrix} 1 & 1 & \cdots & 1 \\ 1 & W & \cdots & W^{N-1} \\ \vdots & \vdots & & \vdots \\ 1 & W^{N-1} & \cdots & W^{(N-1)^2} \end{bmatrix} \begin{bmatrix} x(0) \\ x(1) \\ \vdots \\ x(N-1) \end{bmatrix} \tag{2.19}$$

with $W = e^{-j(2\pi/N)}$.

Following equation (2.3), with $\alpha = 0$, we express the z-transform of the periodic signal by

$$X(z) = \sum_{k=0}^{N-1} \frac{S_k}{1 - e^{j(2\pi/N)k}z^{-1}} \tag{2.20}$$

and its poles are uniformly distributed on the unit circle as shown in Figure 2.4 for N even. Therefore, the signal $x(n)$ satisfies the recursion

$$\sum_{i=0}^{N} a_i x(n-i) = 0 \tag{2.21}$$

where the a_i are the coefficients of the polynomial $P(z)$:

$$P(z) = \sum_{i=0}^{N} a_i z^{-i} = \prod_{k=1}^{N} (1 - e^{j(2\pi/N)k}z^{-1}) \tag{2.22}$$

So $a_0 = 1$, and if all the cisoids are present in the periodic signal, then $a_N = 1$ and $a_i = 0$ for $1 \leqslant i \leqslant N - 1$. The N complex amplitudes, or the real amplitudes and phases, are defined by the N initial conditions. If some of the N possible cisoids are missing, then the coefficients take on values according to the factors in the product (2.22).

The ACF of the periodic signal $x(n)$ is calculated from the following expression, valid for complex data:

$$r(p) = \frac{1}{N} \sum_{n=0}^{N-1} x(n)\bar{x}(n-p) \tag{2.23}$$

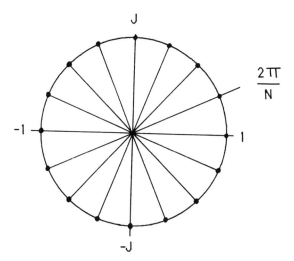

Figure 2.4 Poles of a signal with period N.

where $\bar{x}(n)$ is the complex conjugate of $x(n)$. According to the inverse DFT, $x(n)$ can be expressed from its frequency components by

$$x(n) = \sum_{k=0}^{N-1} S_k e^{j(2\pi/N)kn} \qquad (2.24)$$

Now, combining (2.24) and (2.23) gives

$$r(p) = \sum_{k=0}^{N-1} |S_k|^2 e^{j(2\pi/N)kp} \qquad (2.25)$$

and, for $x(n)$ a real signal and for the configuration of poles shown in Figure 2.4 with N even,

$$r(p) = S_0^2 + S_{N/2}^2 + 2 \sum_{k=1}^{N/2-1} |S_k|^2 \cos\left(\frac{2\pi}{N} kp\right) \qquad (2.26)$$

The corresponding spectrum is made of lines at frequencies which are integer multiples of $1/N$.

The same analysis as above can be carried out for a signal composed of a sum of sinusoids with arbitrary frequencies, which just implies that the period N may grow to infinity. In that case, the roots of the polynomial $P(z)$ take on arbitrary positions on the unit circle. Such a signal is said to be deterministic because it is completely determined by the recurrence relationship (2.21) and the set of initial conditions; in other words, a signal value at time n can be exactly calculated from the N preceding values; there is no innovation in the process; hence, it is also said to be predictable.

The importance of $P(z)$ is worth emphasizing, because it directly determines the signal recurrence relation. Several methods of analysis primarily aim at finding out that polynomial for a start.

The above deterministic or predictable signals have discrete power spectra. To obtain continuous spectra, one must introduce random signals. They bring innovation in the processes.

2.3 RANDOM SIGNALS

A random real signal $x(n)$ is defined by a probability law for its amplitude at each time n. The law can be expressed as a probability density $p(x, n)$ defined by

$$p(x, n) = \lim_{\Delta x \to 0} \frac{\text{Prob}[x \leqslant x(n) \leqslant x + \Delta x]}{\Delta x} \tag{2.27}$$

It is used to calculate, by ensemble averages, the statistics of the signal or process [5].

The signal is second order if it possesses a first-order moment $m_1(n)$ called the mean value or expectation of $x(n)$, denoted $E[x(n)]$ and defined by

$$m_1(n) = E[x(n)] = \int_{-\infty}^{\infty} xp(x, n)\, dx \tag{2.28}$$

and a second-order moment, called the covariance:

$$E[x(n_1)x(n_2)] = m_2(n_1, n_2) = \int_{-\infty}^{\infty} \int_{-\infty}^{\infty} x_1 x_2 p(x_1, x_2; n_1, n_2)\, dx_1\, dx_2 \tag{2.29}$$

where $p(x_1, x_2; n_1, n_2)$ is the joint probability density of the pair of random variables $[x(n_1), x(n_2)]$.

The signal is *stationary* if its statistical properties are independent of the time index n–that is, if the probability density is independent of time n:

$$p(x, n) = p(x) \tag{2.30}$$

The stationarity can be limited to the moments of first and second order. Then the signal is *wide-sense stationary*, and it is characterized by the following equations:

$$E[x(n)] = \int_{-\infty}^{\infty} xp(x)\, dx = m_1 \tag{2.31}$$

$$E[x(n)x(n - p)] = r(p) \tag{2.32}$$

The function $r(p)$ is the (ACF) of the signal.

The statistical parameters are, in general, difficult to estimate or measure directly, because of the ensemble averages involved. A reasonably accurate measurement of an ensemble average requires that many process realizations be available or that the experiment be repeated many times, which is often impractical. On the contrary, time averages are much easier to come by, for time series. Therefore the ergodicity property is of great practical importance; it states that, for a stationary signal, ensemble and time averages are equivalent:

$$m_1 = E[x(n)] = \lim_{N \to \infty} \frac{1}{2N + 1} \sum_{n=-N}^{N} x(n) \tag{2.33}$$

$$r(p) = E[x(n)x(n - p)] = \lim_{N \to \infty} \frac{1}{2N + 1} \sum_{n=-N}^{N} x(n)x(n - p) \tag{2.34a}$$

For complex signals, the ACF is

$$r(p) = E[x(n)\bar{x}(n - p)] = \lim_{N \to \infty} \frac{1}{2N + 1} \sum_{-N}^{N} x(n)\bar{x}(n - p) \tag{2.34b}$$

The factor $x(n - p)$ is replaced by its complex conjugate $\bar{x}(n - p)$; note that $r(0)$ is the signal power and is always a real number.

In the literature, the factor $x(n + p)$ is generally taken to define $r(p)$; however, we use $x(n - p)$ throughout this book because it comes naturally in adaptive filtering.

In some circumstances, moments of order $k > 2$ might be needed. They are defined by

$$m_k = \int_{-\infty}^{\infty} x^k p(x) \, dx \tag{2.35}$$

and they can be calculated efficiently through the introduction of a function $F(u)$, called the characteristic function of the random variable x and defined by

$$F(u) = \int_{\infty}^{\infty} e^{jux} p(x) \, dx \tag{2.36}$$

Using definition (2.35), we obtain the series expansion

$$F(u) = \sum_{k=0}^{\infty} \frac{(ju)^k}{k!} m_k \tag{2.37}$$

Since $F(u)$ is the inverse Fourier transform of the probability density $p(x)$, it can be easy to calculate and can provide the high-order moments of the signal.

An important concept is that of statistical independence of random variables. Two random variables, x_1 and x_2, are independent if and only if their joint density $p(x_1, x_2)$ is the product of the individual probability densities:

$$p(x_1, x_2) = p(x_1)p(x_2) \tag{2.38}$$

which implies the same relationship for the characteristic functions:

$$F(u_1, u_2) = \int\!\!\!\int_{-\infty}^{\infty} e^{j(u_1 x_1 + u_2 x_2)} p(x_1, x_2)\, dx_1\, dx_2 \tag{2.39}$$

and

$$F(u_1, u_2) = F(u_1)F(u_2) \tag{2.40}$$

The correlation concept is related to linear dependency. Two noncorrelated variables, such that $E[x_1 x_2] = 0$, have no linear dependency. But, in general, that does not mean statistical independency, since higher-order dependency can exist.

Among the probability laws, the Gaussian law has special importance in signal processing.

2.4 GAUSSIAN SIGNALS

A random variable x is said to be normally distributed or Gaussian if its probability law has a density $p(x)$ which follows the normal or Gaussian law:

$$p(x) = \frac{1}{\sigma_x \sqrt{2\pi}} e^{-(x-m)^2/2\sigma_x^2} \tag{2.41}$$

The parameter m is the mean of the variable x; the variance σ_x^2 is the second-order moment of the centered random variable $(x - m)$; σ_x is also called the standard deviation.

The characteristic function of the centered Gaussian variable is

$$F(u) = e^{-\sigma_x^2 u^2/2} \tag{2.42}$$

Now, using the series expansion (2.37), the moments are

$$m_{2k+1} = 0$$

$$m_2 = \sigma_x^2, \quad m_4 = 3\sigma_x^4, \quad m_{2k} = \frac{2k!}{2^k k!} \sigma_x^{2k} \tag{2.43}$$

The normal law can be generalized to multidimensional random variables. The characteristic function of a k-dimensional Gaussian variable

$x(x_1, x_2, \ldots, x_k)$ is

$$F(u_1, u_2, \ldots, u_k) = \exp\left(-\frac{1}{2}\sum_{i=1}^{k}\sum_{j=1}^{k} r_{ij}u_i u_j\right) \tag{2.44}$$

with $r_{ij} = E[x_i x_j]$.

If the variables are not correlated, then they are independent, because $r_{ij} = 0$ for $i \neq j$ and $F(u_1, u_2, \ldots, u_k)$ is the product of the characteristic functions. So noncorrelation means independence for Gaussian variables.

A random signal $x(n)$ is said to be Gaussian if, for any set of k time values $n_i(1 \leq i \leq k)$, the k-dimensional random variable $x = [x(n_1), x(n_2), \ldots, x(n_k)]$ is Gaussian. According to (2.44), the probability law of that variable is completely defined by the ACF $r(p)$ of $x(n)$. The power spectral density $S(f)$ is obtained as the Fourier transform of the ACF:

$$S(f) = \sum_{p=-\infty}^{\infty} r(p)e^{-j2\pi pf} \tag{2.45}$$

or, since $r(p)$ is an even function,

$$S(f) = r(0) + 2\sum_{p=1}^{\infty} r(p)\cos(2\pi pf) \tag{2.46}$$

If the data in the sequence $x(n)$ are independent, then $r(p)$ reduces to $r(0)$ and the spectrum $S(f)$ is flat; the signal is then said to be white.

An important aspect of the Gaussian probability laws is that they preserve their character under any linear operation, such as convolution, filtering, differentiation, or integration.

Therefore, if a Gaussian signal is fed to a linear system, the output is also Gaussian. Moreover, there is a natural trend toward Gaussian probability densities, because of the so-called central limit theorem, which states that the random variable

$$x = \frac{1}{\sqrt{N}}\sum_{i=1}^{N} x_i \tag{2.47}$$

where the x_i are N independent identically distributed (i.i.d.) second-order random variables, becomes Gaussian when N grows to infinity.

The Gaussian approximation can reasonably be made as soon as N exceeds a few units, and the importance of Gaussian densities becomes apparent because in nature many signal sources and, particularly, noise sources at the micro- or macroscopic levels add up to make the sequence to be processed. So Gaussian noise is present in virtually every signal processing application.

2.5 SYNTHETIC MOVING AVERAGE AND AUTOREGRESSIVE SIGNALS

In simulation, evaluation, transmission, test and measurement, the data sequences used are often not natural but synthetic signals. They appear also in some analysis techniques, namely analysis by synthesis techniques.

Deterministic signals can be generated in a straightforward manner as isolated or recurring pulses or as sums of sinusoids. A diagram to produce a single sinusoid is shown in Figure 2.3. Note that the sinusoids in a sum must have different phases; otherwise an impulse shape waveform is obtained.

Flat spectrum signals arc characterized by the fact that their energy is uniformly distributed over the entire frequency band. Therefore an approach to produce a deterministic white-noise-like waveform is to generate a set of sinusoids uniformly distributed in frequency with the same amplitude but different phases.

Random signals can be obtained from sequences of statistically independent real numbers generated by standard computer subroutines through a rounding process. The magnitudes of these numbers are uniformly distributed in the interval $(0, 1)$, and the sequences obtained have a flat spectrum.

Several probability densities can be derived from the uniform distribution. Let the Gaussian, Rayleigh, and uniform densities be $p(x)$, $p(y)$, and $p(z)$, respectively. The Rayleigh density is

$$p(y) = \frac{y}{\sigma^2} \exp\left[-\frac{y^2}{2\sigma^2} \right] \tag{2.48}$$

where the variance of the corresponding random variable is $2\sigma^2$. It is a density associated with the peak values of a narrowband Gaussian signal. The change of variables

$$p(z)\,dz = dz = p(y)\,dy$$

leads to

$$\frac{dz}{dy} = \frac{y}{\sigma^2} \exp\left[-\frac{y^2}{2\sigma^2} \right]$$

Hence,

$$z = \exp\left[-\frac{y^2}{2\sigma^2} \right]$$

and a Rayleigh sequence $y(n)$ is obtained from a uniform sequence $z(n)$ in the magnitude interval $(0, 1)$ by the following operation:

$$y(n) = \sigma\sqrt{2\ln[1/z(n)]} \tag{2.49}$$

Now, independent Rayleigh and uniform sequences can be used to derive a Gaussian sequence $x(n)$:

$$x(n) = y(n) \cos[2\pi z(n)] \tag{2.50}$$

Correlated random signals can be obtained by filtering a white sequence with either uniform or Gaussian amplitude probability density, as shown in Figure 2.5. The filter $H(z)$ can take on different structures, corresponding to different models for the output signal [6].

The simplest type is the finite impulse response (FIR) filter, corresponding to the so-called moving average (MA) model, and defined by

$$H(z) = \sum_{i=0}^{N} h_i z^{-i} \tag{2.51}$$

and, in the time domain,

$$x(n) = \sum_{i=0}^{N} h_i e(n - i) \tag{2.52}$$

where the h_i are the filter impulse response.

The output signal ACF is obtained by direct application of definition (2.34), considering that

$$E[e^2(n)] = \sigma_e^2, \qquad E[e(n)e(n - i)] = 0 \quad \text{for } i \neq 0$$

The result is

$$r(p) = \begin{cases} \sigma_e^2 \sum_{i=0}^{N-p} h_i h_{i+p}, & |p| \leqslant N \\ 0, & |p| > N \end{cases} \tag{2.53}$$

Several remarks are necessary. First, the ACF has a finite length in accordance with the filter impulse response. Second, the output signal power σ_x^2 is related to the input signal power by

$$\sigma_x^2 = r(0) = \sigma_e^2 \sum_{i=0}^{N} h_i^2 \tag{2.54}$$

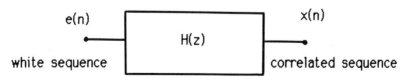

Figure 2.5 Generation of a correlated random signal.

Equation (2.54) is frequently used in subsequent sections. The power spectrum can be computed from the ACF $r(p)$ by using equation (2.46), but another approach is to use $H(z)$, since it is available, via the equation

$$S(f) = \sigma_e^2 \left| \sum_{i=0}^{N} h_i e^{-j2\pi if} \right|^2 \tag{2.55}$$

An infinite impulse response (IIR) filter corresponds to an autoregressive (AR) model. The equations are

$$H(z) = \frac{1}{1 - \sum_{i=1}^{N} a_i z^{-i}} \tag{2.56}$$

and, in the time domain,

$$x(n) = e(n) + \sum_{i=1}^{N} a_i x(n-i) \tag{2.57}$$

The ACF can be derived from the corresponding filter impulse response coefficients h_i:

$$H(z) = \sum_{i=0}^{\infty} h_i z^{-i} \tag{2.58}$$

and, accordingly, it is an infinite sequence:

$$r(p) = \sigma_e^2 \sum_{i=0}^{\infty} h_i h_{i+p} \tag{2.59}$$

The power spectrum is

$$S(f) = \frac{\sigma_e^2}{\left| 1 - \sum_{i=1}^{N} a_i e^{-j2\pi if} \right|^2} \tag{2.60}$$

An example is shown in Figure 2.6 for the filter transfer function:

$$H(z) = \frac{1}{(1 + 0.80z^{-1} + 0.64z^{-2})(1 - 1.23z^{-1} + 0.64z^{-2})}$$

Since the spectrum of a real signal is symmetric about the zero frequency, only the band $[0, f_{s/2}]$, where f_s is the sampling frequency, is represented.

For MA signals, the direct relation (2.53) has been derived between the ACF and filter coefficients. A direct relation can also be obtained here by multiplying both sides of the recursion definition (2.57) by $x(n-p)$ and

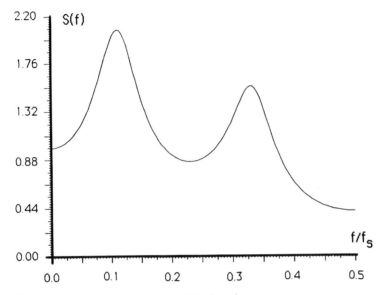

Figure 2.6   Spectrum of an AR signal.

taking the expectation, which leads to

$$r(0) = \sigma_e^2 + \sum_{i=1}^{N} a_i r(i) \qquad (2.61)$$

$$r(p) = \sum_{i=1}^{N} a_i r(p - i), \qquad p \geqslant 1 \qquad (2.62)$$

For $p \geqslant N$, the sequence $r(p)$ is generated recursively from the N preceding terms. For $0 \leqslant p \leqslant N - 1$, the above equations establish a linear dependence between the two sets of filter coefficients and the first ACF values.

They can be expressed in matrix form to derive the coefficients from the ACF terms:

$$\begin{bmatrix} r(0) & r(1) & \cdots & r(N) \\ r(1) & r(1) & \cdots & r(N-1) \\ \vdots & \vdots & \ddots & \vdots \\ r(N) & r(N-1) & \cdots & r(0) \end{bmatrix} \begin{bmatrix} 1 \\ -a_1 \\ \vdots \\ -a_N \end{bmatrix} = \begin{bmatrix} \sigma_e^2 \\ 0 \\ \vdots \\ 0 \end{bmatrix} \qquad (2.63)$$

Equation (2.63) is a normal equation, called the order N forward linear prediction equation, studied in a later chapter.

To complete the AR signal analysis, note that the generating filter impulse response is

$$h_p = r(p) - \sum_{i=1}^{N} a_i h(p + i) \tag{2.64}$$

This equation is a direct consequence of definition relations (2.57) and (2.58), if we notice that

$$h_p = E[x(n)e(n - p)]$$

Since $r(p) = r(-p)$, equation (2.62) shows that the impulse response h_p is zero for negative p, which reflects the filter causality.

A limitation of AR spectra is that they do not take on zero values, whereas MA spectra do. So it may be useful to combine both [7].

2.6 ARMA SIGNALS

An ARMA signal is obtained through a filter with a rational z-transfer function:

$$H(z) = \frac{\sum_{i=0}^{N} b_i z^{-1}}{1 - \sum_{i=1}^{N} a_i z^{-i}} \tag{2.65a}$$

In the time domain,

$$x(n) = \sum_{i=0}^{N} b_i e(n - i) + \sum_{i=1}^{N} a_i x(n - i) \tag{2.65b}$$

The denominator and numerator polynomials of $H(z)$ can always be assumed to have the same order; if necessary, zero coefficients can be added.

The power spectral density is

$$S(f) = \sigma_e^2 \frac{\left| \sum_{i=0}^{N} b_i e^{-j2\pi i f} \right|^2}{\left| 1 - \sum_{i=1}^{N} a_i e^{-j2\pi i f} \right|^2} \tag{2.66}$$

A direct relation between the ACF and the coefficients is obtained by multiplying both sides of the time recursion (2.65) by $x(n - p)$ and taking the expectation:

$$r(p) = \sum_{i=1}^{N} a_i r(p - i) + \sum_{i=0}^{N} b_i E[e(n - i)x(n - p)] \tag{2.67}$$

Now the relationships between ACF and filter coefficients become nonlinear, due to the second term in (2.67). However, that nonlinear term vanishes for $p > N$ because $x(n - p)$ is related to the input signal value with the same index and the preceding values only, not future ones. Hence, a matrix equation can again be derived involving the AR coefficients of the ARMA signal:

$$\begin{bmatrix} r(N) & r(N - 1) & \cdots & r(0) \\ r(N + 1) & r(N) & \cdots & r(1) \\ \vdots & \vdots & \ddots & \vdots \\ r(2N) & r(2N - 1) & \cdots & r(N) \end{bmatrix} \begin{bmatrix} 1 \\ -a_1 \\ \vdots \\ -a_N \end{bmatrix} = b_0 b_N \begin{bmatrix} \sigma_e^2 \\ 0 \\ \vdots \\ 0 \end{bmatrix} \tag{2.68}$$

For $p > N$, the sequence $r(p)$ is again generated recursively from the N preceding terms.

The relationship between the first $(N + 1)$ ACF terms and the filter coefficients can be established through the filter impulse response, whose coefficients h_i satisfy, by definition,

$$x(n) = \sum_{i=0}^{\infty} h_i e(n - i) \tag{2.69}$$

Now replacing $x(n - i)$ in (2.65) gives

$$x(n) = \sum_{i=0}^{N} b_i e(n - i) + \sum_{i=1}^{N} a_i \sum_{j=0}^{\infty} h_j e(n - i - j)$$

and

$$x(n) = \sum_{i=0}^{N} b_i e(n - i) + \sum_{k=1}^{\infty} e(n - k) \sum_{i=1}^{N} a_i h_{k-i} \tag{2.70}$$

Clearly, the impulse response coefficients can be computed recursively:

$$h_0 = b_0, \qquad h_k = 0 \quad \text{for } k < 0$$

$$h_k = b_k + \sum_{i=1}^{N} a_i h_{k-i}, \qquad k \geqslant 1 \tag{2.71}$$

In matrix form, for the $N + 1$ first terms we have

$$
\begin{bmatrix}
1 & 0 & 0 & \cdots & 0 \\
-a_1 & 1 & 0 & \cdots & 0 \\
-a_2 & -a_1 & 1 & \cdots & 0 \\
\vdots & \vdots & \vdots & \ddots & \vdots \\
-a_N & -a_{N-1} & -a_{N-2} & \cdots & 1
\end{bmatrix}
\begin{bmatrix}
h_0 & 0 & 0 & \cdots & 0 \\
h_1 & h_0 & 0 & \cdots & 0 \\
h_2 & h_1 & h_0 & \cdots & 0 \\
\vdots & \vdots & \vdots & \ddots & \vdots \\
h_N & h_{N-1} & h_{N-2} & \cdots & h_0
\end{bmatrix}
$$

$$
=
\begin{bmatrix}
b_0 & 0 & 0 & \cdots & 0 \\
b_1 & b_0 & 0 & \cdots & 0 \\
b_2 & b_1 & b_0 & \cdots & 0 \\
\vdots & \vdots & \vdots & \ddots & \vdots \\
b_N & b_{N-1} & b_{N-2} & \cdots & b_0
\end{bmatrix}
\tag{2.72}
$$

Coming back to the ACF and (2.67), we have

$$
\sum_{i=0}^{N} b_i E[e(n - i)x(n - p)] = \sigma_e^2 \sum_{i=0}^{N} b_i h_{i-p}
$$

and, after simple manipulations,

$$
r(p) = \sum_{i=1}^{N} a_i r(p - i) + \sigma_e^2 \sum_{j=0}^{N-p} b_{j+p} h_j
\tag{2.73}
$$

Now, introducing the variable

$$
d(p) = \sum_{j=0}^{N-p} b_{j+p} h_j
\tag{2.74}
$$

we obtain the matrix equation

$$
\mathscr{A}
\begin{bmatrix}
r(0) \\
r(1) \\
\vdots \\
r(N)
\end{bmatrix}
+ \mathscr{A}'
\begin{bmatrix}
r(0) \\
r(-1) \\
\vdots \\
r(-N)
\end{bmatrix}
= \sigma_e^2
\begin{bmatrix}
d(0) \\
d(1) \\
\vdots \\
d(N)
\end{bmatrix}
\tag{2.75}
$$

where

$$
\mathscr{A} =
\begin{bmatrix}
1 & 0 & \cdots & 0 \\
-a_1 & 1 & \cdots & 0 \\
\vdots & \vdots & \ddots & \vdots \\
-a_N & -a_{N-1} & \cdots & 1
\end{bmatrix}
$$

$$
\mathscr{A}' = \begin{bmatrix} 0 & -a_1 & \cdots & & -a_N \\ 0 & -a_2 & \cdots & & 0 \\ \vdots & \vdots & & & \vdots \\ 0 & -a_N & & & \\ 0 & 0 & \cdots & & 0 \end{bmatrix}
$$

For real signals, the first $(N + 1)$ ACF terms are obtained from the equation

$$
\begin{bmatrix} r(0) \\ r(1) \\ \vdots \\ r(N) \end{bmatrix} = \sigma_e^2 [\mathscr{A} + \mathscr{A}']^{-1} \begin{bmatrix} d(0) \\ d(1) \\ \vdots \\ d(N) \end{bmatrix} \tag{2.76}
$$

In summary, the procedure to calculate the ACF of an ARMA signal from the generating filter coefficients is as follows:

1. Compute the first $(N + 1)$ terms of the filter impulse response through recursion (2.71).
2. Compute the auxiliary variables $d(p)$ for $0 \leqslant p \leqslant N$.
3. Compute the first $(N + 1)$ ACF terms from matrix equation (2.76).
4. Use recursion (2.62) to derive $r(p)$ when $p \geqslant N + 1$.

Obviously, finding the ACF is not a simple task, particularly for large filter orders N. Conversely, the filter coefficients and input noise power can be retrieved from the ACF. First the AR coefficients a_i and the scalar $b_0 b_N \sigma_e^2$ can be obtained from matrix equation (2.68). Next, from the time domain definition (2.65), the following auxiliary signal can be introduced:

$$
u(n) = x(n) - \sum_{i=1}^{N} a_i x(n - i) = e(n) + \sum_{i=1}^{N} b_i e(n - i) \tag{2.77}
$$

where $b_0 = 1$ is assumed.

The ACF $r_u(p)$ of the auxiliary signal $u(n)$ is derived from the ACF of $x(n)$ by the equation

$$
\begin{aligned} r_u(p) &= E[u(n)u(n - p)] \\ &= r(p) - \sum_{i=1}^{N} a_i r(p + i) - \sum_{i=1}^{N} a_i r(p - i) + \sum_{i=1}^{N} \sum_{j=1}^{N} a_i a_j r(p + j - i) \end{aligned}
$$

or, more concisely by

$$
r_u(p) = \sum_{i=-N}^{N} c_i r(p - i) \tag{2.78}
$$

where

$$c_i = c_{-i}, \qquad c_0 = 1 + \sum_{j=1}^{N} a_j^2$$

$$c_i = -a_i + \sum_{j=i+1}^{N} a_j a_{j-i} \tag{2.79}$$

But $r_u(p)$ can also be expressed in terms of MA coefficients, because of the second equation in (2.77). The corresponding expressions, already given in the previous section, are

$$r_u(p) = \begin{cases} \sigma_e^2 \sum_{i=0}^{N-p} b_i b_{i+p}, & |p| \leqslant N \\ 0, & |p| > N \end{cases}$$

From these $N + 1$ equations, the input noise power σ_e^2 and the MA coefficients $b_i (1 \leqslant i \leqslant N; \ b_0 = 1)$ can be derived from iterative Newton–Raphson algorithms. It can be verified that $b_0 b_N \sigma_e^2$ equals the value we previously found when solving matrix equation (2.68) for AR coefficients.

The spectral density $S(f)$ can be computed with the help of the auxiliary signal $u(n)$ by considering the filtering operation

$$x(n) = u(n) + \sum_{i=1}^{N} a_i x(n - i) \tag{2.80}$$

which, in the spectral domain, corresponds to

$$S(f) = \frac{r_u(0) + \sum_{p=1}^{N} r_u(p)\cos(2\pi pf)}{\left|1 - \sum_{i=1}^{N} a_i e^{-j2\pi if}\right|^2} \tag{2.81}$$

This expression is useful in spectral analysis.

Until now, only real signals have been considered in this section. Similar results can be obtained with complex signals by making appropriate complex conjugations in equations. An important difference is that the ACF is no longer symmetrical, which can complicate some procedures. For example, the matrix equation (2.75) to obtain the first $(N + 1)$ ACF terms becomes

$$\mathscr{A}r + \mathscr{A}'\bar{r} = \sigma_e^2 d \tag{2.82a}$$

where r is the correlation vector, $\bar{r}$ the vector with complex conjugate entries, and d the auxiliary variable vector. The conjugate expression of (2.75) is

$$\bar{\mathscr{A}}\bar{r} + \bar{\mathscr{A}}'r = \sigma_e^2 \bar{d} \tag{2.82b}$$

The above equations, after some algebraic manipulations, lead to

$$[\mathscr{A} - \mathscr{A}'(\bar{\mathscr{A}})^{-1}\bar{\mathscr{A}}']r = \sigma_e^2[d - \mathscr{A}'(\bar{\mathscr{A}})^{-1}\bar{d}] \tag{2.83}$$

Now two matrix inversions are needed to get the correlation vector. Note that $\mathscr{A}^{-1}$ is readily obtained from (2.72) by calculating the first $N + 1$ values of the impulse response of the AR filter through the recursion (2.71).

Next, more general signals of the types often encountered in control systems are introduced.

2.7 MARKOV SIGNALS

Markov signals are produced by state variable systems whose evolution from time n to time $n + 1$ is governed by a constant transition matrix [8].

The state of a system of order N at time n is defined by a set of N internal variables represented by a vector $X(n)$ called the state vector. The block diagram of a typical system is shown in Figure 2.7, and the equations are

$$X(n + 1) = AX(n) + Bw(n) \tag{2.84a}$$

$$y(n) = C'X(n) + v(n) \tag{2.84b}$$

The matrix A is the $N \times N$ transition matrix, B is the control vector, and C is the observation vector [9]. The input sequence is $w(n)$; $v(n)$ can be a measurement noise contaminating the output $y(n)$.

The state of the system at time n is obtained from the initial state at time zero by the equation

$$X(n) = A^n X(0) + \sum_{i=1}^{n} A^{n-i} Bw(i - 1) \tag{2.85}$$

Consequently, the behavior of such a system depends on successive powers of the transition matrix A.

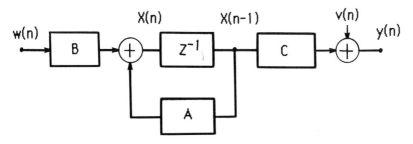

Figure 2.7 State variable system.

The z-transfer function of the system $H(z)$, obtained by taking the z-transform of the state equations, is

$$H(z) = C^t(ZI_N - A)^{-1}B \tag{2.86}$$

with I_N the $N \times N$ unity matrix.

The poles of the transfer function are the values of z for which the determinant of the matrix $(ZI_N - A)$ is zero. That is also the definition of the eigenvalues of A.

The system is stable if and only if the poles are inside the unit circle in the complex plane or, equivalently, if and only if the absolute values of the eigenvalues are less than unity, which can be seen directly from equation (2.85).

Let us assume that $w(n)$ is a centered white noise with power σ_w^2. The state variables are also centered, and their covariance matrix can be calculated. Multiplying state equation (2.84a) on the right by its transpose yields

$$X(n + 1)X^t(n + 1) = AX(n)X^t(n)A + Bw^2(n)B^t$$
$$+ AX(n)w(n)B^t + Bw(n)X^t(n)A^t$$

The expected values of the last two terms of this expression are zero, because $x(n)$ depends only on the past input values. Hence, the covariance matrix $R_{xx}(n + 1)$ is

$$R_{xx}(n + 1) = E[X(n + 1)X^t(n + 1)] = AR_{xx}(n)A^t + \sigma_w^2BB^t \tag{2.87}$$

It can be computed recursively once the covariance of the initial conditions $R_{xx}(0)$ is known. If the elements of the $w(n)$ sequence are Gaussian random variables, the state variables themselves are Gaussian, since they are linear combinations of past input values.

The Markovian representation applies to ARMA signals. Several sets of state variables can be envisaged. For example, in linear prediction, a representation corresponding to the following state equations is used:

$$x(n) = C^t\hat{X}(n) + e(n) \tag{2.88a}$$

$$\hat{X}(n) = A\hat{X}(n - 1) + Be(n - 1) \tag{2.88b}$$

with

$$A = \begin{bmatrix} 0 & 1 & 0 & \cdots & 0 \\ 0 & 0 & 1 & \cdots & 0 \\ \vdots & \vdots & \vdots & \ddots & \vdots \\ 0 & 0 & 0 & \cdots & 1 \\ a_N & a_{N-1} & a_{N-2} & \cdots & a_1 \end{bmatrix}, \quad B = \begin{bmatrix} h_1 \\ h_2 \\ \vdots \\ h_N \end{bmatrix}$$

$$C = \begin{bmatrix} 1 \\ 0 \\ \vdots \\ 0 \end{bmatrix}, \qquad \hat{X}(n) = \begin{bmatrix} \hat{x}_0(n) \\ \hat{x}_1(n) \\ \vdots \\ \hat{x}_{N-1}(n) \end{bmatrix}$$

The elements of vector B are the filter impulse response coefficients of equation (2.69), and those of the state vector, $\hat{x}_i(n)$ are the i-step linear predictions of $x(n)$, defined, for the ARMA signal and as shown later, by

$$\hat{x}_i(n) = \sum_{k=1}^{i} a_k \hat{x}(n-k) + \sum_{j=1}^{N-i} a_{i+j} x(n-i-j) + \sum_{j=1}^{N} b_{i+j} e(n-i-j) \quad (2.89)$$

It can be verified that the characteristic polynomial of the matrix A, whose roots are the eigenvalues, is the denominator of the filter transfer function $H(z)$ in (2.65).

Having presented methods for generating signals, we now turn to analysis techniques. First we introduce some important definitions and concepts [10].

2.8 LINEAR PREDICTION AND INTERPOLATION

The operation which produces a sequence $e(n)$ from a data sequence $x(n)$, assumed centered and wide-sense stationary, by the convolution

$$e(n) = x(n) - \sum_{i=1}^{\infty} a_i x(n-i) \qquad (2.90)$$

is called one-step linear prediction error filtering, if the coefficients are calculated to minimize the variance of the output $e(n)$. The minimization is equivalent, through derivation, to making $e(n)$ orthogonal to all previous data, because it leads to:

$$E[e(n)x(n-i)] = 0, \qquad i \geqslant 1 \qquad (2.91)$$

Since $e(n)$ is a linear combination of past data, the following equations are also valid:

$$E[e(n)e(n-i)] = 0, \qquad i \geqslant 1 \qquad (2.92)$$

and the sequence $e(n)$, called the prediction error or the innovation, is a white noise. Therefore the one-step prediction error filter is also called the whitening filter. The data $x(n)$ can be obtained from the innovations by the inverse filter, assumed realizable, which is called the model or innovation filter. The operations are shown in Figure 2.8.

The prediction error variance $E_a = E[e^2(n)]$ can be calculated from the data power spectrum density $S(e^{j\omega})$ by the conventional expressions for digital filtering:

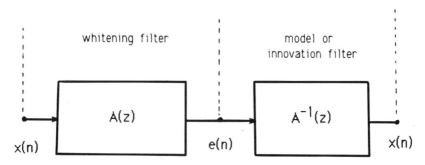

Figure 2.8 Linear prediction filter and inverse filter.

$$E_a = \frac{1}{2\pi} \int_{-\pi}^{\pi} |A(e^{j\omega})|^2 S(e^{j\omega}) \, d\omega \tag{2.93}$$

or, in terms of z-transforms,

$$E_a = \frac{1}{j2\pi} \int_{|z|=1} A(z)A(z^{-1})S(z) \frac{dz}{z} \tag{2.94}$$

where $A(z)$ is the transfer function of the prediction error filter. The prediction filter coefficients depend only on the input signal, and the error power can be expressed as a function of $S(e^{j\omega})$ only. To derive that expression, we must first show that the prediction error filter is minimum phase; in other words, all its zeros are inside or on the unit circle in the complex z-plane.

Let us assume that a zero of $A(z)$, say z_0, is outside the unit circle, which means $|z_0| > 1$, and consider the filter $A'(z)$ given by

$$A'(z) = A(z) \frac{z - \bar{z}_0^{-1}}{z - z_0} \frac{z - z_0^{-1}}{z - \bar{z}_0} \tag{2.95}$$

As Figure 2.9 shows,

$$\left| \frac{z - \bar{z}_0^{-1}}{z - z_0} \right|_{z=e^{j\omega}} \left| \frac{z - z_0^{-1}}{z - \bar{z}_0} \right|_{z=e^{j\omega}} = \frac{1}{|z_0|^2} \tag{2.96}$$

and the corresponding error variance is

$$E'_a = \frac{1}{|z_0|^2} E_a < E_a \tag{2.97}$$

which contradicts the definition of the prediction filter. Consequently, the prediction filter $A(z)$ is minimum phase.

In (2.94) for E_a, we can remove the filter transfer function with the help of logarithms, taking into account that the innovation sequence has a constant

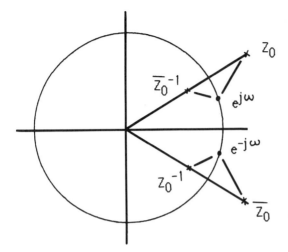

Figure 2.9   Reflection of external zero in the unit circle.

power spectrum density; thus,

$$2\pi j \ln E_a = \int_{|z|=1} \ln A(z)\, \frac{dz}{z} + \int_{|z|=1} \ln A(z^{-1})\, \frac{dz}{z} + \int_{|z|=1} \ln S(z)\, \frac{dz}{z} \quad (2.98)$$

Now, since $A(z)$ is minimum phase, $\ln A(z)$ is analytic for $z \geqslant 1$ and the unit circle can be replaced in the above integral with a circle whose radius is arbitrarily large, and since

$$\lim_{z \to \infty} A(z) = a_0 = 1$$

the first integral vanishes on the right side of (2.98). The second integral also vanishes because it can be shown, by a change of variables from z^{-1} to z that it is equal to the first one.

Finally, the prediction error power is expressed in terms of the signal power spectrum density by

$$E_a = \exp \left\{ \frac{1}{2\pi} \int_{-\pi}^{\pi} \ln S(e^{j\omega})\, d\omega \right\} \quad (2.99)$$

This very important result is known as the Kolmogoroff–Szegö formula.

A useful signal parameter is the prediction gain G, defined as the signal-to-prediction-error ratio:

$$G = \frac{1}{2\pi} \int_{-\pi}^{\pi} S(e^{j\omega})\, d\omega \bigg/ \exp \left\{ \frac{1}{2\pi} \int_{-\pi}^{\pi} \ln S(e^{j\omega})\, d\omega \right\} \quad (2.100)$$

Clearly, for a white noise $G = 1$.

At this stage, it is interesting to compare linear prediction and interpolation. Interpolation is the filtering operation which produces from the data $x(n)$ the sequence

$$e_i(n) = \sum_{j=-\infty}^{\infty} h_j x(n-j), \quad h_0 = 1 \tag{2.101}$$

with coefficients calculated to minimize the output power. Hence, $e_i(n)$ is orthogonal to past and future data:

$$E[e_i(n)x(n-k)] = E_i \delta(k) \tag{2.102}$$

where $\delta(k)$ is the Dirac distribution and

$$E_i = E[e_i^2(n)] \tag{2.103}$$

Clearly, the interpolation error $e_i(n)$ is not necessarily a white noise. Taking the z-transform of both sides of the orthogonal relationship (2.102) leads to

$$H(z)S(z) = E_i \tag{2.104}$$

Also

$$E_i = \frac{1}{j2\pi} \int_{|z|=1} H(z)H(z^{-1})S(z) \frac{dz}{z} \tag{2.105}$$

Combining equations (2.104) and (2.105) gives

$$E_i = 1 / \frac{1}{2\pi} \int_{-\pi}^{\pi} \frac{d\omega}{S(e^{j\omega})} \tag{2.106}$$

Now, it is known from linear prediction that

$$S(e^{j\omega}) = \frac{E_a}{|A(e^{j\omega})|^2} \tag{2.107}$$

and

$$E_i = E_a / \frac{1}{2\pi} \int_{-\pi}^{\pi} |A(e^{j\omega})|^2 \, d\omega = E_a / \sum_{i=0}^{\infty} a_i^2 \tag{2.108}$$

Since $a_0 = 1$, we can conclude that $E_i \leqslant E_a$; the interpolation error power is less than or equal to the prediction error power, which is a not unexpected result.

Linear prediction is useful for classifying signals and, particular, distinguishing between deterministic and random processes.

2.9 PREDICTABLE SIGNALS

A signal $x(n)$ is predictable if and only if its prediction error power is null:

$$E_a = \frac{1}{2\pi} \int_{-\pi}^{\pi} |A(e^{j\omega})|^2 S(e^{j\omega})\, d\omega = 0 \tag{2.109}$$

or, in the time domain,

$$x(n) = \sum_{i=1}^{\infty} a_i x(n - i) \tag{2.110}$$

which means that the present value $x(n)$ of the signal can be expressed in terms of its past values. The only signals which satisfy the above equations are those whose spectrum consists of lines:

$$S(e^{j\omega}) = \sum_{i=1}^{N} |S_i|^2 \delta(\omega - \omega_i) \tag{2.111}$$

The scalars $|S_i|^2$ are the powers of individual lines. The integer N can be arbitrarily large. The minimum degree prediction filter is

$$A_m(z) = \prod_{i=1}^{N} (1 - e^{j\omega_i} z^{-1}) \tag{2.112}$$

However all the filters $A(z)$ with

$$A(z) = 1 - \sum_{i=1}^{\infty} a_i z^{-i} \tag{2.113}$$

and such that $A(e^{j\omega_i}) = 0$ for $1 \leqslant i \leqslant N$ satisfy the definition and are prediction filters.

Conversely, since $A(z)$ is a power series, $A(e^{j\omega})$ cannot equal zero for every ω in an interval, and equations (2.109) and (2.110) can hold only if $S(e^{j\omega}) = 0$ everywhere except at a countable set of points. It follows that $S(e^{j\omega})$ must be a sum of impulses as in (2.111), and $A(z)$ has corresponding zeros on the unit circle.

Finally, a signal $x(n)$ is predictable if and only if its spectrum consists of lines.

The line spectrum signals are an extreme case of the more general class of bandlimited signals. A signal $x(n)$ is said to be bandlimited if $S(e^{j\omega}) = 0$ in one or more frequency intervals. Then a filter $H(\omega)$ exists such that

$$H(\omega)S(e^{j\omega}) \equiv 0 \tag{2.114}$$

and, in the time domain,

$$\sum_{i=-\infty}^{\infty} h_i x(n - i) = 0$$

With proper scaling, we have

$$x(n) = - \sum_{i=1}^{\infty} h_i x(n - i) - \sum_{i=1}^{\infty} h_{-i} x(n + i) \qquad (2.115)$$

Thus the present value can be expressed in terms of past and future values. Again the representation is not unique, because the function $H(\omega)$ is arbitrary, subject only to condition (2.114). It can be shown that a band-limited signal can be approximated arbitrarily closely by a sum involving only its past values. Equality is obtained if $S(e^{j\omega})$ consists of lines only.

The above sections are mainly intended to serve as a gradual preparation for the introduction of one of the most important results in signal analysis, the fundamental decomposition.

2.10 THE FUNDAMENTAL (WOLD) DECOMPOSITION

Any signal is the sum of two orthogonal components, an AR signal and a predictable signal. More specifically:

Decomposition Theorem

An arbitrary unpredictable signal $x(n)$ can be written as a sum of two orthogonal signals:

$$x(n) = x_p(n) + x_r(n) \qquad (2.116)$$

where $x_p(n)$ is predictable and $x_r(n)$ is such that its spectrum $S_r(e^{j\omega})$ can be factored as

$$S_r(e^{j\omega}) = |H(e^{j\omega})|^2, \qquad H(z) = \sum_{i=0}^{\infty} h_i z^{-i} \qquad (2.117)$$

and $H(z)$ is a function analytic for $|z| > 1$.

The component $x_r(n)$ is sometimes said to be regular. Following the development in [10], the proof of the theorem begins with the computation of the prediction error sequence

$$e(n) = x(n) - \sum_{i=1}^{\infty} a_i x(n - i) \qquad (2.118)$$

As previously mentioned, the prediction coefficients are computed so as to make $e(n)$ orthogonal to all past data values, and the error sequence is a white noise with variance E_a.

Conversely, the least squares estimate of $x(n)$ in terms of the sequence $e(n)$ and its past is the sum

$$x_r(n) = \sum_{i=0}^{\infty} h_i e(n - i) \qquad (2.119)$$

and the corresponding error signal

$$x_p(n) = x(n) - x_r(n)$$

is orthogonal to $e(n - i)$ for $i \geqslant 0$. In other words, $e(n)$ is orthogonal to $x_p(n + k)$ for $k \geqslant 0$.

Now, $e(n)$ is also orthogonal to $x_r(n - k)$ for $k \geqslant 1$, because $x_r(n - k)$ depends linearly on $e(n - k)$ and its past and $e(n)$ is white noise. Hence,

$$E[e(n)[x(n - k) - x_r(n - k)]] = 0 = E[e(n)x_p(n - k)], \qquad k \geqslant 1$$

and

$$E[e(n)x_p(n - k)] = 0, \qquad \text{all } k \tag{2.120a}$$

Expression (2.119) yields

$$E[x_r(n)x_p(n - k)] = 0, \qquad \text{all } k \tag{2.120b}$$

The signals $x_r(n)$ and $x_p(n)$ are orthogonal, and their powers add up to give the input signal power:

$$E[x^2(n)] = E[x_p^2(n)] + E[x_r^2(n)] \tag{2.121}$$

Now (2.119) also yields

$$E[x_r^2(n)] = E_a \sum_{i=0}^{\infty} h_i^2 \leqslant E[x^2(n)] \tag{2.122}$$

Therefore,

$$H(z) = \sum_{i=0}^{\infty} h_i z^{-i}$$

converges for $|z| > 1$ and defines a linear causal system which produces $x_r(n)$ when fed with $e(n)$.

In these conditions, the power spectrum of $x_r(n)$ is

$$S_r(e^{j\omega}) = E_a |H(e^{j\omega})|^2 \tag{2.123}$$

The filtering operations which have produced $x_r(n)$ from $x(n)$ are shown in Figure 2.10. If instead of $x(n)$ the component in a signal sequence $x(n) - x_r(n) = x_p(n)$ is fed to the system, the error $e_p(n)$, instead of $e(n)$, is obtained. The sequence

$$e_p(n) = e(n) - \left[x_r(n) - \sum_{i=1}^{\infty} a_i x_r(n - i) \right] \tag{2.124}$$

is a linear combination of $e(n)$ and its past, via equation (2.119). But, by definition,

$$e_p(n) = x_p(n) - \sum_{i=1}^{\infty} a_i x_p(n - i) \tag{2.125}$$

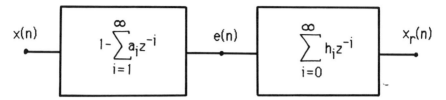

Figure 2.10   Extraction of the regular component in a signal.

which, using equations (2.120), yields

$$E[e_p^2(n)] = E\left\{\left[e(n) - \left(x_r(n) - \sum_{i=1}^{\infty} a_i x_r(n-i)\right)\right]\right.$$
$$\left.\times \left[x_p(n) - \sum_{i=1}^{\infty} a_i x_p(n-i)\right]\right\}$$
$$= 0$$

Therefore $x_p(n)$ is a predictable signal and the whitening filter $A(z)$ is a prediction error filter, although not necessarily the minimum degree filter, which is given by (2.112). On the contrary, $A(z)$ is the unique prediction error filter of $x(n)$.

Finally, the spectrum $S(e^{j\omega})$ of the unpredictable signal $x(n)$ is a sum

$$S(e^{j\omega}) = S_r(e^{j\omega}) + S_p(e^{j\omega}) \tag{2.126}$$

where $S_r(e^{j\omega})$ is the continuous spectrum of the regular signal $x_r(n)$, and $S_p(e^{j\omega})$ is the line spectrum of the deterministic component, the two components being uncorrelated.

2.11 HARMONIC DECOMPOSITION

The fundamental decomposition is used in signal analysis as a reference for selecting a strategy [11]. As an illustration let us consider the case, frequently occurring in practice, where the signal to be analyzed is given as a set of $2N + 1$ autocorrelation coefficients $r(p)$ with $-N \leqslant p \leqslant N$, available from a measuring procedure. To perform the analysis, we have two extreme hypotheses. The first one consists of assuming that the signal has no deterministic component; then a set of N prediction coefficients can be calculated as indicated in the section dealing with AR signals by (2.63), and the power spectrum is obtained from (2.60).

But another hypothesis is that the signal is essentially deterministic and consists of N sinusoids in noise. The associated ACF for real data is

$$r(p) = 2 \sum_{k=1}^{N} |S_k|^2 \cos(p\omega_k) + \sigma_e^2 \delta(p) \tag{2.127}$$

where ω_k are the radial frequencies of the sinusoids and S_k are the amplitudes. In matrix form,

$$
\begin{bmatrix} r(0) - \sigma_e^2 \\ r(1) \\ r(2) \\ \vdots \\ r(N) \end{bmatrix} = 2 \begin{bmatrix} 1 & 1 & \cdots & 1 \\ \cos \omega_1 & \cos \omega_2 & \cdots & \cos \omega_N \\ \cos 2\omega_1 & \cos 2\omega_2 & \cdots & \cos 2\omega_N \\ \vdots & \vdots & & \vdots \\ \cos N\omega_1 & \cos N\omega_2 & \cdots & \cos N\omega_N \end{bmatrix}
$$

$$
\times \begin{bmatrix} |S_1|^2 \\ |S_2|^2 \\ \vdots \\ |S_N|^2 \end{bmatrix} \tag{2.128}
$$

The analysis of the signal consists of finding out the sinusoid frequencies and amplitudes and the noise power σ_e^2. To perform that task, we use the signal sequence $x(n)$. According to the above hypothesis, it can be expressed by

$$
x(n) = x_p(n) + e(n) \tag{2.129}
$$

with

$$
x_p(n) = \sum_{i=1}^{N} a_i x_p(n - i)
$$

Now, the data signal satisfies the recursion

$$
x(n) = \sum_{i=1}^{N} a_i x(n - i) + e(n) - \sum_{i=1}^{N} a_i e(n - i) \tag{2.130}
$$

which is just a special kind of ARMA signal, with $b_0 = 1$ and $b_i = -a_i$ in time domain relation (2.65). Therefore results derived in Section 2.6 can be applied.

The impulse response can be computed recursively, and relations (2.71) yield $h_k = \delta(k)$. The auxiliary variable in (2.74) is $d(p) = -a_p (1 \leqslant p \leqslant N)$. Rewriting the equations giving the autocorrelation values (2.73) leads to

$$
r(p) = \sum_{i=1}^{N} a_i r(p - i) + \sigma_e^2 (-a_p), \qquad 1 \leqslant p \leqslant N \tag{2.131}
$$

or, in matrix form for real data,

$$
\begin{bmatrix} r(0) & r(1) & \cdots & r(N) \\ r(1) & r(0) & \cdots & r(N-1) \\ \vdots & \vdots & \ddots & \vdots \\ r(N) & r(N-1) & \cdots & r(0) \end{bmatrix} \begin{bmatrix} 1 \\ -a_1 \\ \vdots \\ -a_N \end{bmatrix} = \sigma_e^2 \begin{bmatrix} 1 \\ -a_1 \\ \vdots \\ -a_N \end{bmatrix} \tag{2.132}
$$

This is an eigenvalue equation. The signal autocorrelation matrix is symmetric, and therefore all eigenvalues are greater than or equal to zero. For N sinusoids without noise, the $(N + 1) \times (N + 1)$ autocorrelation matrix has one eigenvalue equal to zero; adding to the signal a white noise component of power σ_e^2 results in adding σ_e^2 to all eigenvalues of the autocorrelation matrix. Thus, the noise power σ_e^2 is the smallest eigenvalue of the signal, and the recursion coefficients are the entries of the associated eigenvector. As shown in the next chapter, the roots of the filter

$$A(z) = 1 - \sum_{i=1}^{N} a_i z^{-i} \tag{2.133}$$

called the minimum eigenvalue filter, are located on the unit circle in the complex plane and give the frequencies of the sinusoids. The analysis is then completed by solving the linear system (2.128) for the individual sinusoid powers. The complete procedure, called the Pisarenko method, is presented in more detail in a subsequent chapter [12].

So, it is very important to notice that a signal given by a limited set of correlation coefficients can always be viewed as a set of sinusoids in noise. That explains why the study of sinusoids in noise is so important for signal analysis and, more generally, for processing.

In practice, the selection of an analysis strategy is guided by a priori information on the signal and its generation process.

2.12 MULTIDIMENSIONAL SIGNALS

Most of the algorithms and analysis techniques presented in this book are for monodimensional real or complex sequences, which make up the bulk of the applications. However, the extension to multidimensional signals can be quite straightforward and useful in some important cases—for example, those involving multiple sources and receivers, as in geophysics, underwater acoustics, and multiple-antenna transmission systems [13].

A multidimensional signal is defined as a vector of N sequences

$$X(n) = \begin{bmatrix} x_1(n) \\ x_2(n) \\ \vdots \\ x_N(n) \end{bmatrix}$$

For example, the source and receiver vectors in Figure 1.1 are multidimensional signals. The N sequences are assumed to be dependent; otherwise they could be treated as N different scalar signals. They are characterized by the joint density function between them.

A second-order stationary multidimensional random signal is characterized by a mean vector M_X and a covariance matrix R_{xx}:

$$M_x = \begin{bmatrix} E[x_1(n)] \\ E[x_2(n)] \\ \vdots \\ E[x_N(n)] \end{bmatrix}, \qquad R_{xx} = E[(X(n) - M_x)(X(n) - M_x)^t] \qquad (2.134)$$

The diagonal terms of R_{xx} are the variances of the signal elements. If the elements in the vector are each Gaussian, then they are jointly Gaussian and have a joint density:

$$p(X) = \frac{1}{(2\pi)^{N/2}[\det R_{xx}]^{1/2}} \exp[-\tfrac{1}{2}(X - M_x)^t R_{xx}^{-1}(X - M_x)] \qquad (2.135)$$

For the special case $N = 2$,

$$R_{xx} = \begin{bmatrix} \sigma_{x_1}^2 & \rho\sigma_{x_1}\sigma_{x_2} \\ \rho\sigma_{x_1}\sigma_{x_2} & \sigma_{x_2}^2 \end{bmatrix} \qquad (2.136a)$$

with ρ the correlation coefficient defined by

$$\rho = \frac{1}{\sigma_{x_1}\sigma_{x_2}} E[(x_1 - m_1)(x_2 - m_2)] \qquad (2.136b)$$

If the signal elements are independent, R_{xx} is a diagonal matrix and

$$p(X) = \prod_{i=1}^{N} \frac{1}{\sigma_i^2 \sqrt{2\pi}} \exp\left[-\frac{(x_i - m_i)^2}{2\sigma_i^2} \right] \qquad (2.137)$$

Furthermore, if all the variances are equal, then

$$R_{xx} = \sigma^2 I_N \qquad (2.138)$$

This situation is frequently encountered in roundoff noise analysis in implementations.

For complex data, the Gaussian joint density (2.135) takes a slightly different form:

$$p(X) = \frac{1}{\pi^N} \frac{1}{\det R_{xx}} \exp[-(X - M_x)^{*t} R_{xx}^{-1}(X - M_x)] \qquad (2.139)$$

Multidimensional signals appear naturally in state variable systems, as shown in Section 2.7.

2.13 NONSTATIONARY SIGNALS

A signal is nonstationary if its statistical character changes with time. The fundamental decomposition can be extended to such a signal, and the regular component is

$$x_r(n) = \sum_{i=0}^{\infty} h_i(n)e(n - i) \tag{2.140}$$

where $e(n)$ is a stationary white noise. The generating filter impulse response coefficients are time dependent. An instantaneous spectrum can be defined as

$$S(f, n) = \sigma_e^2 \left| \sum_{i=0}^{\infty} h_i(n)e^{-j2\pi f i} \right|^2 \tag{2.141}$$

So, nonstationary signals can be generated or modeled by the techniques developed for stationary signals, but with additional means to make the system coefficients time varying [14]. For example, the ARMA signal is

$$x(n) = \sum_{i=0}^{N} b_i(n)e(n - i) + \sum_{i=1}^{N} a_i(n)x(n - i) \tag{2.142}$$

The coefficients can be generated in various ways. For example, they can be produced as weighted sums of K given time functions $f_k(n)$:

$$a_i(n) = \sum_{k=1}^{K} a_{ik}f_k(n) \tag{2.143}$$

These time functions may be periodic functions or polynomials; a simple case is the one-degree polynomial, which corresponds to a drift of the coefficients. The signal depends on $(2N + 1)K$ time-independent parameters.

The set of coefficients can also be a multidimensional signal. A realistic example in that class is shown in Figure 2.11. The N time-varying filter coefficients $a_i(n)$ are obtained as the outputs of N fixed-coefficient filters fed by independent white noises with same variances. A typical choice for the coefficient filter transfer function is the first-order low-pass function

$$H_i(z) = \frac{1}{1 - \gamma z^{-1}}, \qquad 0 \ll \gamma < 1 \tag{2.144}$$

whose time constant is

$$\tau = \frac{1}{1 - \gamma} \tag{2.145}$$

For γ close to unity, the time constant is large and the filter coefficients are subject to slow variations.

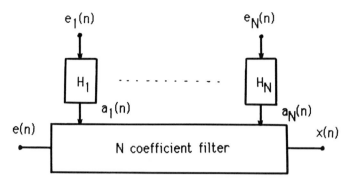

Figure 2.11 Generation of a nonstationary signal.

The analysis of nonstationary signals is complicated because the ergodicity assumption can no longer be used and statistical parameters cannot be computed through time averages. Natural signals are nonstationary. However, they are often slowly time varying and can then be assumed stationary for short periods of time.

2.14 NATURAL SIGNALS

To illustrate the preceding developments, we give several signals from different application fields in this section.

Speech is probably the most commonly processed natural signal through digital communication networks. The waveform for the word "FATHER" is shown in Figure 2.12. The sampling rate is 8 kHz, and the duration is about 0.5 s. Clearly, it is nonstationary. Speech consists of phonemes and can be considered as stationary on durations ranging from 10 to 25 ms.

It can be modeled as the output of a time-varying purely recursive filter (AR model) fed by either a string of periodic pulses for voiced sections or a string of random pulses for unvoiced sections [15].

The output of the demodulator of a frequency-modulated continuous wave (FMCW) radar is shown in Figure 2.13. It is basically a distorted sinusoid corrupted by noise and echoes. The main component frequency is representative of the distance to be measured.

An image can be represented as a one-dimensional signal through scanning. In Figure 2.14, three lines of a black-and-white contrasted picture are shown; a line has 256 samples. The similarities between consecutive lines can be observed, and the amplitude varies quickly within every line. The picture represents a house.

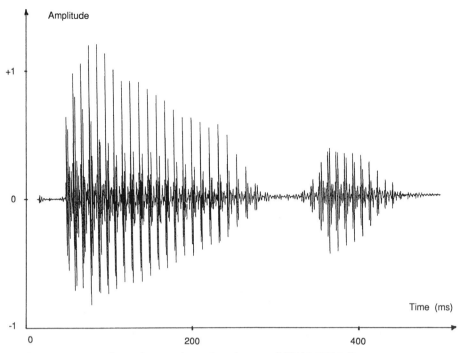

Figure 2.12 Speech waveform for the word "FATHER."

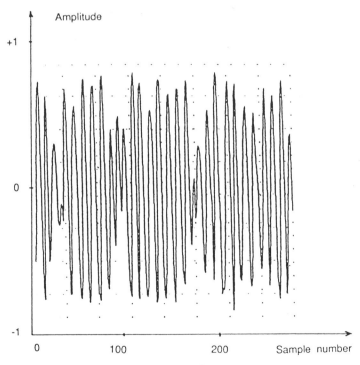

Figure 2.13 FMCW radar signal.

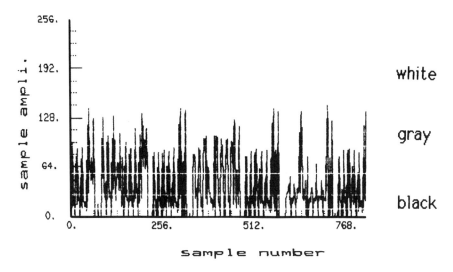

Figure 2.14 Image Signal: Three lines of a black-and-white picture.

2.15 SUMMARY

Any stationary signal can be decomposed into periodic and random components. The characteristics of both classes can be studied by considering as main parameters, the ACF, the spectrum, and the generating model. Periodic signals have been analyzed first. Then random signals have been defined, with attention being focused on wide-sense stationary signals; they have second-order statistics which are independent of time. Synthetic random signals can be generated by a filter fed with white noise. The Gaussian amplitude distribution is especially important because of its nice statistical properties, but also because it is a model adequate for many real situations. The generating filter structures correspond to various output signal classes: MA, AR, and ARMA. The concept of linear prediction is related to a generating filter model, and the class of predictable signals has been defined. A proof of the fundamental Wold decomposition has been presented, and, as an application, it has been shown that a signal specified by a limited set of correlation coefficients can be viewed as a set of sinusoids in noise. That is the harmonic decomposition.

In practice, signals are nonstationary, and, in general, short-term stationarity or slow variations have to be assumed. Several natural signal examples, namely speech, radar, and image samples, have been selected to illustrate the theory.

EXERCISES

1. Calculate the z-transform $Y_R(z)$ of the damped cosinusoid

$$y_R(n) = \begin{cases} 0, & n < 0 \\ e^{-0.1n} \cos \dfrac{n\pi}{2}, & n \geq 0 \end{cases}$$

and show the poles in the complex plane.
    Give the signal energy spectrum and verify the energy relationship

$$E_y = \sum_{n=0}^{\infty} y_R^2(n) = \frac{1}{2\pi j} \int_{|z|=1} Y_R(z) Y_R(z^{-1}) z^{-1} \, dz$$

Give the coefficients, initial conditions, and diagram of the second-order section which generates $y_R(n)$.

2. Find the ACF of the signal

$$x(n) = \cos n\frac{\pi}{3} + \frac{1}{2}\sin n\frac{\pi}{4}$$

Determine the recurrence equation satisfied by $x(n)$ and give the initial conditions.

3. Evaluate the mean and variance associated with the uniform probability density function on the interval $[x_1, x_2]$. Comment on the results.

4. Consider the signal

$$x(n) = \begin{cases} 0 & n < 0 \\ 0.8x(n-1) + e(n), & n \geq 1 \end{cases}$$

assuming $e(n)$ is a stationary zero mean random sequence with power $\sigma_e^2 = 0.5$. The initial condition is deterministic with value $x(0) = 1$.
    Calculate the mean sequence $m_n = E[x(n)]$. Give the recursion, for the variance sequence. What is the stationary solution. Calculate the ACF of the stationary signal.

5. Find the first three terms of the ACF of the AR signal

$$x(n) = 1.27x(n-1) - 0.81x(n-2) + e(n)$$

where $e(n)$ is a unit power centered white noise.

6. An ARMA signal is defined by the recursion

$$x(n) = e(n) + 0.5e(n-1) + 0.9e(n-2) + x(n-1) - 0.5x(n-2)$$

where $e(n)$ is a unit variance centered white noise. Calculate the generating filter z-transfer function and its impulse response. Derive the signal ACF.

7. A two-dimensional signal is defined by

$$X(n) = \begin{cases} \begin{bmatrix} x_1(n) \\ x_2(n) \end{bmatrix} = \begin{bmatrix} 0 \\ 0 \end{bmatrix}, & n \leqslant 0 \\ \begin{bmatrix} 0.63 & 0.36 \\ 0.09 & 0.86 \end{bmatrix} X(n-1) + \begin{bmatrix} 0.01 \\ 0.06 \end{bmatrix} e(n), & n \geqslant 1 \end{cases}$$

where $e(n)$ is a unit power centered white noise. Find the covariance propagation equation and calculate the stationary solution.

8. A measurement has supplied the signal autocorrelation values $r(0) = 5.75$, $r(1) = 4.03$, $r(2) = 0.46$. Calculate the two coefficients of the second-order linear predictor and the prediction error power. Give the corresponding signal power spectrum.

9. Find the eigenvalues of the matrix

$$R_3 = \begin{bmatrix} 1.00 & 0.70 & 0.08 \\ 0.70 & 1.00 & 0.70 \\ 0.08 & 0.70 & 1.00 \end{bmatrix}$$

and the coefficients of the minimum eigenvalue filter. Locate the zeros of that filter and give the harmonic spectrum. Compare with the prediction spectrum obtained in the previous exercise.

REFERENCES

1. T. W. Anderson, *The Statistical Analysis of Time Series*, Wiley, New York, 1971.
2. G. E. P. Box and G. M. Jenkins, *Time Series Analysis: Forecasting and Control*, Holden-Day, San Francisco, 1976.
3. J. E. Cadzow and H. Van Landingham, *Signals, Systems and Transforms*, Prentice-Hall, Englewood Cliffs, N.J., 1985.
4. A. V. Oppenheim, A. S. Willsky, and I. T. Young, *Signals and Systems*, Prentice-Hall, Englewood Cliffs, N.J., 1983.
5. W. B. Davenport, *Probability and Random Processes*, McGraw-Hill, New York, 1970.
6. T. J. Terrel, *Introduction to Digital Filters*, Wiley, New York, 1980.
7. D. Graupe, D. J. Krause, and J. B. Moore, "Identification of ARMA Parameters of Time Series." *IEEE Transactions* **AC-20**, 104–107 (February 1975).
8. J. Lamperti, *Stochastic Processes*, Springer, New York, 1977.
9. R. G. Jacquot, *Modern Digital Control Systems*, Marcel Dekker, New York, 1981.

10. A. Papoulis, "Predictable Processes and Wold's Decomposition: A Review," *IEEE Transactions* **ASSP-33**, 933–938 (August 1985).

11. S. M. Kay and S. L. Marple, "Spectrum Analysis: A Modern Perspective," *Proc. IEEE* **69**, 1380–1419 (November 1981).

12. V. F. Pisarenko, "The Retrieval of Harmonics from a Covariance Function," *Geophysical J. Royal Astronomical Soc.* **33**, 347–366 (1973).

13. D. E. Dudgeon and R. M. Mersereau, *Multidimensional Digital Signal Processing*, Prentice-Hall, Englewood Cliffs, N.J., 1984.

14. Y. Grenier, "Time Dependent ARMA Modeling of Non Stationary Signals," *IEEE Transactions* **ASSP-31**, 899–911 (August 1983).

15. L. R. Rabiner and R. W. Schafer, *Digital Processing of Speech Signals*, Prentice-Hall, Englewood Cliffs, N.J., 1978.

3

Correlation Function and Matrix

The operation and performance of adaptive filters are tightly related to the statistical parameters of the signals involved. Among these parameters, the correlation functions take a significant place. In fact, they are crucial because of their own value for signal analysis but also because their terms are used to form correlation matrices. These matrices are exploited directly in some analysis techniques. However, in the efficient algorithms for adaptive filtering considered here, they do not, in general, really show up, but they are implied and actually govern the efficiency of the processing. Therefore an in-depth knowledge of their properties is necessary. Unfortunately it is not easy to figure out their characteristics and establish relations with more accessible and familiar signal features, such as the spectrum.

This chapter presents correlation functions and matrices, discusses their most useful properties, and, through examples and applications, makes the reader accustomed to them and ready to exploit them. To begin with, the correlation functions, which have already been introduced, are presented in more detail.

3.1 CROSS-CORRELATION AND AUTOCORRELATION

Assume that two sets of N real data, $x(n)$ and $y(n)$, have to be compared, and consider the scalar a which minimizes the cost function

$$J(N) = \sum_{n=1}^{N} [y(n) - ax(n)]^2 \tag{3.1}$$

Setting to zero the derivative of $J(N)$ with respect to a yields

$$a = \frac{\sum\limits_{n=1}^{N} x(n)y(n)}{\sum\limits_{n=1}^{N} x^2(n)} \tag{3.2}$$

The minimum of the cost function is

$$J_{\min}(N) = [1 - k^2(N)] \sum_{n=1}^{N} y^2(n) \tag{3.3}$$

with

$$k(N) = \frac{\sum\limits_{n=1}^{N} x(n)y(n)}{\sqrt{\sum\limits_{n=1}^{N} x^2(n)} \sqrt{\sum\limits_{n=1}^{N} y^2(n)}} \tag{3.4}$$

The quantity $k(N)$, cross-correlation coefficient, is a measure of the degree of similarity between the two sets of N data. To point out the practical significance of that coefficient, we mention that it is the basic parameter of an important class of prediction filters and adaptive systems—the least squares (LS) lattice structures in which it is computed in real time recursively.

From equations (3.2) and (3.4), the correlation coefficient $k(N)$ is bounded by

$$|k(N)| \leqslant 1 \tag{3.5}$$

and it is independent of the signal energies; it is said to be normalized.

If instead of $x(n)$ we consider a delayed version of the signal in the above derivation, a cross-correlation function can be obtained. The general, unnormalized form of the cross-correlation function between two real sequences $x(n)$ and $y(n)$ is defined by

$$r_{yx}(p) = E[y(n)x(n-p)] \tag{3.6}$$

For stationary and ergodic signals we have

$$r_{yx}(p) = \lim_{N \to \infty} \frac{1}{2N+1} \sum_{n=-N}^{N} y(n)x(n-p) \tag{3.7}$$

Several properties result from the above definitions. For example:

$$r_{yx}(-p) = E\{x(n+p)y[(n+p)-p]\} = r_{xy}(p) \tag{3.8}$$

If two random zero mean signals are independent, their cross-correlation functions are zero. In any case, when p approaches infinity the cross-correlation approaches zero. The magnitudes of $r_{yx}(p)$ are not, in general, maximum at the origin, but they are bounded. The inequality

$$[y(n) - x(n - p)]^2 \geqslant 0 \tag{3.9}$$

yields the bound

$$|r_{yx}(p)| \leqslant \tfrac{1}{2}[r_{xx}(0) + r_{yy}(0)] \tag{3.10}$$

If the signals involved are the input and output of a filter

$$y(n) = \sum_{i=0}^{\infty} h_i x(n - i) \tag{3.11}$$

and

$$r_{yx}(p) = E[y(n)x(n - p)] = \sum_{i=0}^{\infty} h_i r_{xx}(p - i) \tag{3.12}$$

the following relationships, in which the convolution operator is denoted $*$, can be derived:

$$\begin{aligned} r_{yx}(p) &= r_{xx}(p) * h(p) \\ r_{xy}(p) &= r_{xx}(p) * h(-p) \\ r_{yy}(p) &= r_{xx}(p) * h(p) * h(-p) \end{aligned} \tag{3.13}$$

When $y(n) = x(n)$, the autocorrelation function (ACF) is obtained; it is denoted $r_{xx}(p)$ or, more simply, $r(p)$, if there is no ambiguity. The following properties hold:

$$r(p) = r(-p), \qquad |r(p)| \leqslant r(0) \tag{3.14}$$

For $x(n)$ a zero mean white noise with power σ_x^2,

$$r(p) = \sigma_x^2 \delta(p) \tag{3.15}$$

and for a sine wave with amplitude S and radial frequency ω_0,

$$r(p) = \frac{S^2}{2} \cos p\omega_0 \tag{3.16}$$

The ACF is periodic with the same period. Note that from (3.15) and (3.16) a simple and efficient noise-elimination technique can be worked out to retrieve periodic components, by just dropping the terms $r(p)$ for small p in the noisy signal ACF.

The Fourier transform of the ACF is the signal spectrum. For the cross-correlation $r_{yx}(p)$ it is the cross spectrum $S_{yx}(f)$.

Considering the Fourier transform $X(f)$ and $Y(f)$ of the sequences $x(n)$, and $y(n)$, equation (3.7) yields

$$S_{yx}(f) = Y(f)\bar{X}(f) \tag{3.17}$$

where $\bar{X}(f)$ is the complex conjugate of $X(f)$.

The frequency domain correspondence for the set of relationships (3.13) is found by introduction of the filter transfer function:

$$H(f) = \frac{Y(f)}{X(f)} = \frac{Y(f)\bar{X}(f)}{|X(f)|^2} \tag{3.18}$$

Now

$$S_{yx}(f) = S_{xx}(f)H(f)$$
$$S_{xy}(f) = S_{xx}(f)\bar{H}(f)$$
$$S_{yy}(f) = S_{xx}(f)|H(f)|^2 \tag{3.19}$$

The spectra and cross spectra can be used to compute ACF and cross-correlation function, through Fourier series development, although it is often the other way round in practice.

Most of the above definitions and properties can be extended to complex signals. In that case the cross-correlation function (3.6) becomes

$$r_{yx}(p) = E[y(n)\bar{x}(n - p)] \tag{3.20}$$

In the preceding chapter the relations between correlation functions and model coefficients have been established for MA, AR, and ARMA stationary signals. In practice, the correlation coefficients must be estimated from available data.

3.2 ESTIMATION OF CORRELATION FUNCTIONS

The signal data may be available as a finite-length sequence or as an infinite sequence, as for stationary signals. In any case, due to the limitations in processing means, the estimations have to be restricted to a finite time window. Therefore a finite set of N_0 data is assumed to be used in estimations.

A first method to estimate the ACF $r(p)$ is to calculate $r_1(p)$ by

$$r_1(p) = \frac{1}{N_0} \sum_{n=p+1}^{N_0} x(n)x(n - p) \tag{3.21}$$

The estimator is biased because

$$E[r_1(p)] = \frac{N_0 - p}{N_0} r(p) \tag{3.22}$$

However, the bias approaches zero as N_0 approaches infinity, and $r_1(p)$ is asymptotically unbiased.

An unbiased estimator is

$$r_2(p) = \frac{1}{N_0 - p} \sum_{n=p+1}^{N_0} x(n)x(n-p) \tag{3.23}$$

In order to limit the range of the estimations, which are exploited subsequently, we introduce a normalized form, given for the unbiased estimator by

$$r_{n2}(p) = \frac{\displaystyle\sum_{n=p+1}^{N_0} x(n)x(n-p)}{\left[\displaystyle\sum_{n=p+1}^{N_0} x^2(n) \sum_{n=p+1}^{N_0} x^2(n-p)\right]^{1/2}} \tag{3.24}$$

The variance is

$$\mathrm{var}\{r_{n2}(p)\} = E[r_{n2}^2(p)] - E^2[r_{n2}(p)] \tag{3.25}$$

and it is not easily evaluated in the general case because of the nonlinear functions involved. However, a linearization method, based on the first derivatives of Taylor expansions, can be applied [1]. For uncorrelated pairs in equation (3.24), we obtain

$$\mathrm{var}\{r_{n2}(p)\} \approx \frac{[1 - r_n^2(p)]^2}{N_0 - p} \tag{3.26}$$

$$r_n(p) = \frac{E[x(n)x(n-p)]}{[E[x^2(n)]E[x^2(n-p)]]^{1/2}} \tag{3.27}$$

is the theoretical normalized ACF.

Thus, the variance also approaches zero as the number of samples approaches infinity, and $r_{n2}(p)$ is a consistent estimate.

The calculation of the estimator according to (3.24) is a demanding operation for large N_0. In a number of applications, like radiocommunications, the correlation calculation may be the first processing operation, and it has to be carried out on high-speed data. Therefore it is useful to have less costly methods available. Such methods exist for Gaussian random signals, and they can be applied as well to many other signals.

The following property is valid for a zero mean Gaussian signal $x(n)$:

$$r(p) = \frac{\pi}{2} r_{yx}(p)r_{yx}(0) \tag{3.28}$$

where

$$y(n) = \mathrm{sign}\{x(n)\}, \qquad y(n) = \pm 1$$

Hence the ACF estimate is

$$r_3(p) = c \frac{1}{N_0 - p} \sum_{n=p+1}^{N_0} x(n - p)\mathrm{sign}\{x(n)\} \tag{3.29}$$

where

$$c = \frac{\pi}{2} r_{yx}(0) = \frac{\pi}{2N_0} \sum_{n=1}^{N_0} |x(n)|$$

In normalized form, we have

$$r_{n3}(p) = \frac{N_0}{N_0 - p} \frac{\displaystyle\sum_{n=p+1}^{N_0} x(n - p)\mathrm{sign}\{x(n)\}}{\displaystyle\sum_{n=1}^{N_0} |x(n)|} \tag{3.29a}$$

A multiplication-free estimate is obtained [2], which is sometimes called the hybrid sign correlation or relay correlation. For uncorrelated pairs and p small with respect to N_0, the variance is approximately [3]

$$\mathrm{var}\{r_{n3}(p)\} \approx \frac{1}{N_0}\left[\frac{\pi}{2} - 2r_n(p)\mathrm{Arcsin}[r_n(p)]\right.$$

$$\left. + \frac{\pi}{2} r_n^2(p) - 2r_n^2(p)\sqrt{1 - r_n^2(p)}\right] \tag{3.30}$$

This estimator is also consistent.

The simplification process can be carried one step further, through the polarity coincidence technique, which relies on the following property of zero mean Gaussian signals:

$$r(p) = r(0)\sin\left[\frac{\pi}{2} E[\mathrm{sign}\{x(n)x(n - 1)\}]\right] \tag{3.31}$$

The property reflects the fact that a Gaussian function is determined by its zero crossings, except for a constant factor. Hence we have the simple estimate

$$r_{n4}(p) = \sin\left(\frac{\pi}{2} \frac{1}{N_0 - p} \sum_{n=p+1}^{N_0} \mathrm{sign}\{x(n)x(n - p)\}\right) \tag{3.32}$$

which is called the sign or polarity coincidence correlator. Its variance can be approximated for N_0 large by [4]

$$\mathrm{var}\{r_{n4}(p)\} \approx \frac{1}{N_0} \frac{\pi^2}{4} [1 - r_n^2(p)]\left[1 - \left(\frac{2}{\pi} \mathrm{Arcsin}\, r(p)\right)^2\right] \tag{3.32a}$$

The variances of the three normalized estimators r_{n2}, r_{n3}, and r_{n4} are shown in Figure 3.1 versus the theoretical autocorrelation (AC) $r(p)$. Clearly the lower computational cost of the hybrid sign and polarity coincidence correlators is paid for by a lower accuracy.

The performance evaluation of the estimators has been carried out under the assumption of uncorrelated sample pairs, which is no longer valid when the estimate is extracted on the basis of a single realization of a correlated process, i.e., a single data record. The evaluation can be carried out by considering the correlation between pairs of samples; it shows a degradation in performance [5].

For example, if the sequence $x(n)$ is a bandlimited noise with bandwidth B, the following bound can be derived for a large number of data N_0 [6]:

$$\text{var}\{r_2(p)\} \leqslant \frac{r^2(0)}{B(N_0 - p)} \tag{3.33}$$

The worst case occurs when the bandwidth B is half the sampling frequency; then $x(n)$ is a white noise, and the data are independent, which

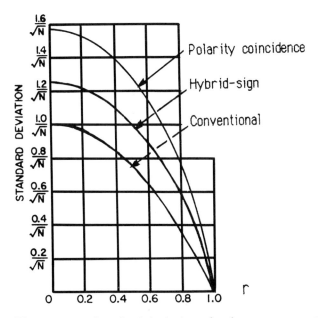

Figure 3.1 Standard deviation of estimators versus theoretical autocorrelation for large number of data N.

leads to

$$\operatorname{var}\{r_2(p)\} \leqslant \frac{2r^2(0)}{N_0 - p} \tag{3.33a}$$

This bound is compatible with estimation (3.26). Anyway the estimator for correlated data is still consistent for fixed p.

Furthermore, the Gaussian hypothesis is also needed for the hybrid sign and polarity coincidence estimators. So, these estimators have to be used with care in practice. An example of performance comparison is presented in Figure 3.2 for a speech sentence of 1.25 s corresponding to $N_0 = 10,000$ samples.

In spite of noticeable differences between conventional and polarity coincidence estimators for small AC values, the general shape of the function is the same for both.

Concerning correlated data, an important aspect of simplified correlators applied to real-life data is that they may attenuate or even cancel small useful components. Therefore, if small critical components in the signal have to be kept, the correlation operation accuracy in equipment must be determined to ensure that they are kept. Otherwise, reduced word lengths, such as 8 bits or 4 bits or even less, can be employed.

The first estimator introduced, $r_1(p)$, is just a weighted version of $r_2(p)$; hence its variance is

$$\operatorname{var}\{r_1(p)\} = \operatorname{var}\left\{\frac{N_0 - p}{N_0} r_2(p)\right\} = \left(\frac{N_0 - p}{N_0}\right)^2 \operatorname{var}\{r_2(p)\} \tag{3.34}$$

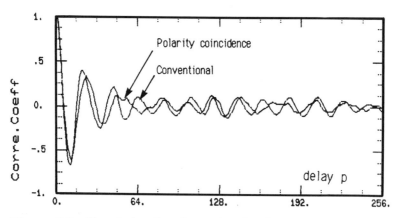

Figure 3.2 Correlation function estimation for a speech sentence.

The estimator $r_1(p)$ is biased, but it has a smaller variance than $r_2(p)$. It is widely used in practice.

So far, stationarity has been assumed. However, when the signal is just short-term stationary, the estimation has to be carried out on a compatible short-time window. An updated estimation is obtained every time if the window slides on the time axis; it is a sliding window technique, in which the oldest datum is discarded as a new datum enters the summation.

An alternative, more convenient, and widely used approach is recursive estimation.

3.3 RECURSIVE ESTIMATION

The time window estimation, according to (3.21) or (3.23), is a finite impulse response (FIR) filtering, which can be approximated by an infinite impulse response (IIR) filtering method. The simplest IIR filter is the first-order low-pass section, defined by

$$y(n) = x(n) + by(n - 1), \qquad 0 < b < 1 \tag{3.35}$$

Before investigating the properties of the recursive estimator, let us consider the simple case where the input sequence $x(n)$ is the sum of a constant m and a zero mean white noise $e(n)$ with power σ_e^2. Furthermore, if $y(n) = 0$ for $n < 0$, then

$$y(n) = m\frac{1 - b^{n+1}}{1 - b} + \sum_{i=0}^{n} b^i e(n - i) \tag{3.36}$$

Taking the expectation gives

$$E[y(n)] = m\frac{1 - b^{n+1}}{1 - b} \tag{3.37}$$

Therefore, an estimation of the input mean m is provided by the product $(1 - b)y(n)$, that is by the first-order section with z-transfer function:

$$H(z) = \frac{1 - b}{1 - bz^{-1}} \tag{3.38}$$

The noise power σ_0^2 at the output of such a filter is

$$\sigma_0^2 = \sigma_e^2\frac{1 - b}{1 + b} \tag{3.39}$$

Consequently, the input noise is all the more attenuated than b is close to unity. Taking $b = 1 - \delta,\ 0 < \delta \ll 1$ yields

$$\sigma_0^2 \approx \sigma_e^2\frac{\delta}{2} \tag{3.40}$$

The diagram of the recursive estimator is shown in Figure 3.3. The corresponding recursive equation is

$$M(n) = (1 - \delta)M(n - 1) + \delta x(n) \tag{3.41}$$

According to equation (3.37) the estimation is biased and the duration needed to reach a good estimation is inversely proportional to δ. In digital filter theory, a time constant τ can be defined by

$$e^{-1/\tau} = b \tag{3.42}$$

which for b close to 1, leads to

$$\tau \approx \frac{1}{1-b} = \frac{1}{\delta} \tag{3.43}$$

In order to relate recursive and window estimations, we define an equivalence. The FIR estimator

$$y(n) = \frac{1}{N_0} \sum_{i=0}^{N_0-1} x(n - i) \tag{3.44}$$

which is unbiased, yields the output noise power

$$(\sigma_0')^2 = \frac{\sigma_e^2}{N_0} \tag{3.45}$$

Comparing with (3.40), we get

$$2\tau \approx N_0 \tag{3.46}$$

The recursive estimator can be considered equivalent to a window estimator whose width is twice the time constant.

For example, consider the recursive estimation of the power of a white Gaussian signal $x(n)$, the true value being σ_x^2. The input to the recursive

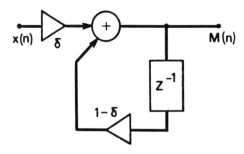

Figure 3.3 Recursive estimator.

estimator, $x^2(n)$, can be viewed as the sum of the constant $m = \sigma_x^2$ and a zero mean white noise, with variance

$$\sigma_e^2 = E[x^4(n)] - \sigma_x^4 = 2\sigma_x^4 \tag{3.47}$$

The standard deviation of the output, ΔP, is

$$\Delta P = \sigma_x^2 \sqrt{\delta} \tag{3.48}$$

and the relative error on the estimated power is $\sqrt{\delta}$.

Recursive estimation techniques can be applied to the ACF and to cross-correlation coefficients; a typical example is the lattice adaptive filter.

Once the ACF has been estimated, it can be used for analysis or any further processing.

3.4 THE AUTOCORRELATION MATRIX

Often in signal analysis or adaptive filtering, the ACF appears under the form of a square matrix, called the autocorrelation matrix.

The $N \times N$ AC matrix R_{xx} of the real sequence $x(n)$ is defined by

$$R_{xx} = \begin{bmatrix} r(0) & r(1) & \cdots & r(N-1) \\ r(1) & r(0) & \cdots & r(N-2) \\ \vdots & \vdots & \ddots & \vdots \\ r(N-1) & r(N-2) & \cdots & r(0) \end{bmatrix} \tag{3.49}$$

It is a symmetric matrix and $R_{xx}^t = R_{xx}$. For complex data the definition is slightly different:

$$R_{xx} = \begin{bmatrix} r(0) & r(1) & \cdots & r(N-1) \\ r(-1) & r(0) & \cdots & r(N-2) \\ \vdots & \vdots & \ddots & \vdots \\ r[-(N-1)] & r[-(N-2)] & \cdots & r(0) \end{bmatrix} \tag{3.50}$$

Since $r(-p)$ is the complex conjugate of $r(p)$, the matrix is Hermitian; that is,

$$R_{xx}^* = R_{xx} \tag{3.51}$$

where "*" denotes transposition and complex conjugation.

To illustrate how naturally the AC matrix appears, let us consider an FIR filtering operation with N coefficients:

$$y(n) = \sum_{i=0}^{N-1} h_i x(n-i) \tag{3.52}$$

In vector notation (3.52) is

$$y(n) = H^t X(n) = X^t(n)H$$

The output power is

$$E[y^2(n)] = E[H^t X(n)X^t(n)H] = H^t R_{xx} H \tag{3.53}$$

The inequality

$$H^t R_{xx} H \geqslant 0 \tag{3.54}$$

is valid for any coefficient vector H and characterizes positive semidefinite or nonnegative definite matrices [7]. A matrix is positive definite if

$$H^t R_{xx} H > 0 \tag{3.55}$$

The matrix R_{xx} is also symmetrical about the secondary diagonal; hence it is said to be doubly symmetric or persymmetric. Define by J_N the $N \times N$ co-identity matrix, which acts as a reversing operator on vectors and shares a number of properties with the identity matrix I_N:

$$I_N = \begin{bmatrix} 1 & 0 & \cdots & 0 & 0 \\ 0 & 1 & \cdots & 0 & 0 \\ \vdots & \vdots & & \vdots & \vdots \\ 0 & 0 & \cdots & 1 & 0 \\ 0 & 0 & \cdots & 0 & 1 \end{bmatrix}, \quad J_N = \begin{bmatrix} 0 & 0 & \cdots & 0 & 1 \\ 0 & 0 & \cdots & 1 & 0 \\ \vdots & \vdots & & \vdots & \vdots \\ 0 & 1 & \cdots & 0 & 0 \\ 1 & 0 & \cdots & 0 & 0 \end{bmatrix} \tag{3.56}$$

The double symmetry property is expressed by

$$R_{xx} J_N = J_N R_{xx} \tag{3.57}$$

Autocorrelation matrices have an additional property with respect to doubly symmetric matrices, namely their diagonal entries are identical; they are said to have a Toeplitz form or, in short, to be Toeplitz. This property is crucial and leads to drastic simplifications in some operations and particularly the inverse calculation, needed in the normal equations introduced in Section 1.4, for example. Examples of AC matrices can be given for MA and AR signals. If $x(n)$ is an MA signal, generated by filtering a white noise with power σ_e^2 by an FIR filter having $P < N/2$ coefficients, then R_{xx} is a band matrix. For $P = 2$,

$$x(n) = h_0 e(n) + h_1 e(n-1) \tag{3.58}$$

Using the results of Section 2.5 yields

$$R_{MA1} = \sigma_e^2 \begin{bmatrix} h_0^2 + h_1^2 & h_0 h_1 & 0 & \cdots & 0 & 0 \\ h_0 h_1 & h_0^2 + h_1^2 & h_0 h_1 & \cdots & 0 & 0 \\ 0 & h_0 h_1 & h_0^2 + h_1^2 & \cdots & 0 & 0 \\ \vdots & \vdots & \vdots & & \vdots & \vdots \\ 0 & 0 & 0 & \cdots & h_0^2 + h_1^2 & h_0 h_1 \\ 0 & 0 & 0 & \cdots & h_0 h_1 & h_0^2 + h_1^2 \end{bmatrix} \quad (3.59)$$

Similarly, for a first-order AR process, we have

$$x(n) = ax(n - 1) + e(n)$$

The matrix takes the form

$$R_{AR1} = \frac{\sigma_e^2}{1 - a^2} \begin{bmatrix} 1 & a & a^2 & \cdots & a^{N-1} \\ a & 1 & a & \cdots & a^{N-2} \\ a^2 & a & 1 & \cdots & a^{N-3} \\ \vdots & \vdots & \vdots & & \vdots \\ a^{N-1} & a^{N-2} & a^{N-3} & \cdots & 1 \end{bmatrix} \quad (3.60)$$

The inverse of the AR signal AC matrix is a band matrix because the inverse of the filter used to generate the AR sequence is an FIR filter. In fact, except for edge effects, it is an MA matrix.

Adjusting the first entry gives for the first-order case

$$R_{AR1}^{-1} = \frac{1}{\sigma_e^2} \begin{bmatrix} 1 & -a & 0 & \cdots & 0 \\ -a & 1 + a^2 & -a & \cdots & 0 \\ 0 & -a & 1 + a^2 & \cdots & 0 \\ \vdots & \vdots & \vdots & 1 + a^2 & -a \\ 0 & 0 & 0 & -a & 1 \end{bmatrix} \quad (3.61)$$

This is an important result, which is extended and exploited in subsequent sections.

Since AC matrices often appear in linear systems, it is useful, before further exploring their properties, to briefly review linear systems.

3.5 SOLVING LINEAR EQUATION SYSTEMS

Let us consider a set of N_0 linear equations represented by the matrix equation

$$MH = Y \quad (3.62)$$

The column vector Y has N_0 elements. The unknown column vector H has N elements, and the matrix M has N_0 rows and N columns. Depending on the respective values of N_0 and N, three cases can be distinguished. First, when $N_0 = N$, the system is exactly determined and the solution is

$$H = M^{-1}Y \qquad (3.63)$$

Second, when $N_0 > N$, the system is overdetermined because there are more equations than unknowns. A typical example is the filtering of a set of N_0 data $x(n)$ by an FIR filter whose N coefficients must be calculated so as to make the output set equal to the given vector Y:

$$\begin{bmatrix} x(0) & 0 & \cdots & 0 \\ x(1) & x(0) & \cdots & 0 \\ \vdots & \vdots & & \vdots \\ x(N-1) & x(N-2) & \cdots & x(0) \\ \vdots & \vdots & & \vdots \\ x(N_0-1) & x(N_0-2) & \cdots & x(N_0-N) \end{bmatrix} \begin{bmatrix} h_0 \\ h_1 \\ \vdots \\ h_{N-1} \end{bmatrix} = \begin{bmatrix} y(0) \\ y(1) \\ \vdots \\ y(N_0-1) \end{bmatrix} \qquad (3.64)$$

A solution in the LS sense is found by minimizing the scalar J:

$$J = (Y - MH)^t(Y - MH)$$

Through derivation with respect to the entries of the vector H, the solution is found to be

$$H = (M^tM)^{-1}M^tY \qquad (3.65)$$

Third, when $N_0 < N$, the system is underdetermined and there are more unknowns than equations. The solution is then

$$H = M^t(MM^t)^{-1}Y \qquad (3.66)$$

The solution of an exactly determined system must be found in all cases. The matrix (M^tM) is symmetrical, and standard algorithms exist to solve equation systems based on such matrices, which are assumed positive definite. The Cholesky method uses a triangular factorization of the matrix and needs about $N^3/3$ multiplications; the subroutine is given in Annex 3.1.

Iterative techniques can also be used to solve equation (3.62). The matrix M can be decomposed as

$$M = D + E$$

where D is a diagonal matrix and E is a matrix with zeros on the main diagonal. Now

$$H = D^{-1}Y - D^{-1}EH$$

and an iterative procedure is as follows:

$$H_0 = D^{-1}Y$$
$$H_1 = D^{-1}Y - D^{-1}EH_0$$
$$\cdots\cdots\cdots\cdots\cdots\cdots\cdots\cdots\cdots\cdots \tag{3.67}$$
$$H_{n+1} = D^{-1}Y - D^{-1}EH_n$$

The decrement after n iterations is

$$H_{n+1} - H_n = -(D^{-1}E)^{n+1}D^{-1}Y$$

The procedure may be stopped when the norm of the vector $H_{n+1} - H_n$ falls below a specified value.

3.6 EIGENVALUE DECOMPOSITION

The eigenvalue decomposition of an AC matrix leads to the extraction of the basic components of the corresponding signal [8–11]—hence its significance.

The eigenvalues λ_i and eigenvectors V_i of the $N \times N$ matrix R are defined by

$$RV_i = \lambda_i V_i, \qquad 0 \leqslant i \leqslant N - 1 \tag{3.68}$$

If the matrix R now denotes the AC matrix R_{xx}, it is symmetric for real signals and Hermitian for complex signals because

$$\overline{\lambda} V_i^* V_i = (V_i^* R V_i)^* = \lambda V_i^* V_i \tag{3.69}$$

The eigenvalues are the real solutions of the characteristic equation

$$\det(R - \lambda I_N) = 0 \tag{3.70}$$

The identity matrix I_N has $+1$ as single eigenvalue with multiplicity N, and the co-identity matrix J_N has ± 1.

The relations between the zeros and coefficients of polynomials yield the following important results:

$$\det R = \prod_{i=0}^{N-1} \lambda_i \tag{3.71}$$

$$Nr(0) = N\sigma_x^2 = \sum_{i=0}^{N-1} \lambda_i \tag{3.72}$$

That is, if the determinant of the matrix is nonzero, each eigenvalue is nonzero and the sum of the eigenvalues is equal to N times the signal power. Furthermore, since the AC matrix is nonnegative definite, all the eigenvalues are nonnegative:

$$\lambda_i \geqslant 0, \qquad 0 \leqslant i \leqslant N - 1 \tag{3.73}$$

Once the eigenvalues have been found, the eigenvectors are obtained by solving equations (3.68). The eigenvectors associated with different eigenvalues of a symmetric matrix are orthogonal because of the equality

$$V_i^t V_j = \frac{1}{\lambda_i} V_i^t R V_j = \frac{\lambda_j}{\lambda_i} V_i^t V_j \tag{3.74}$$

When all the eigenvalues are distinct, the eigenvectors make an orthonormal base and the matrix can be diagonalized as

$$R = M^t \Lambda M \tag{3.75}$$

with M the $N \times N$ orthonormal modal matrix made of the N eigenvectors, and Λ the diagonal matrix of the eigenvalues:

$$M = [V_0 \quad V_1 \quad \cdots \quad V_{N-1}], \qquad M^t = M^{-1},$$
$$\Lambda = \text{diag}(\lambda_0, \lambda_1, \ldots, \lambda_{N-1}) \tag{3.76}$$

For example, take a periodic signal $x(n)$ with period N. The AC function is also periodic with the same period and is symmetrical. The AC matrix is a circulant matrix, in which each row is derived from the preceding one by shifting. Now, if $|S(k)|^2$ denotes the signal power spectrum and T_N the discrete Fourier transform (DFT) matrix of order N:

$$T_N = \begin{bmatrix} 1 & 1 & \cdots & 1 \\ 1 & w & \cdots & w^{N-1} \\ \vdots & \vdots & & \vdots \\ 1 & w^{N-1} & \cdots & w^{(N-1)(N-1)} \end{bmatrix}, \qquad w = e^{-j2\pi/N} \tag{3.77}$$

it can be directly verified that

$$R T_N = T_N \, \text{diag}(|S(k)|^2) \tag{3.78}$$

Due to the periodicity assumed for the AC function, the same is also true for the discrete cosine Fourier transform matrix, which is real and defined by

$$T_{cN} = \tfrac{1}{2}[T_N + T_N^*] \tag{3.77a}$$

Thus

$$R T_{cN} = T_{cN} \, \text{diag}(|S(k)|^2) \tag{3.78a}$$

and the N column vectors of T_{cN} are the N orthogonal eigenvectors of the matrix R. Then

$$R = T_{cN} \, \text{diag}(|S(k)|^2) T_{cN} \tag{3.79}$$

So, it appears that the eigenvalues of the AC matrix of a periodic signal are the power spectrum; and the eigenvector matrix is the discrete cosine Fourier transform matrix.

However, the diagonalization of an AC matrix is not always unique. Let us assume that the N cisoids in the signal $x(n)$ have frequencies ω_i which are no longer multiples of $2\pi/N$:

$$x(n) = \sum_{i=1}^{N} S_i e^{jn\omega_i} \tag{3.80}$$

The ACF is

$$r(p) = \sum_{i=1}^{N} |S_i|^2 e^{jp\omega_i} \tag{3.81}$$

and the AC matrix can be expressed as

$$R = M^* \operatorname{diag}(|S_i|^2)M \tag{3.82}$$

with

$$M = \begin{bmatrix} 1 & e^{j\omega_1} & \cdots & e^{j(N-1)\omega_1} \\ 1 & e^{j\omega_2} & \cdots & e^{j(N-1)\omega_2} \\ \vdots & \vdots & & \vdots \\ 1 & e^{j\omega_N} & \cdots & e^{j(N-1)\omega_N} \end{bmatrix}$$

But the column vectors in M^* are neither orthogonal nor eigenvectors of R, as can be verified. If there are K cisoids with $K < N$, M becomes a $K \times N$ rectangular matrix and factorization (3.82) is still valid. But then the signal space dimension is restricted to the number of cisoids K, and $N - K$ eigenvalues are zero.

The white noise is a particularly simple case because $R = \sigma_e^2 I_N$ and all the eigenvalues are equal. If that noise is added to the useful signal, the matrix $\sigma_e^2 I_N$ is added to the AC matrix and all the eigenvalues are increased by σ_e^2.

Example

Consider the sinusoid in white noise

$$x(n) = \sqrt{2} \sin(n\omega) + e(n)$$

The AC function is

$$r(p) = \cos(p\omega) + \sigma_e^2 \delta(p)$$

The eigenvalues of the 3×3 AC matrix are

$$R = \begin{bmatrix} r(0) & r(1) & r(2) \\ r(1) & r(0) & r(1) \\ r(2) & r(1) & r(0) \end{bmatrix}, \qquad \begin{aligned} \lambda_1 &= \sigma_e^2 + 1 - \cos 2\omega \\ \lambda_2 &= \sigma_e^2 + 2 + \cos 2\omega \\ \lambda_3 &= \sigma_e^2 \end{aligned}$$

and the unit norm eigenvectors are

$$U_1 = \frac{1}{\sqrt{2}} \begin{bmatrix} 1 \\ 0 \\ -1 \end{bmatrix}, \qquad U_2 = \frac{1}{(1 + 2\cos^2 \omega)^{1/2}} \begin{bmatrix} \cos \omega \\ 1 \\ \cos \omega \end{bmatrix},$$

$$U_3 = \frac{1}{(2 + 4\cos^2 \omega)^{1/2}} \begin{bmatrix} 1 \\ -2\cos \omega \\ 1 \end{bmatrix}$$

The variations of the eigenvalues with frequency are shown in Figure 3.4 [12].

Once a set of N orthogonal eigenvectors has been obtained, any signal vector $X(n)$ can be expressed as a linear combination of these vectors, which can be scaled to have a unit norm denoted by U_i:

$$X(n) = \sum_{i=0}^{N-1} \alpha_i(n)U_i \tag{3.83}$$

The coefficients $\alpha_i(n)$ are the projections of $X(n)$ on the vectors U_i. Another expression of the AC matrix can then be obtained, assuming real signals:

$$R = E[X(n)X^t(n)] = \sum_{i=0}^{N-1} E[\alpha_i^2(n)]U_i U_i^t \tag{3.84}$$

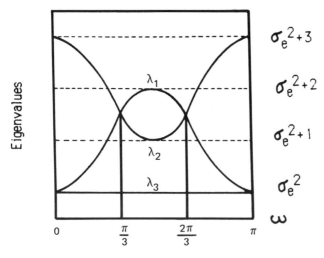

Figure 3.4   Variation of eigenvalues with frequency.

The definition of the eigenvalues yields

$$E[\alpha_i^2(n)] = \lambda_i \qquad (3.85)$$

Equation (3.85) provides an important interpretation of the eigenvalues: they can be considered as the powers of the projections of the signal vectors on the eigenvectors. The subspace spanned by the eigenvectors corresponding to nonzero eigenvalues is called the signal subspace.

The eigenvalue or spectral decomposition is derived from (3.84):

$$R = \sum_{i=0}^{N-1} \lambda_i U_i U_i^t \qquad (3.86)$$

It is a fundamental result which shows the actual constitution of the signal and is exploited in subsequent sections. For signals in noise, expression (3.86) can serve to separate signal subspace and noise subspace.

Among the eigenparameters the minimum and maximum eigenvalues have special properties.

3.7 EIGENFILTERS

The maximization of the signal-to-noise ratio (SNR) through FIR filtering leads to an eigenvalue problem [13].

The output power of an FIR filter is given in terms of the input AC matrix and filter coefficients by equation (3.53):

$$E[y^2(n)] = H^t R H$$

If a white noise with power σ_e^2 is added to the input signal, the output SNR is

$$SNR = \frac{H^t R H}{H^t H \sigma_e^2} \qquad (3.87)$$

It is maximized by the coefficient vector H, which maximizes $H^t R H$, subject to the constraint $H^t H = 1$. Using a Lagrange multiplier, one has to maximize $H^t R H + \lambda(1 - H^t H)$ with respect to H, and the solution is $RH = \lambda H$. Therefore the optimum filter is the signal AC matrix eigenvector associated with the largest eigenvalue, and is called the maximum eigenfilter. Similarly, the minimum eigenfilter gives the smallest output signal power. These filters are characterized by their zeros in the complex plane.

The investigation of the eigenfilter properties begins with the case of distinct maximum or minimum eigenvalues; then it will be shown that the filter zeros are on the unit circle.

Let us assume that the smallest eigenvalue λ_{min} is zero. The corresponding eigenvector U_{min} is orthogonal to the other eigenvectors, which span the signal space. According to the harmonic decomposition of Section 3.11, the matrix R is the AC matrix of a set of $N - 1$ cisoids, and the signal space is also spanned by $N - 1$ vectors V_i:

$$V_i = \begin{bmatrix} 1 \\ e^{j\omega_i} \\ \vdots \\ e^{j(N-1)\omega_i} \end{bmatrix}, \qquad 1 \leqslant i \leqslant N - 1$$

Therefore U_{min} is orthogonal to all the vectors V_i, and the $N - 1$ zeros of the corresponding filter are $e^{j\omega_i}(1 \leqslant i \leqslant N - 1)$, and they are on the unit circle in the complex plane.

Now, if λ_{min} is not zero, the above development applies to the matrix $(R - \lambda_{min} \ I_N)$, which has the same eigenvectors as R, as can be readily verified.

For the maximum eigenvector U_{max} corresponding to λ_{max}, it is sufficient to consider the matrix $(\lambda_{max} \ I_N - R)$, which has all the characteristics of an AC matrix. Thus the maximum eigenfilter also has its zeros on the unit circle in the z-plane as soon as λ_{max} is distinct.

The above properties can be checked for the example in the preceding section, which shows, in particular, that the zeros for U_{min} are $e^{\pm j\omega}$.

Next, if the minimum (or maximum) eigenvalue is multiple, for example $N - K$, it means that the dimension of the signal space is K and that of the noise space is $N - K$. The minimum eigenfilters, which are orthogonal to the signal space, have K zeros on the unit circle, but the remaining $N - 1 - K$ zeros may or may not be on the unit circle.

We give an example for two simple cases of sinusoidal signals in noise. The AC matrix of a single cisoid, with power S^2, in noise is

$$R = \begin{bmatrix} S^2 + \sigma_e^2 & S^2 e^{j\omega} & \cdots & S^2 e^{j(N-1)\omega} \\ S^2 e^{-j\omega} & S^2 + \sigma_e^2 & \cdots & S^2 e^{j(N-2)\omega} \\ \vdots & \vdots & \ddots & \vdots \\ S^2 e^{-j(N-1)\omega} & S^2 e^{-j(N-2)\omega} & \cdots & S^2 + \sigma_e^2 \end{bmatrix} \qquad (3.88)$$

The eigenvalues are

$$\lambda_1 = NS^2 + \sigma_e^2, \qquad \lambda_i = \sigma_e^2, \qquad 2 \leqslant i \leqslant N$$

and the maximum eigenfilter is

$$U_{max} = \frac{1}{\sqrt{N}} \begin{bmatrix} 1 \\ e^{-j\omega} \\ \vdots \\ e^{-j(N-1)\omega} \end{bmatrix} \qquad (3.89)$$

The corresponding filter z-transfer function is

$$H_M(z) = \frac{1}{\sqrt{N}} \frac{z^N - e^{jN\omega}}{z - e^{j\omega}} e^{-j(N-1)\omega} \tag{3.90}$$

and the $N - 1$ roots

$$z_i = e^{j(\omega + 2\pi i/N)}, \qquad 1 \leqslant i \leqslant N - 1$$

are spread on the unit circle, except at the frequency ω. $H_M(z)$ is the conventional matched filter for a sine wave in noise.

Because the minimum eigenvalue is multiple, the unnormalized eigenvector $V_{\min}$ is

$$V_{\min} = \begin{bmatrix} -\sum_{i=2}^{N} v_i e^{j(i-1)\omega} \\ v_2 \\ \vdots \\ v_N \end{bmatrix} \tag{3.91}$$

where $N - 2$ arbitrary scalars v_i are introduced.

Obviously there are $N - 1$ linearly independent minimum eigenvectors which span the noise subspace. The associated filter z-transfer function is

$$H_m(z) = (z - e^{j\omega}) \sum_{i=2}^{N} v_i [z^{i-2} + z^{i-3}e^{j\omega} + \cdots + e^{j(i-2)\omega}] \tag{3.92}$$

One zero is at the cisoid frequency on the unit circle; the others may or may not be on that circle.

The case of two cisoids, with powers S_1^2 and S_2^2 in noise leads to more complicated calculations. The correlation matrix

$R =$

$$\begin{bmatrix} S_1^2 + S_2^2 + \sigma_e^2 & S_1^2 e^{j\omega_1} + S_2^2 e^{j\omega_2} & \vdots & S_1^2 e^{j(N-1)\omega_1} + S_2^2 e^{j(N-1)\omega_2} \\ S_1^2 e^{-j\omega_1} + S_2^2 e^{-j\omega_2} & S_1^2 + S_2^2 + \sigma_e^2 & \vdots & S_1^2 e^{j(N-2)\omega_1} + S_2^2 e^{j(N-2)\omega_2} \\ \vdots & \vdots & \ddots & \vdots \\ S_1^2 e^{-j(N-1)\omega_1} + S_2^2 e^{-j(N-1)\omega_2} & S_1^2 e^{-j(N-2)\omega_1} + S_2^2 e^{-j(N-2)\omega_2} & \cdots & S_1^2 + S_2^2 + \sigma_e^2 \end{bmatrix}$$

has eigenvalues [14]

$$\lambda_1 = \sigma_e^2 + \frac{N}{2} [S_1^2 + S_2^2] + \sqrt{\frac{N^2}{4}(S_1^2 - S_2^2)^2 + N^2 S_1^2 S_2^2 F^2(\omega_1 - \omega_2)}$$

$$\lambda_2 = \sigma_e^2 + \frac{N}{2}(S_1^2 + S_1^2) - \sqrt{\frac{N^2}{4}(S_1^2 - S_2^2)^2 + N^2 S_1^2 S_2^2 F^2(\omega_1 - \omega_2)}$$

$$\lambda_i = \sigma_e^2, \qquad 3 \leqslant i \leqslant N \tag{3.93}$$

$F(\omega)$ is the familiar function

$$F(\omega) = \frac{\sin(N\omega/2)}{N \sin(\omega/2)} \tag{3.94}$$

These results, when applied to a sinusoid amplitude A, $x(n) = A \sin(n\omega)$, yield

$$\lambda_{1,2} = \sigma_e^2 + \frac{A^2}{4}\left(N \pm \frac{\sin(N\omega)}{\sin \omega}\right) \tag{3.93a}$$

The extent to which λ_1 and λ_2 reflect the powers of the two cisoids depends on their respective frequencies, through the function $F(\omega)$, which corresponds to a length-N rectangular time window. For N large and frequencies far apart enough,

$$F(\omega_1 - \omega_2) \approx 0, \quad \lambda_1 = NS_1^2 + \sigma_e^2, \quad \lambda_2 = NS_2^2 + \sigma_e^2 \tag{3.95}$$

and the largest eigenvalues represent the cisoid powers.

The z-transfer function of the minimum eigenfilters is

$$H_m(z) = (z - e^{j\omega_1})(z - e^{j\omega_2})P(z) \tag{3.96}$$

with $P(z)$ a polynomial of degree less than $N - 2$. Two zeros are on the unit circle at the cisoid frequencies; the other zeros may or may not be on that circle.

To conclude: for a given signal the maximum eigenfilter indicates where the power is in the frequency domain, and the zeros of the minimum eigenvalue filter give the exact frequencies associated with the harmonic decomposition of that signal.

Together, the maximum and minimum eigenfilters constitute a powerful tool for signal analysis. However, in practice, the appeal of that technique is somewhat moderated by the computation load needed to extract the eigenparameters, which becomes enormous for large matrix dimensions. Savings can be obtained by careful exploitation of the properties of AC matrices [15]. For example, the persymmetry relation (3.57) yields, for any eigenvector V_i,

$$J_N R V_i = \lambda_i J_N V_i = R J_N V_i$$

Now, if λ_i is a distinct eigenvalue, the vectors V_i and $J_N V_i$ are colinear, which means that V_i is also an eigenvector of the co-identity matrix J_N, whose eigenvalues are ± 1. Hence the relation

$$J_N V_i = \pm V_i \tag{3.97}$$

holds.

The corresponding property of the AC matrix can be stated as follows: the eigenvectors associated with distinct eigenvalues are either symmetric or skew symmetric; that is, they verify (3.97).

Iterative techniques help manage the computation load. Before presenting such techniques, we give additional properties of extremal eigenvalues.

3.8 PROPERTIES OF EXTREMAL EIGENVALUES

In the design process of an adaptive filter it is sometimes enough to have simple evaluations of the extremal eigenvalues λ_{max} and λ_{min}. A loose bound for the maximum eigenvalue of an AC matrix, derived from (3.72), is

$$\lambda_{max} \leqslant N\sigma_x^2 \tag{3.98}$$

with σ_x^2 the signal power and $N \times N$ the matrix dimension. A tighter bound, valid for any square matrix R with entries r_{ij}, is known from matrix theory to be

$$\lambda_{max} \leqslant \max_j \sum_{i=0}^{N-1} |r_{ij}| \tag{3.99}$$

or

$$\lambda_{max} \leqslant \max_i \sum_{j=0}^{N-1} |r_{ij}|$$

To prove the inequality, single out the entry with largest magnitude in the eigenvector V_{max} and bound the elements of the vector RV_{max}.

In matrix theory, λ_{max} is called the spectral radius. It serves as a matrix norm as well as the right side of (3.99).

The Rayleigh quotient of R is defined by

$$R_a(V) = \frac{V^t R V}{V^t V}, \qquad V \neq 0 \tag{3.100}$$

As shown in the preceding section,

$$\lambda_{max} = \max_V R_a(V) \tag{3.101}$$

The diagonalization of R yields

$$R = M^{-1} \operatorname{diag}(\lambda_i) M \tag{3.102}$$

It is readily verified that

$$R^{-1} = M^{-1} \operatorname{diag}\left(\frac{1}{\lambda_i}\right) M \tag{3.103}$$

Therefore $\lambda_{\min}^{-1}$ is the maximum eigenvalue of R^{-1}. The condition number of R is defined by

$$\text{cond}(R) = \|R\| \, \|R^{-1}\| \tag{3.104}$$

If the matrix norm $\|R\|$ is $\lambda_{\max}$, then

$$\text{cond}(R) = \frac{\lambda_{\max}}{\lambda_{\min}} \tag{3.105}$$

The condition number is a matrix parameter which impacts the accuracy of the operations, particularly inversion [7]. It is critical in solving linear systems, and it is directly related to some stability conditions in LS adaptive filters.

In adaptive filters, sequences of AC matrices with increasing dimensions are sometimes encountered, and it is useful to know how the extremal eigenvalues vary with matrix dimensions for a given signal. Let us denote by $U_{\max,N}$ the maximum unit-norm eigenvector of the $N \times N$ AC matrix R_N. The maximum eigenvalue is

$$\lambda_{\max,N} = U_{\max,N}^t R_N U_{\max,N} \tag{3.106}$$

Now, because of the structure of the $(N+1) \times (N+1)$ AC matrix, the following equation is valid:

$$\lambda_{\max,N} = [U_{\max,N}^t, 0]$$

$$\times \begin{bmatrix} r(0) & r(1) & \cdots & r(N-1) & r(N) \\ r(1) & r(0) & \cdots & r(N-2) & r(N-1) \\ \vdots & \vdots & R_N & \vdots & \vdots \\ r(N-1) & r(N-2) & \cdots & r(0) & r(1) \\ \hline r(N) & r(N-1) & \cdots & r(1) & r(0) \end{bmatrix} \begin{bmatrix} U_{\max,N} \\ \hline 0 \end{bmatrix} \tag{3.107}$$

At the dimension $N+1$, $\lambda_{\max,N+1}$ is defined as the maximum of the product $U_{N+1}^t R_{N+1} U_{N+1}$ for any unit-norm vector U_{N+1}. The vector obtained by appending a zero to $U_{\max,N}$ is such a vector, and the following inequality is proven:

$$\lambda_{\max,N} \leqslant \lambda_{\max,N+1} \tag{3.108}$$

Also, considering the minimization procedure, we have

$$\lambda_{\min,N} \geqslant \lambda_{\min,N+1} \tag{3.109}$$

When N approaches infinity, $\lambda_{\max}$ and $\lambda_{\min}$ approach the maximum and the minimum, respectively, of the signal power spectrum, as shown in the next section.

3.9 SIGNAL SPECTRUM AND EIGENVALUES

According to relation (3.72), the eigenvalue extraction can be viewed as an energy decomposition of the signal. In order to make comparisons with the spectrum, we choose the following definition for the Fourier transform $Y(f)$ of the signal $x(n)$:

$$Y(f) = \lim_{n \to \infty} \frac{1}{\sqrt{2N+1}} \sum_{-N}^{N} x(n)e^{-j2\pi fn} \tag{3.110}$$

The spectrum is the square of the modulus of $Y(f)$:

$$S(f) = Y(f)\bar{Y}(f) = |Y(f)|^2 \tag{3.111}$$

When the summations in the above definition of $S(f)$ are rearranged, the correlation function $r(p)$ shows up, and the following expression is obtained:

$$S(f) = \sum_{p=-\infty}^{\infty} r(p)e^{-j2\pi fp} \tag{3.112}$$

Equation (3.112) is appropriate for random signals with statistics that are known or that can be measured or estimated.

Conversely, the spectrum $S(f)$ is a periodic function whose period is the reciprocal of the sampling frequency, and the correlation coefficients are the coefficients of the Fourier series expansion of $S(f)$:

$$r(p) = \int_{-1/2}^{1/2} S(f)e^{j2\pi pf} \, df \tag{3.113}$$

In practice, signals are time limited, and often a finite-duration record of N_0 data representing a single realization of the process is available. Then it is sufficient to compute the spectrum at frequencies which are integer multiples of $1/N_0$, since intermediate values can be interpolated, and the DFT with appropriate scaling factor

$$Y(k) = \frac{1}{\sqrt{N_0}} \sum_{n=0}^{N_0-1} x(n)e^{-j(2\pi/N)nk} \tag{3.114}$$

is employed to complete that task. The operation is equivalent to making the signal periodic with period N_0; the corresponding AC function is also periodic, with the same period, and the eigenvalues of the AC matrix are $|Y(k)|^2$, $0 \leqslant k \leqslant N_0 - 1$.

Now, the N eigenvalues λ_i of the $N \times N$ AC matrix R_N and their associated eigenvectors V_i are related by

$$\lambda_i V_i^* V_i = V_i^* R_N V_i \tag{3.115}$$

The right side is the power of the output of the eigenfilter; it can be expressed in terms of the frequency response by

$$V_i^* R_N V_i = \int_{-1/2}^{1/2} |H_i(f)|^2 S(f)\, df \tag{3.116}$$

The left side of (3.115) can be treated similarly, which leads to

$$\min_{-1/2 \leqslant f \leqslant 1/2} S(f) \leqslant \lambda_i \leqslant \max_{-1/2 \leqslant f \leqslant 1/2} S(f) \tag{3.117}$$

It is also interesting to relate the eigenvalues of the order N AC matrix to the DFT of a set of N data, which is easily obtained and familiar to practitioners. If we denote the set of N data by the vector X_N, the DFT, expressed by the matrix T_N (3.77), yields the vector Y_N:

$$Y_N = \frac{1}{\sqrt{N}} T_N X_N$$

The energy conservation relation is verified by taking the Euclidean norm of the complex vector Y_N:

$$\|Y_N\|^2 = Y_N^* Y_N = X_N^* X_N$$

Or, explicitly, we can write

$$\sum_{k=0}^{N-1} |Y(k)|^2 = \sum_{n=0}^{N-1} |x(n)|^2$$

The covariance matrix of the DFT output is

$$E[Y_N Y_N^*] = \frac{1}{N} T_N R T_N \tag{3.118}$$

The entries of the main diagonal are

$$E[|Y(k)|^2] = \frac{1}{N} V_k^* R_N V_k \tag{3.119}$$

with

$$V_k^* = [1, e^{j2\pi/N}, \ldots, e^{j(2\pi/N)(N-1)}]$$

From the properties of the eigenvalues, the following inequalities are derived:

$$\lambda_{\max} \geqslant \max_{0 \leqslant f \leqslant N-1} E[|Y(k)|^2]$$

$$\lambda_{\min} \leqslant \min_{0 \leqslant k \leqslant N-1} E[|Y(k)|^2] \tag{3.120}$$

These relations state that the DFT is a filtering operation and the output signal power is bounded by the extreme eigenvalues.

When the data vector length N approaches infinity, the DFT provides the exact spectrum, and, due to relations (3.117) and (3.120), the extreme eigenvalues λ_{min} and λ_{max} approach the extreme values of the signal spectrum [16].

3.10 ITERATIVE DETERMINATION OF EXTREMAL EIGENPARAMETERS

The eigenvalues and eigenvectors of an AC matrix can be computed by classical algebraic methods [7]. However, the computation load can be enormous, and it is useful to have simple and efficient methods to derive the extremal eigenparameters, particularly if real-time operation is envisaged.

A first, gradient-type approach is the unit-norm constrained algorithm [17]. It is based on minimization or maximization of the output power of a filter with coefficient vector $H(n)$, as shown in Figure 3.5, using the eigenfilter properties presented in Section 3.7. The output of the unit-norm filter is

$$e(n) = \frac{H^t(n)X(n)}{[H^t(n)H(n)]^{1/2}} \tag{3.121}$$

The gradient of $e(n)$ with respect to $H(n)$ is the vector

$$\nabla e(n) = \frac{1}{[H^t(n)H(n)]^{1/2}}\left[X(n) - e(n)\frac{H(n)}{[H^t(n)H(n)]^{1/2}} \right] \tag{3.122}$$

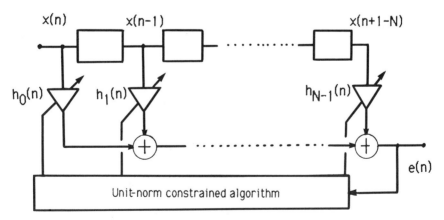

Figure 3.5 Unit-norm constrained adaptive filter.

Now, the power of the sequence $e(k)$ is minimized if the coefficient vector at time $n + 1$ is taken as

$$H(n + 1) = H(n) - \delta e(n)\nabla e(n) \tag{3.123}$$

where δ, the adaptation step size, is a positive constant. After normalization, the unit-norm filter coefficient vector is

$$\frac{H(n + 1)}{\|H(n + 1)\|} = \frac{1}{\|H(n + 1)\|}\left[H(n) - \frac{\delta e(n)}{\|H(n)\|}\left(X(n) - e(n)\frac{H(n)}{\|H(n)\|} \right)\right] \tag{3.124}$$

with

$$\|H(n)\| = [H^t(n)H(n)]^{1/2}$$

In the implementation, the expression contained in the brackets is computed first and the resulting coefficient vector is then normalized to unit norm. In that way there is no roundoff error propagation. The gradient-type approach leads to the eigenequation, as can be verified by rewriting equation (3.123):

$$H(n + 1) = H(n) - \frac{\delta}{\|H(n)\|}\left[X(n)X^t(n)\frac{H(n)}{\|H(n)\|} - e^2(n)\frac{H(n)}{\|H(n)\|} \right] \tag{3.125}$$

Taking the expectation of both sides, after convergence, yields

$$R\frac{H(\infty)}{\|H(\infty)\|} = E[e^2(n)]\frac{H(\infty)}{\|H(\infty)\|} \tag{3.126}$$

The output signal power is the minimum eigenvalue, and $H(\infty)$ is the corresponding eigenvector. Changing the sign in equation (3.123) leads to the maximum eigenvalue instead.

The step size δ controls the adaptation process. Its impact is analyzed in-depth in the next chapter.

Faster convergence can be obtained by minimizing the conventional cost function

$$J(n) = \sum_{p=1}^{n} W^{n-p}e^2(p), \qquad 0 \ll W \leqslant 1 \tag{3.127}$$

using a recursive LS algorithm [18]. The improvement in speed and accuracy is paid for by a significant increase in computation load. Furthermore, because of approximations made in the derivation, an initial guess for the coefficient vector sufficiently close to the exact solution is needed to achieve convergence. In contrast, a method based on the conjugate gradient technique converges for any initial guess in approximately M steps, where M is the number of independent eigenvalues of the AC matrix [19].

The method assumes that the AC matrix R is known, and it begins with an initial guess of the minimum eigenvector $H_{min}(0)$ and with an initial direction vector. The minimum eigenvalue is computed as $U^t_{min}(0)RU_{min}(0)$, and then successive approximations $U_{min}(k)$ are developed to minimize the cost function U^tRU in successive directions, which are R-conjugates,, until the desired minimum eigenvalue is found.

The FORTRAN subroutine is given in Annex 3.2.

3.11 ESTIMATION OF THE AC MATRIX

The AC matrix can be formed with the estimated values of the AC function. The bias and variance of the estimators impact the eigenparameters. The bias can be viewed as a modification of the signal. For example, windowing effects, as in (3.21), smear the signal spectrum and increase the dimension of the signal subspace, giving rise to spurious eigenvalues [20]. The effects of the estimator variance can be investigated by considering small random perturbations on the elements of the AC matrix. In adaptive filters using the AC matrix, explicitly or implicitly as in fast least squares (FLS) algorithms,, random perturbations come from roundoff errors and can affect, more or less independently, all the matrix entries.

Let us assume that the matrix R has all its eigenvalues distinct and is affected by a small perturbation matrix ΔR. The eigenvalues and vectors are explicit functions of the matrix elements, and their alteration can be developed in series; considering only the first term in the series, the eigenvalue equation with unit-norm vectors is

$$(R + \Delta R)(U_i + \Delta U_i) = (\lambda_i + \Delta\lambda_i)(U_i + \Delta U_i), \quad 0 \leqslant i \leqslant N - 1$$

$$(3.128)$$

Neglecting the second-order terms and premultiplying by U^t_i yields

$$\Delta\lambda_i = U^t_i\Delta RU_i \tag{3.129}$$

Due to the summing operation in the right side, the perturbation of the eigenvalue is very small, if the error matrix elements are i.i.d. random variables.

In order to investigate the eigenvector deviation, we introduce the normalized error matrix ΔE, associated with the diagonalization (3.75) of the matrix R:

$$\Delta E = \Lambda^{1/2}M\Delta RM^t\Lambda^{-1/2} \tag{3.130}$$

We can write (3.128), without the second-order terms and taking (3.129) into account,

$$(R - \lambda_iI_N)\Delta U_i = (U_iU^t_i - I_N)\Delta RU_i \tag{3.131}$$

After some algebraic manipulations, we get

$$\Delta U_i = \sum_{\substack{k=0 \\ k \neq i}}^{N-1} \frac{\sqrt{\lambda_i \lambda_k}}{\lambda_i - \lambda_k} \Delta E(k, i) U_k \tag{3.132}$$

where the $\Delta E(k, i)$ are the elements of the normalized error matrix.

Clearly, the deviation of the unit-norm eigenvectors U_i depends on the spread of the eigenvalues, and large deviations can be expected to affect eigenvectors corresponding to close eigenvalues [21].

Overall, the bias of the AC function estimator affects the AC matrix eigenvalues, and the variance of errors on the AC matrix elements affects the eigenvector directions.

In recursive algorithms, the following estimation appears:

$$R_N(n) = \sum_{p=1}^{n} W^{n-p} X(n) X^t(n) \tag{3.133}$$

where W is a weighting factor $(0 \ll W \leqslant 1)$ and $X(n)$ is the vector of the N most recent data. In explicit form, assuming $X(0) = 0$, we can write

$$R_N(n) = \begin{bmatrix} \sum\limits_{i=1}^{n} W^{n-i} x^2(i) & \sum\limits_{i=2}^{n} W^{n-i} x(i) x(i-1) & \cdots & \sum\limits_{i=N}^{n} W^{n-i} x(i) x(i-N+1) \\ \sum\limits_{i=2}^{n} W^{n-i} x(i-1) x(i) & \sum\limits_{i=2}^{n} W^{n-i} x^2(i-1) & \cdots & \\ \vdots & \vdots & \ddots & \vdots \\ \sum\limits_{i=N}^{n} W^{n-i} x(i) x(i-N+1) & \cdots & \cdots & \sum\limits_{i=N}^{n} W^{n-i} x^2(i-N+1) \end{bmatrix} \tag{3.134}$$

The matrix is symmetric. For large n it is almost doubly symmetric. Its expectation is

$$E[R_N(n)] = \frac{1}{1-W}$$

$$\times \begin{bmatrix} (1-W^n) r(0) & (1-W^{n-1}) r(1) & \cdots & (1-W)^{n-N+1} r(N-1) \\ (1-W^{n-1}) r(1) & (1-W^{n-1}) r(0) & \cdots & \vdots \\ \vdots & \vdots & \ddots & \\ (1-W^{n-N+1}) r(N-1) & \cdots & \cdots & (1-W^{n-N+1}) r(0) \end{bmatrix} \tag{3.135}$$

For large n

$$E[R_N(n)] \approx \frac{1}{1 - W} R \qquad (3.136)$$

In these conditions, the eigenvectors of $R_N(n)$ are those of R, and the eigenvalues are multiplied by $(1 - W)^{-1}$.

Example

$$x(n) = \sin\left(n\frac{\pi}{4}\right), \qquad n > 0$$

$$x(n) = 0, \qquad\qquad n \leqslant 0$$

The eigenvalues of the 8×8 AC matrix can be found from (3.93), in which

$$S_1 = S_2 = \frac{1}{2}, \qquad \omega_1 - \omega_2 = \frac{\pi}{2}$$

so that the term in the square root vanishes. Expression (3.93a) can be used as well, with $A = 1$:

$$\lambda_1 = \lambda_2 = 2, \qquad \lambda_3 = \cdots = \lambda_8 = 0$$

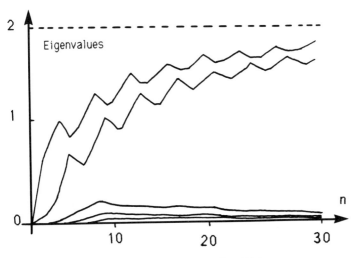

Figure 3.6 Eigenvalues of the matrix $R'(n)$.

The eigenvalues of the matrix $R'(n)$

$$R'(n) = \frac{R_N(n)}{2 \sum\limits_{i=1}^{n} W^{n-i} x^2(i)}, \qquad W = 0.95$$

are shown in Figure 3.6 for the first values of n. They approach the theoretical values as n increases.

3.12 EIGEN (KL) TRANSFORM AND APPROXIMATIONS

The projections of a signal vector X on the eigenvectors of the AC matrix form a vector

$$[\alpha] = M^t X \tag{3.137}$$

where M is the $N \times N$ orthonormal modal matrix defined in Section 3.6. The transform is unitary $(M^t M = I_N)$ and called the Karhunen-Loève (KL) transform. It is optimal for the class of all signals having the same second-order statistics [22]. Optimality means the efficiency of a transform in achieving data compression: the KL transform provides the optimum sets of data to represent signal vectors within a specified mean square error. For example, if M out of the N eigenvalues are zero or negligible, the N element data vectors can be represented by $N - M$ numbers only.

To prove that property we assume that the elements of the vector X are N centered random variables and look for the unitary transform I which best compresses the N elements of X into $M(M < N)$ elements out of the N elements y_i of the vector Y given by

$$Y = TX$$

The mean square error is

$$\text{MSE} = \sum_{i=M+1}^{N} E(y_i^2)$$

If the new vectors of T are designated by V_{Ti}^i. then

$$\text{MSE} = \sum_{i=M+1}^{N} V_{Ti}^t E[XX^t] V_{Ti}$$

The minimization of the above expression under the constraint of unit norm vectors, using Lagrange multipliers, leads to:

$$E[XX^t]V_{Ti} = \lambda_i V_{Ti}, \qquad M + 1 \leqslant i \leqslant N$$

The minimum is obtained if the scalars λ_i are the $N - M$ smallest eigenvalues of the matrix $E[XX^t]$ and V_{Ti} the corresponding unit norm eigenvectors. The minimum mean square error is

$$(MSE)_{min} = \sum_{i=M+1}^{N} \lambda_i$$

and, in fact, referring to Section 3.6, it is the amount of signal energy which is lost in the compression process.

However, compared with other unitary transforms like the DFT, the KL transform suffers from several drawbacks in practice. First, it has to be adjusted when the signal second-order statistics change. Second, as seen in the preceding sections, it requires a computation load proportional to N^2. Therefore it is helpful to find approximations which are sufficiently close for some signal classes and amenable to easy calculation through fast algorithms. Such approximations can be found for the first-order AR signal.

Because of the dual diagonalization relation

$$R^{-1} = M^t \Lambda^{-1} M \qquad (3.138)$$

the KL transform coefficients can be found from the inverse AC matrix as well. For the first-order unity-variance AR signal, the AC matrix is given by (3.60). The inverse (3.61) is a tridiagonal matrix, and the elements of the KL transform for N even are [23]

$$m_{kn} = c_n \sin\left[\omega_n\left(k - \frac{N+1}{2}\right) + n\frac{\pi}{2}\right] \qquad (3.139)$$

where c_n are normalization constants and ω_n are the positive roots of

$$\tan(N\omega) = -\frac{(1 - a^2)\sin\omega}{\cos\omega - 2a + a^2\cos\omega} \qquad (3.140)$$

The eigenvalues of R are

$$\lambda_i = \frac{1 - a^2}{(1 - 2a\cos\omega_i + a^2)^{1/2}}, \qquad 1 \leqslant i \leqslant N \qquad (3.141)$$

Now, the elements of the KL transform of a data vector are

$$\alpha_k = \sum_{n=1}^{N} c_n x(n)\sin\left[\omega_n\left(k - \frac{N+1}{2}\right) + n\frac{\pi}{2}\right] \qquad (3.142)$$

Due to the nonharmonicity of the sine terms, a fast algorithm is unavailable in calculating the above expressions, and N^2 computations are required. However, if R^{-1} is replaced by

$$R' = \frac{1}{1-a^2} \begin{bmatrix} 1+a^2 & -a & 0 & \cdots & 0 \\ -a & 1+a^2 & -a & \cdots & 0 \\ 0 & -a & 1+a^2 & \cdots & 0 \\ \vdots & \vdots & \vdots & \ddots & -a \\ 0 & 0 & 0 & -a & 1+a^2 \end{bmatrix} \qquad (3.143)$$

where R' differs by just the first and last entries in the main diagonal, the elements of the modal matrix become

$$m'_{kn} = \sqrt{\frac{2}{N+1}} \sin\left(\frac{kn\pi}{N+1}\right) \qquad (3.144)$$

and the eigenvalues are

$$\lambda'_i = 1 - 2\frac{a}{1+a^2} \cos\left(\frac{i\pi}{N+1}\right), \qquad i = 1,\ldots,N \qquad (3.145)$$

The elements of the corresponding transform of a data vector are

$$\alpha'_k = \sqrt{\frac{2}{N+1}} \sum_{n=1}^{N} x(n)\sin\left(\frac{nk\pi}{N+1}\right) \qquad (3.146)$$

This defines the discrete sine transform (DST), which can be implemented via a fast Fourier transform (FFT) algorithm.

Finally, for an order 1 AR signal, the DST is an efficient approximation of the KL transform.

Another approximation is the discrete cosine transform (DCT), defined as

$$\alpha''_0 = \frac{\sqrt{2}}{N} \sum_{n=1}^{N} x(n)$$

$$\alpha''_k = \frac{2}{N} \sum_{n=1}^{N} x(n)\cos\frac{(2n-1)k\pi}{2N}, \qquad 1 \leqslant k \leqslant N-1 \qquad (3.147)$$

It can be extended to two dimensions and is widely used in image processing [24].

3.13 SUMMARY

Estimating the ACF is often a preliminary step in signal analysis. After definition and basic properties have been introduced, efficient estimation techniques have been compared.

The AC matrix is behind adaptive filtering operations, and it is essential to be familiar with its major characteristics, which have been presented and illustrated by several simple examples. The eigenvalue decomposition has a profound meaning, because it leads to distinguishing between the signal or source space and the noise space, and to extracting the basic components. The filtering aspects help to understand and assess the main properties of eigenvalues and vectors. The extremal eigenparameters are especially crucial not only for the theory but also because they control adaptive filter performance and because they can provide superresolution analysis techniques.

Perturbations of the matrix elements, caused by bias and variance in the estimation process, affect the processing performance and particularly the operation of FLS algorithms. It has been shown that the bias can affect the eigenvalues and the variance causes deviations of eigenvectors. The KL transform is an illustrative application of the theoretical results.

EXERCISES

1. Use the estimators $r_1(p)$ and $r_2(p)$ to calculate the ACF of the sequence

$$x(n) = \sin\left(n\frac{\pi}{5}\right), \qquad 0 \leqslant n \leqslant 15$$

How are the deviations from theoretical values affected by the signal frequency?

2. For the symmetric matrix

$$R = \begin{bmatrix} 1.1 & -0.6 & 0.2 \\ -0.6 & 1.0 & -0.4 \\ 0.2 & -0.4 & 0.6 \end{bmatrix}$$

calculate R^2 and R^3 and the first element $r_{00}^{(4)}$ of the main diagonal of R^4. Compare the ratio $r_{00}^{(4)}/r_{00}^{(3)}$ with the largest eigenvalue λ_{max}.

Show that the following approximation is valid for a symmetric matrix R and N sufficiently large:

$$\left(\frac{R}{\lambda_{max}}\right)^{N+1} \approx \left(\frac{R}{\lambda_{max}}\right)^N$$

This expression can be used for the numerical calculation of the extremal eigenvalues.

3. For the AC matrix

$$R = \begin{bmatrix} 1.0 & 0.7 & 0.0 \\ 0.7 & 1.0 & 0.7 \\ 0.0 & 0.7 & 1.0 \end{bmatrix}$$

calculate its eigenvalues and eigenvectors and check the properties given in Section 3.6. Verify the spectral decomposition (3.86).

4. Find the frequency and amplitude of the sinusoid contained in the signal with AC matrix

$$R = \begin{bmatrix} 1.00 & 0.65 & 0.10 \\ 0.65 & 1.00 & 0.65 \\ 0.10 & 0.65 & 1.00 \end{bmatrix}$$

What is the noise power? Check the results with the curves in Figure 3.3.

5. Find the spectral decomposition of the matrix

$$R = \begin{bmatrix} 1.0 & 0.7 & 0.0 & -0.7 \\ 0.7 & 1.0 & 0.7 & 0.0 \\ 0.0 & 0.7 & 1.0 & 0.7 \\ -0.7 & 0.0 & 0.7 & 1.0 \end{bmatrix}$$

What is the dimension of the signal space? Calculate the projections of the vectors

$$X^t(n) = \left[\cos\left(n\frac{\pi}{4} \right), \cos\left[(n-1)\frac{\pi}{4} \right], \cos\left[(n-2)\frac{\pi}{4} \right], \right.$$
$$\left. \times \cos\left[(n-3)\frac{\pi}{4} \right] \right], \qquad n = 0, 1, 2, 3$$

on the eigenvectors.

6. Consider the order 2 AR signal

$$x(n) = 0.9x(n-1) - 0.5x(n-2) + e(n)$$

with $E[e^2(n)] = \sigma_e^2 = 1$. Calculate its ACF and give its 3×3 AC matrix R_3. Find the minimum eigenvalue and eigenvector. Give the corresponding harmonic decomposition of the signal and compare with the spectrum.

   Calculate the 4×4 AC matrix R_4 and its inverse R_4^{-1}. Comment on the results.

7. Give expressions to calculate the DST (3.131) and the DCT by a standard DFT. Estimate the computational complexity for $N = 2^P$.

ANNEX 3.1 FORTRAN SUBROUTINE TO SOLVE A LINEAR SYSTEM WITH SYMMETRICAL MATRIX

```
          SUBROUTINE CHOL(N,A,X,B)
C
C         SOLVES THE SYSTEM [A]X=B
C         A : SYMMETRIC COVARIANCE MATRIX (N*N)
C         N : SYSTEM ORDER (N > 2)
C         X : SOLUTION VECTOR
C         B : RIGHT SIDE VECTOR
C
          DIMENSION A(20,20),X(1),B(1)
          A(2,1)=A(2,1)/A(1,1)
          A(2,2)=A(2,2)-A(2,1)*A(1,1)*A(2,1)
          DO40I=3,N
          A(I,1)=A(I,1)/A(1,1)
          DO20J=2,I-1
          S=A(I,J)
          DO10K=1,J-1
   10     S=S-A(I,K)*A(K,K)*A(J,K)
   20     A(I,J)=S/A(J,J)
          S=A(I,I)
          DO30K=1,I-1
   30     S=S-A(I,K)*A(K,K)*A(I,K)
   40     A(I,I)=S
          X(1)=B(1)
          DO60I=2,N
          S=B(I)
          DO50J=1,I-1
   50     S=S-A(I,J)*X(J)
   60     X(I)=S
          X(N)=X(N)/A(N,N)
          DO80K=1,N-1
          I=N-K
          S=X(I)/A(I,I)
          DO70J=I+1,N
   70     S=S-A(J,I)*X(J)
   80     X(I)=S
          RETURN
          END
C
```

ANNEX 3.2 FORTRAN SUBROUTINE TO COMPUTE THE EIGENVECTOR CORRESPONDING TO THE MINIMUM EIGENVALUE BY THE CONJUGATE GRADIENT METHOD [19] *(Courtesy of Tapan K. Sarkar, Department of Electrical Engineering, Syracuse University, Syracuse, N.Y. 13244-1240)*

```
      SUBROUTINE GMEVCG(N, X, A, B, U, SML, W, M)
C
C     THIS SUBROUTINE IS USED FOR ITERATIVELY FINDING THE EIGENVECTOR
C     CORRESPONDING TO THE MINIMUM EIGENVALUE OF A GENERALIZED EIGEN-
C     SYSTEM  AX = UBX.
C
C     A    - INPUT REAL SYMMETRIC MATRIX OF ORDER N, WHOSE MINIMUM
C            EIGENVALUE AND THE CORRESPONDING EIGENVECTOR ARE TO
C            BE COMPUTED.
C     B    - INPUT REAL POSITIVE DEFINITE MATRIX OF ORDER N.
C     N    - INPUT ORDER OF THE MATRIX A.
C     X    - OUTPUT EIGENVECTOR OF LENGTH N CORRESPONDING TO THE
C            MINIMUM EIGENVALUE AND ALSO PUT INPUT INITIAL GUESS
C            IN IT.
C     U    - OUTPUT MINIMUM EIGENVALUE.
C     SML  - INPUT UPPER BOUND OF THE MINIMUM EIGENVALUE.
C     W    - INPUT ARBITRARY VECTOR OF LENGTH N.
C     M    - OUTPUT NUMBER OF ITERATIONS.
C
      LOGICAL AAEZ, BBEZ
      REAL A(N,N), B(N,N), X(N), P(5), R(5), W(N), AP(5), BP(5),
     *     AX(5), BX(5)
      NU = 0
      M = 0
      U1 = 0.0
    1 DO 20 I=1,N
      BX(I) = 0.0
      DO 10 J=1,N
      BX(I) = BX(I) + B(I,J)*X(J)
   10 CONTINUE
   20 CONTINUE
      XBX = 0.0
      DO 30 I=1,N
      XBX = XBX + BX(I)*X(I)
   30 CONTINUE
      XBX = SQRT(XBX)
      DO 40 I=1,N
      X(I) = X(I)/XBX
   40 CONTINUE
      DO 60 I=1,N
      AX(I) = 0.0
      DO 50 J=1,N
      AX(I) = AX(I) + A(I,J)*X(J)
   50 CONTINUE
   60 CONTINUE
      U = 0.0
      DO 70 I=1,N
      U = U + AX(I)*X(I)
```

```
 70   CONTINUE
      DO 80 I=1,N
        R(I) = U*BX(I) - AX(I)
        P(I) = R(I)
 80   CONTINUE
  2   DO 100 I=1,N
        AP(I) = 0.0
        DO 90 J=1,N
          AP(I) = AP(I) + A(I,J)*P(J)
 90     CONTINUE
100   CONTINUE
      DO 120 I=1,N
        BP(I) = 0.0
        DO 110 J=1,N
          BP(I) = BP(I) + B(I,J)*P(J)
110     CONTINUE
120   CONTINUE
      PA = 0.0
      PB = 0.0
      PC = 0.0
      PD = 0.0
      DO 130 I=1,N
        PA = PA + AP(I)*X(I)
        PB = PB + AP(I)*P(I)
        PC = PC + BP(I)*X(I)
        PD = PD + BP(I)*P(I)
130   CONTINUE
      AA = PB*PC - PA*PD
      BB = PB - U*PD
      CC = PA - U*PC
      AAEZ = ABS(AA) .LE. 1.0E-75
      BBEZ = ABS(BB) .LE. 1.0E-75
      IF(AAEZ .AND. BBEZ) GO TO 12
      IF(AAEZ) GO TO 11
      DD = -BB + SQRT(BB*BB-4.0*AA*CC)
      T = DD/(2.0*AA)
      GO TO 15
 11   T = -CC/BB
      GO TO 15
 12   T = 0.0
 15   DO 140 I=1,N
        X(I) = X(I) + T*P(I)
140   CONTINUE
      DO 160 I=1,N
        BX(I) = 0.0
        DO 150 J=1,N
          BX(I) = BX(I) + B(I,J)*X(J)
150     CONTINUE
160   CONTINUE
      XBX = 0.0
      DO 170 I=1,N
        XBX = XBX + BX(I)*X(I)
170   CONTINUE
      XBX = SQRT(XBX)
      DO 180 I=1,N
        X(I) = X(I)/XBX
180   CONTINUE
```

```
      DO 200 I=1,N
        AX(I) = 0.0
        DO 190 J=1,N
          AX(I) = AX(I) + A(I,J)*X(J)
 190    CONTINUE
 200  CONTINUE
      U = 0.0
      DO 210 I=1,N
        U = U + AX(I)*X(I)
 210  CONTINUE
      AI = ABS(U1 - U)
      AJ = ABS(U)*1.0E-03
      AK = AI - AJ
      IF(AK .LT. 0.0) GO TO 3
      DO 220 I=1,N
        R(I) = U*BX(I) - AX(I)
 220  CONTINUE
      QN = 0.0
      DO 230 I=1,N
        QN = QN + R(I)*AP(I)
 230  CONTINUE
      Q = -QN/PB
      DO 240 I=1,N
        P(I) = R(I) + Q*P(I)
 240  CONTINUE
      M = M + 1
      U1 = U
C     WRITE (3, 9998) M
9998  FORMAT (/1X, 3HM =, I3)
C     WRITE (3, 9997)
9997  FORMAT (/2H U/)
C     WRITE (3, 9996) U
9996  FORMAT (1X, E14.6)
C     WRITE (3, 9995)
9995  FORMAT (/5H X(I)/)
C     WRITE (3, 9994) X
9994  FORMAT (1X, F11.6)
      GO TO 2
   3  CONTINUE
      IF(U .LT. SML) RETURN
      NU = NU + 1
      CX = 0.0
      DO 250 I=1,N
        CX = CX + W(I)*BX(I)
 250  CONTINUE
      CX = CX/XBX
      DO 260 I=1,N
        W(I) = W(I) - CX*X(I)
        X(I) = W(I)
 260  CONTINUE
      IF(NU .GT. N) GO TO 4
      GO TO 1
   4  WRITE (3, 9999)
9999  FORMAT (28H NO EIGENVALUE LESS THAN SML)
      STOP
      END
```

REFERENCES

1. H. C. Cramer, *Mathematical Methods of Statistics*, Princeton University Press, Princeton, N.J., 1974, pp. 341–359.
2. D. Hertz, "A Fast Digital Method of Estimating the Autocorrelation of a Gaussian Stationary Process," *IEEE Trans.* **ASSP-30**, 329 (April 1982).
3. S. Cacopardi, "Applicability of the Relay Correlator to Radar Signal Processing," *Electronics Lett.* **19**, 722–723 (September 1983).
4. K. J. Gabriel, "Comparison of 3 Correlation Coefficient Estimators for Gaussian Stationary Processes," *IEEE Trans.* **ASSP-31**, 1023–1025 (August 1983).
5. G. Jacovitti and R. Cusani, "Performances of the Hybrid-Sign Correlation Coefficients Estimator for Gaussian Stationary Processes," *IEEE Trans.* **ASSP-33**, 731–733 (June 1985).
6. J. Bendat and A. Piersol, *Measurement and Analysis of Random Data*, Wiley, New York, 1966.
7. G. H. Golub and C. F. Van Loan, *Matrix Computations*, The John Hopkins University Press, Baltimore, 1983.
8. A. R. Gourlay and G. A. Watson, *Computational Methods for Matrix Eigenproblems*, Wiley, New York, 1973.
9. V. Clema and A. Laub, "The Singular Value Decomposition: Its Computation and Some Applications," *IEEE Trans.* **AC-25**, 164–176 (April 1980).
10. S. S. Reddi, "Eigenvector Properties of Toeplitz Matrices and Their Application to Spectral Analysis of Time Series," in *Signal Processing*, vol. 7, North-Holland, 1984, pp. 46–56.
11. J. Makhoul, "On the Eigenvectors of Symmetric Toeplitz Matrices," *IEEE Trans.* **ASSP-29**, 868–872 (August 1981).
12. M. Aktar, B. Sankur, and Y. Istefanopoulos, "Properties of the Maximum Likehood and Pisarenko Spectral Estimates," in *Signal Processing*, vol. 8, North-Holland, 1985, pp. 401–413.
13. J. D. Mathews, J. K. Breakall, and G. K. Karawas, "The Discrete Prolate Spheroidal Filter as a Digital Signal Processing Tool," *IEEE Trans.* **ASSP-33**, 1471–1478 (December 1985).
14. L. Genyuan, X. Xinsheng, and Q. Xiaoyu, "Eigenvalues and Eigenvectors of One or Two Sinusoidal Signals in White Noise," *Proc. IEEE-ASSP Workshop*, Academia Sinica, Beijing. 1986, pp. 310–313.
15. A. Cantoni and P. Butler, "Properties of the Eigenvectors of Persymmetric Matrices with Applications to Communication theory," *IEEE Trans.* **COM-24**, 804–809 (August 1976).
16. R. M. Gray, "On the Asymptotic Eigenvalue Distribution of Toeplitz Matrices," *IEEE Trans* **IT-16**, 725–730 (1972).

17. O. L. Frost, "An Algorithm for Linearly Constrained Adaptive Array Processing," *Proc. IEEE* **60**, 926–935 (August 1972).
18. V. U. Reddy, B. Egardt, and T. Kailath, "Least Squares Type Algorithm for Adaptive Implementation of Pisarenko's Harmonic Retrieval Method," *IEEE Trans.* **ASSP-30**, 399–405 (June 1982).
19. H. Chen, T. K. Sarkar, S. A. Dianat, and J. D. Brule, "Adaptive Spectral Estimation by the Conjugate Gradient Method," *IEEE Trans.* **ASSP-34**, 272–284 (April 1986).
20. B. Lumeau and H. Clergeot, "Spatial Localization—Spectral Matrix Bias and Variance—Effects on the Source Subspace," in *Signal Processing*, no. 4, North-Holland, 1982, pp. 103–123.
21. P. Nicolas and G. Vezzosi, "Location of Sources with an Antenna of Unknown Geometry," *Proc. GRETSI-85*, Nice, France, 1985, pp. 331–337.
22. V. R. Algazi and D. J. Sakrison, "On the Optimality of the Karhunen-Loève Expansion," *IEEE Trans.* **IT-15**, 319–321 (March 1969).
23. A. K. Jain, "A fast Karhunen-Loève Transform for a Class of Random Processes," *IEEE Trans.* **COM-24**, 1023–1029 (1976).
24. N. Ahmed, T. Natarajan, and K. R. Rao, "Discrete Cosine Transform," *IEEE Trans.* **C-23**, 90–93 (1974).

4

Gradient Adaptive Filters

The adaptive filters based on gradient techniques make a class which is highly appreciated in engineering for its simplicity, flexibility, and robustness. Moreover, they are easy to design, and their performance is well characterized. By far, it is the most widely used class in all technical fields, particularly in communications and control [1, 2].

Gradient techniques can be applied to any structure and provide simple equations. However, because of the looped structure, the exact analysis of the filters obtained may be extremely difficult, and it is generally carried out under restrictive hypotheses not verified in practice [3, 4]. However, simplified approximate investigations provide sufficient results in the vast majority of applications.

The emphasis is on engineering aspects in this chapter. Our purpose is to present the results and information necessary to design an adaptive filter and build it successfully, taking into account the variety of options which make the approach flexible.

4.1 THE GRADIENT—LMS ALGORITHM

The diagram of the gradient adaptive filter is shown in Figure 4.1. The error sequence $e(n)$ is obtained by subtracting from the reference signal $y(n)$ the filtered sequence $\tilde{y}(n)$. The coefficients $c_i(n)$, $0 \leqslant i \leqslant N - 1$, are updated by

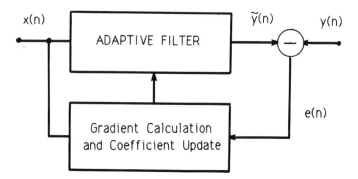

Figure 4.1 Principle of a gradient adaptive filter.

the equation

$$c_i(n + 1) = c_i(n) - \delta \frac{\partial e(n + 1)}{\partial c_i(n)} e(n + 1) \qquad (4.1)$$

The products $[\partial e(n + 1)/\partial c_i(n)]e(n + 1)$ are the elements of the vector V_G, which is the gradient of the function $\frac{1}{2}e^2(n + 1)$. The scalar δ is the adaptation step. In the mean, the operation corresponds to minimizing the error power, hence the denomination least mean squares (LMS) for the algorithm.

The adaptive filter can have any structure. However, the most straightforward and most widely used is the transversal or FIR structure, for which the error gradient is just the input data vector.

The equations of the gradient adaptive transversal filter are

$$e(n + 1) = y(n + 1) - H^t(n)X(n + 1) \qquad (4.2)$$

and

$$H(n + 1) = H(n) + \delta X(n + 1)e(n + 1) \qquad (4.3)$$

where $H^t(n)$ is the transpose of the coefficient vector and $X(n + 1)$ is the vector of the N most recent input data.

The implementation is shown in Figure 4.2. It closely follows the implementation of the fixed FIR filter, a multiplier accumulator circuit being added to produce the time-varying coefficients. Clearly, $2N + 1$ multiplications are needed, as well as $2N$ additions and $2N$ active memories.

Once the number of coefficients N has been chosen, the only filter parameter to be adjusted is the adaptation step δ.

In view of the looped configuration, our first consideration is stability.

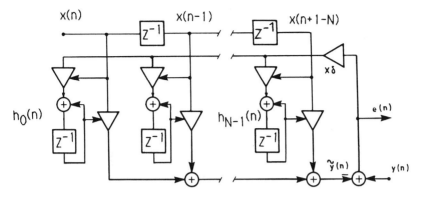

Figure 4.2   Gradient adaptive transversal filter.

4.2 STABILITY CONDITION AND SPECIFICATIONS

The error sequence calculated by equation (4.2) is called "a priori," because it employs the coefficients before updating. The "a posteriori" error is defined as

$$\varepsilon(n + 1) = y(n + 1) - H^t(n + 1)X(n + 1) \tag{4.4}$$

and it can be computed after (4.2) and (4.3) have been completed. Now, from (4.2) and (4.3), (4.4) can be written as

$$\varepsilon(n + 1) = e(n + 1)[1 - \delta X^t(n + 1)X(n + 1)] \tag{4.5}$$

The system can be considered stable if the expectation of the a posteriori error magnitude is smaller than that of the a priori error, which is logical since more information is incorporated in $\varepsilon(n+1)$. If the error $e(n+1)$ is assumed to be independent of the N most recent input data, which is approximately true after convergence, the stability condition is

$$|1 - \delta E[X^t(n + 1)X(n + 1)]| < 1 \tag{4.6}$$

which yields

$$0 < \delta < \frac{2}{N\sigma_x^2} \tag{4.7}$$

where the input signal power σ_x^2 is generally known or easy to estimate.

The stability condition (4.7) is simple and easy to use. However, in practice, to account for the hypotheses made in the derivation, it is wise to take some margin. For example, a detailed analysis for Gaussian signals shows that stability is guaranteed if [5, 6]

$$0 < \delta < \frac{1}{3} \frac{2}{N\sigma_x^2}$$ (4.8)

So, a margin factor of a few units is recommended when using condition (4.7). Once the stability is achieved, the final determination of the step δ in the allowed range is based on performance, compared to specifications.

The two main specifications for gradient adaptive filtering are the system gain and the time constant. The system gain G_S^2 can be defined as the reference to error signal power ratio:

$$G_S^2 = \frac{E[y^2(n)]}{E[e^2(n)]}$$ (4.9)

For example, in adaptive prediction, G_S is the prediction gain. The specification is given as a lower bound for the gain, and the adaptation step and the computation accuracy must be chosen accordingly.

The speed of adaptation is controlled by a time constant specification τ_e, generally imposed on the error sequence. The filter time constant τ can be taken as an effective initial time constant obtained by fitting the sequence $E[e^2(n)]$ to an exponential for $n = 0$ and $n = 1$, which yields

$$(E[e^2(0)] - E[e^2(\infty)])e^{-2/\tau} = E[e^2(1)] - E[e^2(\infty)]$$ (4.10)

Since τ is related to the adaptation step δ, as shown in the following sections, imposing an upper limit τ_e puts a constraint on δ. Indeed the gain and speed specifications must be compatible and lead to a nonempty range of values for δ; otherwise another type of algorithm, like least squares, must be relied upon.

First, the relation between adaptation step and residual error is investigated.

4.3 RESIDUAL ERROR

The gradient adaptive filter equations (4.2) and (4.3) yield

$$H(n + 1) = [I_N - \delta X(n + 1)X'(n + 1)]H(n) + \delta X(n + 1)y(n + 1)$$ (4.11)

When the time index n approaches infinity, the coefficients reach their steady-state values and the average of $H(n + 1)$ becomes equal to the average of $H(n)$. Hence, assuming independence between coefficient variations and input data vectors, we get

$$E[H(\infty)] = R^{-1}r_{yx} = H_{\text{opt}}$$ (4.12)

Using the notation of Section 1.4, we write

$$R = E[X(n)X^t(n)], \qquad r_{yx} = E[X(n+1)y(n+1)] \tag{4.13}$$

Therefore the gradient algorithm provides the optimal coefficient set H_{opt} after convergence and in the mean. The vector r_{yx} is the cross-correlation between the reference and input signals.

The minimum output error power E_{min} can also be expressed as a function of the signals and their cross-correlation.

For the set of coefficients $H(n)$, the mean square output error $E(n)$ is

$$E(n) = E[(y(n) - H^t(n)X(n))^2] \tag{4.14}$$

Now, setting the coefficients to their optimal values gives

$$E_{min} = E[y^2(n)] - H^t_{opt}RH_{opt} \tag{4.15}$$

or

$$E_{min} = E[y^2(n)] - H^t_{opt}r_{yx} \tag{4.16}$$

or

$$E_{min} = E[y^2(n)] - r^t_{yx}R^{-1}r_{yx} \tag{4.17}$$

In these equations the filter order N appears as the dimension of the AC matrix R and of the cross-correlation vector r_{yx}.

For fixed coefficients $H(n)$ the mean square error (MSE) $E(n)$ can be rewritten as a deviation from the minimum:

$$E(n) = E_{min} + [H_{opt} - H(n)]^t R[H_{opt} - H(n)] \tag{4.18}$$

The input data AC matrix R can be diagonalized as

$$R = M^t \text{diag}(\lambda_i)M, \qquad M^tM = I_N \tag{4.19}$$

where, as shown in the preceding chapter, $\lambda_i(0 \leqslant i \leqslant N - 1)$ are the eigenvalues and M the modal unitary matrix.

Letting

$$[\alpha(n)] = M[H_{opt} - H(n)] \tag{4.20}$$

be the coefficient difference vector in the transformed space, we obtain the concise form of (4.18)

$$E(n) = E_{min} + [\alpha(n)]^t \text{diag}(\lambda_i)[\alpha(n)] \tag{4.21}$$

Completing the products, we have

$$E(n) = E_{min} + \sum_{i=0}^{N-1} \lambda_i \alpha_i^2(n) \tag{4.22}$$

If Λ denotes the column vector of the eigenvalues λ_i, and $[\alpha^2(n)]$ denotes the column vector with elements $\alpha_i^2(n)$, then

$$E(n) = E_{\min} + \Lambda^t[\alpha^2(n)] \tag{4.23}$$

The analysis of the gradient algorithm is carried out by following the evolution of the vector $[\alpha(n)]$ according to the recursion

$$[\alpha(n+1)] = [\alpha(n)] - \delta M X(n+1)e(n+1) \tag{4.24}$$

The corresponding covariance matrix is

$$[\alpha(n+1)][\alpha(n+1)]^t = [\alpha(n)][\alpha(n)]^t - 2\delta M X(n+1)e(n+1)[\alpha(n)]^t$$
$$+ \delta^2 e^2(n+1)M X(n+1)X^t(n+1)M^t \tag{4.25}$$

The definition of $e(n+1)$ yields

$$e(n+1) = y(n+1) - H_{\text{opt}}^t X(n+1) + X^t(n+1)M^t[\alpha(n)] \tag{4.26}$$

Equations (4.25) and (4.26) determine the evolution of the system. In order to get useful results, we make simplifying hypotheses, particularly about $e^2(n)$ [7].

It is assumed that the following variables are independent:

The error sequence when the filter coefficients are optimal
The data vector $X(n+1)$
The coefficient deviations $H(n) - H_{\text{opt}}$

Thus

$$E\{[y(n+1) - H_{\text{opt}}^t X(n+1)]X^t(n+1)M^t[\alpha(n)]\} = 0 \tag{4.27}$$

Although not rigorously verified, the above assumptions are reasonable approximations, because the coefficient deviations and optimum output error are noiselike sequences and the objective of the filter is to make them uncorrelated with the N most recent input data. Anyway, the most convincing argument in favor is that the results derived are in good agreement with experiments.

Now, taking the expectation of both sides of (4.25), yields

$$E\{[\alpha(n+1)][\alpha(n+1)]^t\} = [I_N - 2\delta \, \text{diag}(\lambda_i)]E\{[\alpha(n)][\alpha(n)]^t\}$$
$$+ \delta^2 E[e^2(n+1)] \, \text{diag}(\lambda_i) \tag{4.28}$$

For varying coefficients, under the above independence hypotheses, expression (4.23) becomes

$$E[e^2(n+1)] = E_{\min} + \Lambda^t E[\alpha^2(n)] \tag{4.29}$$

Considering the main diagonals of the matrices, and using vector notation and expression (4.29) for the error power, we derive the equation

$$E[\alpha^2(n+1)] = [I_N - 2\delta \operatorname{diag}(\lambda_i) + \delta^2 \Lambda \Lambda^t] E[\alpha^2(n)] + \delta E_{\min} \Lambda \qquad (4.30)$$

A sufficient condition for convergence is that the sum of the absolute values of the elements of any row in the matrix multiplying the vector $E[\alpha^2(n)]$ be less than unity:

$$0 < 1 - 2\delta\lambda_i + \delta^2\lambda_i \left(\sum_{j=0}^{N-1} \lambda_j \right) < 1, \qquad 0 \leqslant i \leqslant N-1 \qquad (4.31)$$

from which we obtain the stability condition

$$0 < \delta < \frac{2}{\displaystyle\sum_{j=0}^{N-1} \lambda_j} = \frac{2}{N\sigma_x^2}$$

which is the condition already found in Section 4.2.

Once the stability conditions are fulfilled, recursion (4.28) yields, as $n \to \infty$,

$$E\{[\alpha(\infty)][\alpha(\infty)]^t\} = \frac{\delta}{2} E(\infty) I_N \qquad (4.32)$$

Due to the definition of the vector $[\alpha(n)]$, equation (4.32) also applies to the coefficient deviations themselves. Thus the coefficient deviations, after convergence, are statistically independent and have the same power.

Now, combining (4.32) and (4.29) yields the residual error E_R:

$$E(\infty) = E_R = \frac{E_{\min}}{1 - (\delta/2)N\sigma_x^2} \qquad (4.33)$$

Finally, the gradient algorithm produces an excess output MSE related to the adaptation step. Indeed when δ approaches the stability limit, the output error power approaches infinity. The ratio of the steady-state MSE to the minimum attainable MSE is called the final misadjustment M_{adj}:

$$M_{\text{adj}} = \frac{E_R}{E_{\min}} = \frac{1}{1 - (\delta/2)N\sigma_x^2} \qquad (4.34)$$

In practical realizations, due to the margin generally taken for the adaptation step size, the approximation

$$E_R \approx E_{\min}\left(1 + \frac{\delta}{2} N\sigma_x^2\right) \qquad (4.35)$$

is often valid, and the excess output MSE is approximately proportional to the step size. In fact, it can be viewed as a gradient noise, due to the approximation of the true cost function gradient by an instantaneous value.

4.4 LEARNING CURVE AND TIME CONSTANT

The adaptive filter starts from an initial state, which often corresponds to zero coefficients. From there, its evolution is controlled by the input and reference signals, and it is possible to define learning curves by parameter averaging.

Recursion (4.30) gives the evolution of the filter coefficients. For small δ it can be simplified to

$$E[\alpha^2(n+1)] \approx \text{diag}(1 - \delta\lambda_i)^2 E[\alpha^2(n)] + \delta E_{\min}\Lambda \tag{4.30}$$

Now, an output error power recursion can be obtained from (4.29):

$$E(n+1) - E_{\min} = \Lambda^t \text{diag}(1 - \delta\lambda_i)^2 E[\alpha^2(n)] + \delta E_{\min}\Lambda^t\Lambda \tag{4.36}$$

Neglecting the last term in that expression and using (4.29) at the time origin, we get

$$E(n) - E_{\min} = \Lambda^t \text{diag}(1 - \delta\lambda_i)^{2n} E[\alpha^2(0)] \tag{4.37}$$

Clearly, the evolution of the output MSE depends on the input signal matrix eigenvalues, which provide as many different modes. In the long run, it is the smallest eigenvalue which controls the convergence.

The filter time constant τ_e obtained from an exponential fitting to the output rms error is obtained by applying definition (4.10) and neglecting the residual error:

$$E(0)e^{-2/\tau_e} = \Lambda^t \text{diag}(1 - \delta\lambda_i)^2 E[\alpha^2(0)] \tag{4.38}$$

We can also obtain it approximately by applying (4.29) at the time origin:

$$\Lambda^t E[\alpha^2(0)]\left[1 - \frac{2}{\tau_e}\right] = \Lambda^t \text{diag}(1 - 2\delta\lambda_i)E[\alpha^2(0)] \tag{4.39}$$

Hence

$$\tau_e = \frac{1}{\delta} \frac{\displaystyle\sum_{i=0}^{N-1} \lambda_i E\{\alpha_i^2(0)\}}{\displaystyle\sum_{i=0}^{N-1} \lambda_i E\{\alpha_i^2(0)\}\lambda_i} \tag{4.40}$$

If the eigenvalues are not too dispersed, we have

$$\tau_e \approx \frac{N}{\delta \displaystyle\sum_{i=0}^{N-1} \lambda_i} = \frac{1}{\delta\sigma_x^2} \tag{4.41}$$

The filter time constant is proportional to the inverses of the adaptation step size and of the input signal power. Therefore, an estimation of the signal power is needed to adjust the adaptation speed. Moreover, if the signal is nonstationary, the power estimation must be carried out in real time to reach a high level of performance.

A limit on the adaptation speed is imposed by the stability condition (4.7).

From equation (4.30), it appears that the rows of the square matrix are quadratic functions of the adaptation step and all take their minimum norm for

$$\delta_m = \frac{1}{\sum\limits_{i=0}^{N-1} \lambda_i} = \frac{1}{N\sigma_x^2} \qquad (4.42)$$

which corresponds to the fastest convergence. Therefore the smallest time constant is

$$\tau_{e,min} = N \qquad (4.43)$$

In these conditions, if the eigenvalues are approximately equal to the signal power, which occurs for noiselike signals in certain modeling applications, the learning curve, taken as the output MSE function, is obtained from (4.36) by

$$E(n) - E_R = (E(0) - E_R)\left(1 - \frac{1}{N}\right)^{2n} \qquad (4.44)$$

For zero initial values of the coefficients, $E(0)$ is just the reference signal power.

Overall, the three expressions (4.7), (4.33), and (4.41) give the basic information to choose the adaptation step δ and evaluate a transversal gradient adaptive filter. They are sufficient in many practical cases.

Example

Consider the second-order adaptive FIR prediction filter in Figure 4.3, with equations

$$e(n + 1) = x(n + 1) - a_1(n)x(n) - a_2(n)x(n - 1)$$
$$\begin{bmatrix} a_1(n + 1) \\ a_2(n + 1) \end{bmatrix} = \begin{bmatrix} a_1(n) \\ a_2(n) \end{bmatrix} + \delta \begin{bmatrix} x(n) \\ x(n - 1) \end{bmatrix} e(n + 1) \qquad (4.45)$$

The input signal is a sinusoid in noise:

$$x(n) = \sin\left(\frac{n\pi}{4}\right) + b(n) \qquad (4.46)$$

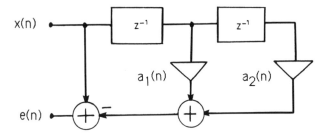

Figure 4.3   Second-order prediction filter.

The noise $b(n)$ has power $\sigma_b^2 = 5 \times 10^{-5}$. The input signal power is $\sigma_x^2 = 0.5$. The step size δ is 0.05. Starting from zero-valued coefficients, the evolution of the output error, the two coefficients, and the corresponding zeros in the complex plane are shown in Figure 4.4. Clearly the output error time constant is in reasonably good agreement with estimation (4.41).

In the filter design process, the next step is the estimation of the coefficient and internal data word lengths needed to meet the adaptive filter specifications.

4.5 WORD-LENGTH LIMITATIONS

Word-length limitations introduce roundoff error sources, which degrade the filter performance. The roundoff process generally takes place at the output of the multipliers, as represented by the quantizers Q in Figure 4.5.

In roundoff noise analysis a number of simplifying hypotheses are generally made concerning the source statistics. The errors are identically distributed and independent; with rounding, the distribution law is uniform in the interval $[-q/2, q/2]$, where q is the quantization step size, the power is $\frac{q^2}{12}$, and the spectrum is flat.

Concerning the adaptive transversal filter, there are two different categories of roundoff errors, corresponding to internal data and coefficients [8].

The quantization processes at each of the N filter multiplication outputs amount to adding N noise sources at the filter output. Therefore, the output MSE is augmented by $Nq_2^2/12$, assuming q_2 is the quantization step.

The quantization with step q_1 of the multiplication result in the coefficient updating section is not so easily analyzed. Recursion (4.28) is modified as follows, taking into account the hypotheses on the roundoff noise sources and

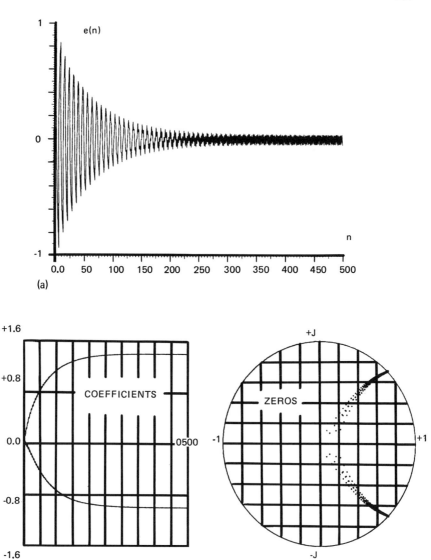

Figure 4.4 The second-order adaptive FIR prediction filter: (a) Output error sequence; (b) Coefficient versus time; (c) Zeros in the complex plane.

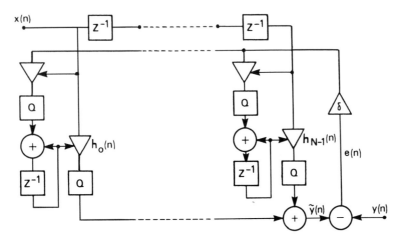

Figure 4.5   Adaptive FIR filter with word-length limitations.

their independence of the other variables:

$$E\{[\alpha(n+1)][\alpha(n+1)]'\} = [I_N - 2\delta \operatorname{diag}(\lambda_i)] E\{[\alpha(n)][\alpha(n)]'\}$$
$$+ \delta^2 E[e^2(n+1)] \operatorname{diag}(\lambda_i) + \frac{q_1^2}{12} I_N \qquad (4.47)$$

An additional gradient noise is introduced.

When $n \to \infty$, equation (4.29) yields, as before,

$$E_{RT}\left(1 - \frac{\delta}{2} N\sigma_x^2\right) = E_{\min} + \frac{q_1^2}{12} \frac{N}{2\delta} \qquad (4.48)$$

Hence, the total residual error, taking into account the quantization of the filter coefficients with step q_1 and the quantization of internal data with step q_2, as shown in Figure 4.5, is

$$E_{RT} = \frac{1}{1 - (\delta/2)N\sigma_x^2} \left[E_{\min} + \frac{N}{2\delta} \frac{q_1^2}{12} + N \frac{q_2^2}{12} \right] \qquad (4.49)$$

or, assuming a small excess output MSE,

$$E_{RT} \approx E_{\min}\left(1 + \frac{\delta}{2} N\sigma_x^2\right) + \frac{N}{2\delta} \frac{q_1^2}{12} + N \frac{q_2^2}{12} \qquad (4.50)$$

This expression shows that the effects of the two kinds of quantizations are different. Because of the factor $\frac{1}{\delta}$, the coefficient quantization and the corresponding word length can be very sensitive. In fact, there is an optimum δ_{opt} for the adaptation step size which minimizes the total residual error;

according to (4.50) it is obtained through derivation as

$$\tfrac{1}{2}E_{\min}N\sigma_x^2 - \frac{N}{2}\frac{q_1^2}{12}\frac{1}{\delta_{\text{opt}}^2} = 0 \tag{4.51}$$

and

$$\delta_{\text{opt}} = \frac{1}{\sqrt{E_{\min}}\sigma_x}\frac{1}{\sqrt{3}}\frac{q_1}{2} \tag{4.52}$$

The curve of the residual error versus the adaptation step size is shown in Figure 4.6. For δ decreasing from the stability limit, the minimum is reached for δ_{opt}; if δ is decreased further, the curve indicates that the total error should grow, which indeed has no physical meaning. The hypotheses which led to (4.50) are no longer valid, and a different phenomenon occurs, namely blocking.

According to the coefficient evolution equation (4.3), the coefficient $h_i(n)$ is frozen if

$$|\delta x(n-i)e(n)| < \frac{q_1}{2} \tag{4.53}$$

Let us assume that the elements of the vector $\delta X(n)e(n)$ are uncorrelated with each other and distribute uniformly in the interval $[-q_{1/2}, q_{1/2}]$. Then

$$\delta^2 E\{e^2(n)X(n)X^t(n)\} = \frac{q_1^2}{12}I_N \tag{4.54}$$

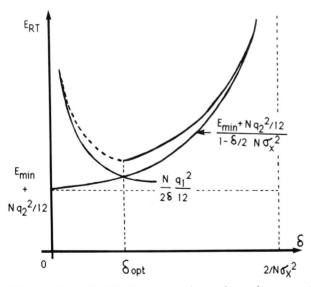

Figure 4.6 Residual error against adaptation step size.

If the coefficients are close to their optimal values and if the input signal can be approximated by a white noise, then equations (4.54) and (4.51) are equivalent. A blocking radius ρ can then be defined for the coefficients by

$$\rho^2 = E\{[H(n) - H_{opt}]^t[H(n) - H_{opt}]\} \tag{4.55}$$

Now, considering that

$$H(n) - H_{opt} = R^{-1}E[e(n)X(n)] \tag{4.56}$$

we have, from (4.54) and the identity $X^tX = \text{trace}(XX^t)$,

$$\rho^2 = \frac{1}{12}\left(\frac{q_1}{\delta}\right)^2 \sum_{i=0}^{N-1} \lambda_i^{-2} \tag{4.57}$$

The blocking radius is a function of the spread of the input AC matrix eigenvalues. Blocking can occur for adaptation step sizes well over δ_{opt}, given by (4.52), if there are small eigenvalues.

In adaptive filter implementations, the adaptation step size is often imposed by system specifications (e.g., the time constant), and the coefficient quantization step size q_1 is chosen small enough to avoid the blocking zone with some margin.

Quantization steps q_1 and q_2 are generally derived from expression (4.50). Considering the crucial advantage of digital processing, which is that operations can be carried out with arbitrary accuracy, the major contribution in the total residual error should be the theoretical minimal error E_{min}. In a balanced realization, the degradations from different origins should be similar. Hence, a reasonable design choice is

$$\frac{1}{2}\left[E_{min}\frac{N\delta\sigma_x^2}{2}\right] = \frac{N}{2\delta}\frac{q_1^2}{12} = N\frac{q_2^2}{12} \tag{4.58}$$

If b_c is the number of bits of the coefficients and h_{max} is the largest coefficient magnitude, then, assuming fixed-point binary representation, we have

$$q_1 = h_{max}2^{1-b_c} \tag{4.59}$$

Under these conditions

$$2^{2b_c} = \frac{2}{3}\frac{h_{max}^2}{\delta^2 E_{min}\sigma_x^2} \tag{4.60}$$

with the assumption that E_{min} is the dominant term in (4.50), that is,

$$G_S^2 E_{min} \approx \sigma_y^2$$

By introducing the time constant specification τ_e, one has approximately

$$b_c \approx \log_2(\tau_e) + \log_2(G_S) + \log_2\left(h_{max}\frac{\sigma_x}{\sigma_y}\right) \tag{4.61}$$

This expression gives an estimation of the coefficient word length necessary to meet the specifications of a gradient adaptive filter. However there is one variable which is not readily available, h_{max}; a simple bound can be derived, if we assume a large system gain and refer to the eigenfilters of Section 3.7:

$$\sigma_y^2 = E[y^2(n)] \approx H^t(n)RH(n) \geqslant \lambda_{min} H^t(n)H(n) \tag{4.62}$$

Now

$$\sigma_y^2 \geqslant \lambda_{min} h_{max}^2$$

and

$$\left(h_{max}\frac{\sigma_x}{\sigma_y}\right)^2 \leqslant \frac{\sigma_x^2}{\lambda_{min}} \tag{4.63}$$

Therefore, the last term on the right side of (4.61) is bounded by zero for input signals whose spectrum is approximately flat, but it can take positive values for narrowband signals.

Estimate (4.61) can produce large values for b_c; that word length is necessary in the coefficient updating accumulator but not in the filter multiplications.

In practice, additional quantizers can be introduced just before the multiplications by $h_i(n)$ in Figure 4.5 in order to avoid multiplications with high precision factors. The effects of the additional roundoff noise sources introduced that way can be investigated as above.

Often, nonstationary signals are handled, and estimate (4.61) is for stationary signals. In this case, a first approach is to incorporate the signal dynamic range in the last term of (4.61).

To complete the filter design, the number of bits b_i of the internal data can be determined by setting

$$q_2 = \max\{|x(n)|, |y(n)|\}2^{1-b_i} \tag{4.64}$$

with the assumption that $\sigma_x^2 \geqslant \sigma_y^2$, which is true in linear prediction and often valid in system modeling, and taking the value 4 as the peak factor of the signal $x(n)$ as in the Gaussian case. Thus

$$q_2 = 4\sigma_x 2^{1-b_i}$$

Now, (4.58) yields

$$2^{2b_i} = 2^4 \frac{4}{3} \frac{1}{E_{min}\delta}$$

By introducing the specifications we obtain

$$b_i \approx 2 + \log_2\left(\frac{\sigma_x}{\sigma_y}\right) + \log_2(G_S) + \tfrac{1}{2}\log_2(\tau_e) \tag{4.65}$$

This completes the implementation parameter estimation for the standard gradient algorithm. However, some modifications can be made to this algorithm, which are either useful or even mandatory.

4.6 LEAKAGE FACTOR

When the input signal vanishes, the driving term in recursion (4.3) becomes zero and the coefficients are locked up. In such conditions, it might be preferable to have them return to zero. This is achieved by the introduction of a leakage factor γ in the updating equation:

$$H(n + 1) = (1 - \gamma)H(n) + \delta X(n + 1)e(n + 1) \tag{4.66}$$

The coefficient recursion is

$$H(n + 1) = [(1 - \gamma)I_N - \delta X(n + 1)X'(n + 1)]H(n)$$
$$+ \delta y(n + 1)X(n + 1) \tag{4.67}$$

After convergence,

$$H_\infty = E[H(\infty)] = \left[R + \frac{\gamma}{\delta} I_N \right]^{-1} r_{yx} \tag{4.68}$$

The leakage factor γ introduces a bias on the filter coefficients, which can be expressed in terms of the optimal values as

$$H_\infty = \left[R + \frac{\gamma}{\delta} I_N \right]^{-1} R H_{\text{opt}} \tag{4.69}$$

The same effect is obtained when a white noise is added to the input signal $x(n)$; a constant equal to the noise power is added to the elements of the main diagonal of the input AC matrix.

To evaluate the impact of the leakage factor, we rewrite the coefficient vector H_∞ as

$$H_\infty = M' \operatorname{diag}\left(\frac{\lambda_i}{\lambda_i + \gamma/\delta} \right) M H_{\text{opt}} \tag{4.70}$$

The significance of the bias depends on the relative values of $\lambda_{\min}$ and $\frac{\gamma}{\delta}$.

Another aspect is that the cost function actually minimized in the whole process is

$$J_\gamma(n) = E\left\{ [y(n) - X'(n)H(n - 1)]^2 + \frac{\gamma}{\delta} H'(n - 1)H(n - 1) \right\} \tag{4.71}$$

. The last term represents a constraint which is imposed on the coefficient magnitudes [9].

The LS solution is given by (4.65), and the coefficient bias is

$$H - H_{\text{opt}} = \left[\left(R + \frac{\gamma}{\delta} I_N \right)^{-1} R - I_N \right] H_{\text{opt}} \tag{4.72}$$

Hence the filter output MSE becomes

$$E_R = E_{\min} + [H - H_{\text{opt}}]' R [H - H_{\text{opt}}] \tag{4.73}$$

The leakage factor is particularly useful for handling nonstationary signals. With such signals, the leakage value can be chosen to reduce the output error power.

If the coefficients are computed by minimizing the above cost function taken on a limited set of data, the coefficient variance can be estimated by

$$E\{[H - H_0][H - H_0]'\} = E_R \left[R + \frac{\gamma}{\delta} I_N \right]^{-1} R \left[R + \frac{\gamma}{\delta} \right]^{-1} \tag{4.74}$$

and the coefficient MSE H_{MSE} is

$$H_{\text{MSE}} = [H - H_{\text{opt}}]' [H - H_{\text{opt}}] + \text{trace}(E\{[H - H_0][H - H_0]'\}) \tag{4.75}$$

When γ increases from zero, H_{MSE} decreases from $E_R \text{trace}(R^{-1})$, then reaches a minimum and increases, because in (4.75) the variance decreases faster than the bias increases at the beginning, as can be seen directly for dimension $N = 1$ [9]. A minimal output MSE corresponds to the minimum of H_{MSE}.

A similar behavior can be observed when the gradient algorithm is applied to nonstationary signals. An illustration is provided by applying a speech signal to an order 8 linear predictor. The prediction gain measured is shown in Figure 4.7 versus the leakage factor for several adaptation step sizes δ. The

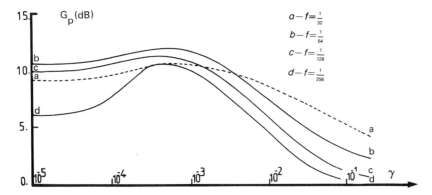

Figure 4.7 Prediction gain vs. leakage factor for a speech sentence.

maximum of the prediction gain is clearly visible. It is also a justification for the values sometimes retained for speech prediction, which are $\delta = 2^{-6}$ and $\gamma = 2^{-8}$.

The leakage factor, which can nicely complement the conventional gradient algorithm, is recommended for the sign algorithm because it bounds the coefficients and thus prevents divergence.

4.7 THE LMAV AND SIGN ALGORITHMS

Instead of the LS, the least absolute value (LAV) criterion can be used to compare variables, vectors, or functions. It has two specific advantages: it does not necessarily lead to minimum phase solutions; it is robust to outliers in a data set. Similarly, the least mean absolute value (LMAV) can replace the LMS in adaptive filters [10].

The gradient of the function $|e(n + 1)|$ is the vector whose elements are

$$\frac{\partial |e(n + 1)|}{\partial h_i} = \frac{\partial}{\partial h_i} |y(n + 1) - X^t(n + 1)H(n)|$$

$$= -x(n + 1 - i)\text{sign}\, e(n + 1) \qquad (4.76)$$

where sign e is $+1$ if e is positive, and -1 otherwise. The LMAV algorithm for the transversal adaptive filter is

$$H(n + 1) = H(n) + \Delta X(n + 1)\text{sign}\, e(n + 1) \qquad (4.77)$$

where Δ, a positive constant, is the adaptation step.

The convergence can be studied by considering the evolution of the coefficient vector toward the optimum H_{opt}. Equation (4.77) can be rewritten as

$$H(n + 1) - H_{\text{opt}} = H(n) - H_{\text{opt}} + \Delta X(n + 1)\text{sign}\, e(n + 1)$$

Taking the norm squared of both sides yields

$$[H(n + 1) - H_{\text{opt}}]^t[H(n + 1) - H_{\text{opt}}] = [H(n) - H_{\text{opt}}]^t[H(n) - H_{\text{opt}}]$$

$$+ 2\Delta\, \text{sign}\, e(n + 1)X^t(n + 1)[H(n) - H_{\text{opt}}] + \Delta^2 X^t(n + 1)X(n + 1) \qquad (4.78)$$

or, with further decomposition,

$$\|H(n + 1) - H_{\text{opt}}\|^2 = \|H(n) - H_{\text{opt}}\|^2 + \Delta^2\|X(n + 1)\|^2 - 2\Delta|e(n + 1)|$$

$$+ 2\Delta\, \text{sign}\, e(n + 1)[y(n + 1) - X^t(n + 1)H_{\text{opt}}]$$

Hence we have the inequality

$$\|H(n + 1) - H_{opt}\|^2 \leqslant \|H(n) - H_{opt}\|^2 + \Delta^2 \|X(n + 1)\|^2 - 2\Delta|e(n + 1)|$$
$$+ 2\Delta|y(n + 1) - X^t(n + 1)H_{opt}|$$

Taking the expectation of both sides gives

$$E\{\|H(n + 1) - H_{opt}\|^2\} \leqslant \|H(n) - H_{opt}\|^2$$
$$+ \Delta^2 N\sigma_x^2 - 2\Delta E\{|e(n + 1)|\} + 2\Delta E_{min} \qquad (4.79)$$

where the minimal error E_{min} is

$$E_{min} = E[|y(n + 1) - X^t(n + 1)H_{opt}|] \qquad (4.80)$$

If the system starts with zero coefficients, then

$$E\{\|H(n + 1) - H_{opt}\|^2\} \leqslant \|H_{opt}\|^2$$
$$+ (n + 1)(\Delta^2 N\sigma_x^2 + 2\Delta E_{min}) - 2\Delta \sum_{p=1}^{n+1} E\{|e(p)|\}$$

Since the left side is nonnegative, the accumulated error is bounded by

$$\frac{1}{n + 1} E\left\{\sum_{p=1}^{n+1} |e(p)|\right\} \leqslant \frac{\Delta}{2} N\sigma_x^2 + E_{min} + \frac{\|H_{opt}\|^2}{2\Delta(n + 1)} \qquad (4.81)$$

This is the basic equation of LMAV adaptive filters. It has the following implications:

Convergence is obtained for any positive step size Δ.
After convergence the residual error E_R is bounded by

$$E_R \leqslant E_{min} + \frac{\Delta}{2} N\sigma_x^2 \qquad (4.82)$$

It is difficult to define a time constant as in Section 4.1. However, an adaptation time τ_A can be defined as the number of iterations needed for the last term in (4.81) to become smaller than E_{min}. Then we have

$$\tau_A = \frac{1}{\Delta} \frac{\|H_{opt}\|^2}{2E_{min}} \qquad (4.83)$$

The performance of the LMAV adaptive filters can be assessed from the above expressions. A comparison with the results given in Sections 4.3 and 4.4 for the standard LMS algorithm clearly shows the price paid for the simplification in the coefficient updating circuitry. The main observation is that, if a small excess output MSE is required, the adaptation time can become very large.

Another way of simplifying gradient adaptive filters is to use the following coefficient updating technique:

$$H(n + 1) = H(n) + \Delta e(n + 1)\text{sign } X(n + 1) \tag{4.84}$$

This algorithm can be viewed as belonging to the LMS family, but with a normalized step size. Since

$$\text{sign } x = \frac{x}{|x|} \tag{4.85}$$

and $|x|$ can be coarsely approximated by the efficient value σ_x, equation (4.84) corresponds to a gradient filter with adaptation step size

$$\delta = \frac{\Delta}{\sigma_x} \tag{4.86}$$

The performance can be assessed by replacing δ in the relevant equations. Pursuing further in that direction, we obtain the sign algorithm

$$H(n + 1) = H(n) + \Delta \text{ sign } e(n + 1)\text{sign } X(n + 1) \tag{4.87}$$

The detailed analysis is rather complicated. However, a coarse but generally sufficient approach consists of assuming a standard gradient algorithm with step size

$$\delta = \frac{\Delta}{\sigma_x \sigma_e} \tag{4.88}$$

where σ_x and σ_e are the efficient values of the input signal and output error, respectively.

In the learning phase, starting with zero-valued coefficients, it can be assumed that $\sigma_e \approx \sigma_y$ and the initial time constant τ_S of the sign algorithm can be roughly estimated by

$$\tau_S \approx \frac{1}{\Delta} \frac{\sigma_y}{\sigma_x} \tag{4.89}$$

After convergence it is reasonable to assume $\sigma_e^2 = E_{\min}$. If the adaptation step is small, the residual error E_{RS} in the sign algorithm can be estimated by

$$E_{RS} \approx E_{\min}\left(1 + \frac{N\Delta}{2} \frac{\sigma_x}{\sqrt{E_{\min}}}\right) \tag{4.90}$$

A stability condition is obtained by combining (4.7) and (4.88), which yields

$$\Delta \leqslant \frac{2}{N} \frac{\sqrt{E_{\min}}}{\sigma_x} \tag{4.91}$$

Also, for stability reasons, a leakage term is generally introduced in the sign algorithm coefficient, giving

$$H(n + 1) = (1 - \gamma)H(n) + \Delta \operatorname{sign} e(n + 1)\operatorname{sign} X(n + 1) \qquad (4.92)$$

Under these conditions, the coefficients are bounded by

$$|h_i(n)| \leqslant \frac{\Delta}{\gamma}, \qquad 0 \leqslant i \leqslant N - 1 \qquad (4.93)$$

Overall, it can be stated that the sign algorithm is slower than the standard gradient algorithm and leads to larger excess output MSE [11–12]. However, it is very simple; moreover it is robust because of the built-in normalization of its adaptation step, and it can handle nonstationary signals. It is one of the most widely used adaptive filter algorithms.

4.8 NORMALIZED ALGORITHMS FOR NONSTATIONARY SIGNALS

When handling nonstationary signals, adaptive filters are expected to trace as closely as possible the evolution of the signal parameters. However, due to the time constant there is a delay which leads to a tracking error. Therefore the excess output MSE has two components: the gradient misadjustment error, and the tracking error.

The efficiency of adaptive filters eepends on the signal characteristics. Clearly, the most favorable situation is that of slow variations, as mentioned in Section 2.13. The detailed analysis of adaptive filter performance is based on nonstationary signal modeling techniques. Nonstationarity can affect the reference signal as well as the filter input signal. In this section a highly simplified example is considered to illustrate the filter behavior.

When only the reference signal is assumed to be nonstationary, the developments of the previous sections can, with adequate modifications, be kept. The nonstationarity of the reference is reflected in the coefficient updating equation (4.3) by the fact that the optimal vector is time dependent:

$$H(n + 1) - H_{\text{opt}}(n + 1) = H(n) - H_{\text{opt}}(n) + \delta e(n + 1)X(n + 1) \qquad (4.94)$$

If it can be assumed that the optimal coefficients are generated by a first-order model whose inputs are zero mean i.i.d. random variables $e_{nS,i}(n)$, with variance σ_{nS}^2, as in Section 2.13, then

$$H_{\text{opt}}(n + 1) = (1 - \gamma)H_{\text{opt}}(n) + [e_{nS,0}(n+1), \ldots, l_{nS(N-1)}(n+1)]^t \qquad (4.95)$$

Furthermore, if the variations are slow, which implies $\gamma \approx 1$, the net effect of the nonstationarity is the introduction of the extra term $\sigma_{nS}I_N$ in recursion

(4.28). As already seen for the coefficient roundoff, the residual error E_{RTnS} is

$$E_{RTnS}\left(1 - \frac{\delta}{2}N\sigma_x^2\right) = E_{\min} + \frac{N}{2\delta}\sigma_{nS}^2 \tag{4.96}$$

or, for small adaptation step size,

$$E_{RTnS} \approx E_{\min}\left(1 + \frac{\delta}{2}N\sigma_x^2\right) + \frac{N}{2\delta}\sigma_{nS}^2 \tag{4.97}$$

In this simplified expression for the residual output error power with a nonstationary reference signal, the contributions of the gradient misadjustment and the tracking error are well characterized. Clearly, there is an optimum for the adaptation step size, δ_{opt}, which is

$$\delta_{opt} = \frac{\sigma_{nS}}{\sigma_x\sqrt{E_{\min}}} \tag{4.98}$$

which corresponds to balanced contributions.

The above model is indeed sketchy, but it provides hints for the filter behavior in more complicated circumstances [13]. For example, an order 12 FIR adaptive predictor is applied to three different speech signals: (a) a male voice, (b) a female voice, and (c) unconnected words. The prediction gain is shown in Figure 4.8(a) for various adaptation step sizes. The existence of an optimal step size is clearly visible in each case.

The performance of adaptive filters can be significantly improved if the most crucial signal parameters can be estimated in real time. For the gradient algorithms the most important parameter is the input signal power, which determines the step size. If the signal power can be estimated, then the normalized LMS algorithm

$$H(n + 1) = H(n) + \frac{\delta}{\sigma_x^2}X(n + 1)e(n + 1) \tag{4.99}$$

can be implemented. The most straightforward estimation σ_x^2 is $P_{x1}(n)$ given by

$$P_{x1}(n) = P_0 + \frac{1}{N_0}\sum_{i=0}^{N_0-1} x^2(n - i) \tag{4.100}$$

where P_0 is a positive constant which prevents division by zero. The parameter N_0, the observation time window, is the duration over which the signal can be assumed to be stationary.

For the prediction filter example mentioned above, the results corresponding to $P_0 = 0.5$ and $N_0 = 100$ (the long-term speech power is unity) are given in Figure 4.8(b). The improvements brought by normalization are clearly

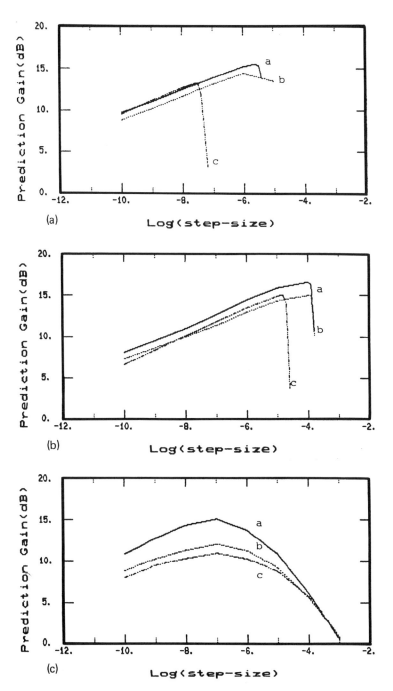

Figure 4.8 Prediction gain vs. adaptation step size for three speech signals: (a) LMS with fixed step; (b) Normalized LMS; (c) Sign algorithm.

visible for all three sentences. The results obtained with the sign algorithm (4.87) are shown in Figure 4.8(c) for comparison purposes. The prediction gain is reduced, particularly for sentences b and c, but the robustness is worth pointing out: there is no steep divergence for too large δ, but a gradual performance degradation instead.

In practice, equation (4.100) is costly to implement, and the recursive estimate of Section 3.3 is preferred:

$$P_{x2}(n + 1) = (1 - \gamma)P_{x2}(n) + \gamma x^2(n + 1) \tag{4.101}$$

Estimates (4.100) and (4.101) are additive. For faster reaction to rapid changes, exponential estimations can be worked out. An efficient and simple method to implement corresponds to a variable adaptation step size $\Delta(n)$ given by

$$\Delta(n) = \frac{\delta}{P_x(n)} = 2^{-I(n)} \tag{4.102}$$

where $I(n)$ is an integer variable, itself updated through an additive process (e.g., a sign algorithm [14]).

The step responses of $P_{x1}(n)$, $P_{x2}(n)$ and the exponential estimate are sketched in Figure 4.9. Better performance can be expected with the exponential technique for rapidly changing signals.

Adaptation step size normalization can also be achieved indirectly by reusing the data at each iteration.

The a posteriori error $\varepsilon(n + 1)$ in equation (4.4) is calculated with the updated coefficients. It can itself be used to update the coefficients a second time, leading to a new error $\varepsilon_1(n + 1)$. After K such iterations, the a posteriori error $\varepsilon_K(n + 1)$ is

$$\varepsilon_K(n + 1) = [1 - \delta X'(n + 1)X(n + 1)]^{K+1}e(n + 1) \tag{4.103}$$

For δ sufficiently small and K large, $\varepsilon_K(n + 1) \approx 0$, which would have been obtained with a step size Δ satisfying

$$1 - \Delta X'(n + 1)X(n + 1) = 0$$

that is

$$\Delta = \frac{1}{X'(n + 1)X(n + 1)} \tag{4.104}$$

The equivalent step size corresponds to the fastest convergence defined in Section 4.4 by equation (4.42). So, the data reusing method can lead to fast convergence, while preserving the stability, in the presence of nonstationary signals.

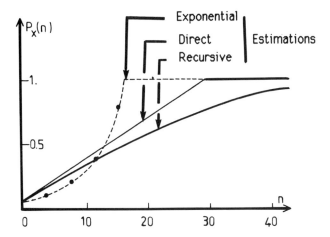

Figure 4.9   Step responses of signal power estimations.

The performance of normalized LMS algorithms can be studied as in the above sections, with the additional complication brought by the variable step size. For example, considering the so-called projection LMS algorithm

$$H(n + 1) = H(n) + \frac{\delta}{X^t(n + 1)X(n + 1)} X(n + 1)e(n + 1) \qquad (4.105)$$

one can show that a bias is introduced on the coefficients, which becomes independent of the step size for small values, while the variance remains proportional to δ [15].

A coarse approach to performance evaluation consists of keeping the results obtained for fixed step algorithms and considering the extreme parameter values.

4.9 DELAYED LMS ALGORITHMS

In the implementation, it can be advantageous to update the coefficients with some delay, say d sampling periods. For example, with integrated signal processors a delay $d = 1$ can ease programming. In these conditions it is interesting to investigate the effects of the updating delay on the adaptive filter performance [16].

The delayed LMS algorithm corresponds to the equation

$$H(n + 1) = H(n) + \delta X(n + 1 - d)e(n + 1 - d) \qquad (4.106)$$

The developments of Section 4.3 can be carried out again based on the above equation. For the sake of brevity and conciseness, a simplified analysis is performed here, starting from equation (4.24), rewritten as

$$[\alpha(n+1)] = [\alpha(n)] - \delta M X(n+1-d)e(n+1-d) \qquad (4.107)$$

Substituting (4.26) in this equation and taking the expectation yields, under the hypotheses of Section 4.3,

$$E\{[\alpha(n+1)]\} = E\{[\alpha(n)]\} - \delta \, \text{diag}(\lambda_i)E\{[\alpha(n-d)]\} \qquad (4.108)$$

The system is stable if the roots of the characteristic equation

$$r^{d+1} - r^d + \delta\lambda_i = 0 \qquad (4.109)$$

are inside the unit circle in the complex plane. Clearly, for $d = 0$, the condition is

$$0 < \delta < \frac{2}{\lambda_{\text{max}}} \qquad (4.110)$$

which a stability condition sometimes used for the conventional LMS algorithms, less stringent than (4.7).

When $d = 1$, the stability condition is

$$0 < \delta < \frac{1}{\lambda_{\text{max}}} \qquad (4.111)$$

which implies that delay makes the stability condition more stringent. If δ is small enough $(\delta < \frac{1}{4}\lambda_{\text{max}})$, the roots of the second-order characteristic equation are real:

$$r_1 \approx 1 - \delta\lambda_i(1 + \delta\lambda_i), \qquad r_2 \approx \delta\lambda_i(1 + \delta\lambda_i) \qquad (4.112)$$

The corresponding digital filter can be viewed as a cascade of two first-order sections, whose time constants can be calculated; its step response is approximately proportional to $1 - (1 + \delta\lambda_i)r_1^n$, where the factor $1 + \delta\lambda_i$ reflects the effect of the root r_2. However, neglecting the root r_2, we can state that, for small adaptation step sizes, the adaptation speed of the delayed algorithm is similar to that of the conventional gradient algorithm. In the context of this simplified analysis, the time constant τ_i for each mode is roughly

$$\tau_i \approx \frac{1}{\delta\lambda_i} \qquad (4.113)$$

When $d = 2$, the largest real root inside the unit circle is r_{max}, such that

$$r_{\text{max}} = 1 - \frac{\delta\lambda_i}{r_{\text{max}}^2} < 1 - \delta\lambda_i \qquad (4.114)$$

Again the corresponding filter step response can be approximated. Overall, the algorithm becomes slower when the delay d is increased.

For the excess output MSE, a first estimation can be obtained by considering only the largest root of the characteristic equation and assuming that the delayed LMS is equivalent to the conventional LMS with a slightly larger adaptation step. For $d = 1$, referring to equation (4.112), we can take the multiplying factor to be $1 + \delta\lambda_{max}$. The most adverse situation for delayed LMS algorithms is the presence of nonstationary signals, because the tracking error can grow substantially.

4.10 FIR FILTERS IN CASCADE FORM

In certain applications it is important to track the roots of the adaptive filter z-transfer function—for instance, for stability control if the inverse system is to be realized. It is then convenient to design the filter as a cascade of L second-order sections $H_l(z)$, $1 \leqslant l \leqslant L$, such that

$$H_l(z) = 1 + h_{1l}z^{-1} + h_{2l}z^{-2}$$

For real coefficients, if the roots z_l are complex, then

$$h_{1l} = -2\,\mathrm{Re}(z_l), \qquad h_{2l} = |z_l|^2 \tag{4.115}$$

The roots are inside the unit circle if

$$|h_{2l}| < 1, \quad |h_{1l}| < 1 + h_{2l}, \qquad 1 \leqslant l \leqslant L \tag{4.116}$$

The filter transfer function is

$$H(z) = \prod_{l=1}^{L} (1 + h_{1l}z^{-1} + h_{2l}z^{-2})$$

The error gradient vector is no longer the input data vector, and it must be calculated.

The filter output sequence can be obtained from the inverse z-transform

$$\tilde{y}(n) = \frac{1}{2\pi j} \int_\Gamma z^{n-1} \prod_{l=1}^{L} (1 + h_{1l}z^{-1} + h_{2l}z^{-2})X(z)\,dz \tag{4.117}$$

where Γ is a suitable integration contour. Hence

$$\frac{\partial e(n+1)}{\partial h_{ki}} = -\frac{\partial \tilde{y}(n+1)}{\partial h_{ki}}$$

$$= -\frac{1}{2\pi j} \int_\Gamma z^n z^{-k} \prod_{\substack{l=1 \\ l \neq i}}^{L} (1 + h_1 z^{-1} + h_{2l}z^{-2})X(z)\,dz$$

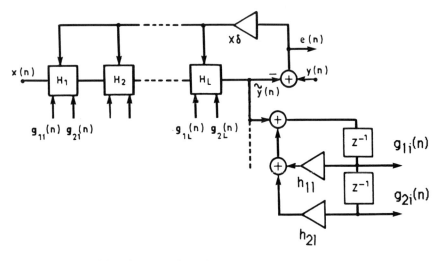

Figure 4.10   Adaptive FIR filter in cascade form.

or, more concisely,

$$\frac{\partial e(n+1)}{\partial h_{ki}} = -\frac{1}{2\pi j}\int_{\Gamma} z^{n}z^{-k}\,\frac{H(z)}{1+h_{1i}z^{-1}+h_{2i}z^{-2}}\,X(z)\,dz \qquad (4.118)$$

Therefore, to form the gradient term $g_{ki}(n)=\partial e(n)/\partial h_{ki}$, it is sufficient to apply the filter output $\tilde{y}(n)$ to a purely recursive second-order section, whose transfer function is just the reciprocal of the section with index i. The recursive section has the same coefficients, but with the opposite sign. The corresponding diagram is given in Figure 4.10.

The coefficients are updated as follows:

$$h_{ki}(n+1)=h_{ki}(n)+\delta e(n+1)g_{ki}(n+1), \qquad k=1,2,\ 1\leqslant i\leqslant L \qquad (4.119)$$

The filter obtained in this way is more complicated than the transversal FIR filter, but it offers a simple method of finding and tracking the roots, which, due to the presence of the recursive part, should be inside the unit circle in the z-plane to ensure stability [17].

However, there are some implementation problems, because the individual sections have to be characterized for the filter to work properly. That can be achieved by imposing different initial conditions or by separating the zero trajectories in the z-plane.

4.11 IIR GRADIENT ADAPTIVE FILTERS

In general, IIR filters achieve given minimum phase functions with fewer coefficients than their FIR counterparts. Moreover, in some applications, it is precisely an IIR function that is looked for. Therefore, IIR adaptive filters are an important class, particularly useful in modeling or identifying systems [18–20].

The output of an IIR filter is

$$\tilde{y}(n) = \sum_{l=0}^{L} a_l x(n - l) + \sum_{k=1}^{K} b_k \tilde{y}(n - k) \tag{4.120}$$

The elements of the error gradient vector are calculated from the derivatives of the filter output:

$$\frac{\partial \tilde{y}(n)}{\partial a_l} = x(n - l) + \sum_{k=1}^{K} b_k \frac{\partial \tilde{y}(n - k)}{\partial a_l}, \qquad 0 \leqslant l \leqslant L \tag{4.121}$$

and

$$\frac{\partial \tilde{y}(n)}{\partial b_k} = \tilde{y}(n - k) + \sum_{i=1}^{K} b_i \frac{\partial \tilde{y}(n - i)}{\partial b_k}, \qquad 1 \leqslant k \leqslant K \tag{4.122}$$

To show the method of realization, let us consider the z-transfer function

$$H(z) = \frac{\displaystyle\sum_{l=0}^{L} a_l z^{-l}}{1 - \displaystyle\sum_{k=1}^{K} b_k z^{-k}} = \frac{N(z)}{D(z)} \tag{4.123}$$

The filter output can be written

$$\tilde{y}(n) = \frac{1}{2\pi j} \int_{\Gamma} z^{n-1} H(z) X(z) \, dz$$

Consequently

$$\frac{\partial \tilde{y}(n)}{\partial a_l} = \frac{1}{2\pi j} \int_{\Gamma} z^{n-1} z^{-l} \frac{X(z)}{D(z)} \, dz \tag{4.124}$$

$$\frac{\partial \tilde{y}(n)}{\partial b_k} = \frac{1}{2\pi j} \int_{\Gamma} z^{n-1} z^{-k} \frac{1}{D(z)} H(z) X(z) \, dz \tag{4.125}$$

The gradient is thus calculated by applying $x(n)$ and $\tilde{y}(n)$ to the circuits corresponding to the transfer function $\frac{1}{D(z)}$.

To simplify the implementation, the second terms in (4.121) and (4.122) can be dropped, which leads to the following set of equations for the adaptive filter (in vector notation):

$$e(n + 1) = y(n + 1) - [A^t(n), B^t(n)] \begin{bmatrix} X(n + 1) \\ \tilde{Y}(n) \end{bmatrix} \tag{4.126}$$

$$\begin{bmatrix} A(n + 1) \\ B(n + 1) \end{bmatrix} = \begin{bmatrix} A(n) \\ B(n) \end{bmatrix} + \delta \begin{bmatrix} X(n + 1) \\ \tilde{Y}(n) \end{bmatrix} e(n + 1) \tag{4.127}$$

The block diagram is shown in Figure 4.11(a). The filter is called a parallel IIR gradient adaptive filter.

The analysis of the performance of such a filter is not simple, because of the vector $\tilde{Y}(n)$ of the most recent filter output data in the system equations [18]. To begin with, the stability can only be ensured if the error sequence $e(n)$ is filtered by a z-transfer function $C(z)$, such that the function $C(z)/D(z)$ be strictly positive real, which means

$$\text{Re} \begin{bmatrix} \frac{C(z)}{D(z)} \end{bmatrix} > 0, \quad |z| = 1 \tag{4.128}$$

An obvious choice is $C(z) = D(z)$.

An alternative approach to get realizable IIR filters is based on the observation that, after convergence, the error signal is generally small and the filter output $\tilde{y}(n)$ is close to the reference $y(n)$. Thus, in the system equations, the filter output vector can be replaced by the reference vector:

$$e(n + 1) = y(n + 1) - [A^t(n), B^t(n)] \begin{bmatrix} X(n + 1 \\ Y(n) \end{bmatrix} \tag{4.129}$$

$$\begin{bmatrix} A(n + 1) \\ B(n + 1) \end{bmatrix} = \begin{bmatrix} A(n) \\ B(n) \end{bmatrix} + \delta \begin{bmatrix} X(n + 1) \\ Y(n) \end{bmatrix} e(n + 1) \tag{4.130}$$

The filter is said to be of the series-parallel type; its diagram is shown in Figure 4.11(b). Now, only FIR filter sections are used, and there is no fundamental stability problem anymore. The performance analysis can be carried out as in the above sections. The stability bound for the adaptation step size is

$$0 < \delta < \frac{2}{L\sigma_x^2 + K\sigma_y^2} \tag{4.131}$$

Overall the performance of the series-parallel IIR gradient adaptive filter can be derived from that of the FIR filter by changing $N\sigma_x^2$ into $L\sigma_x^2 + K\sigma_y^2$.

In order to compare the performance of the parallel type and series-parallel approaches, let us consider the expectation of the recursive coefficient

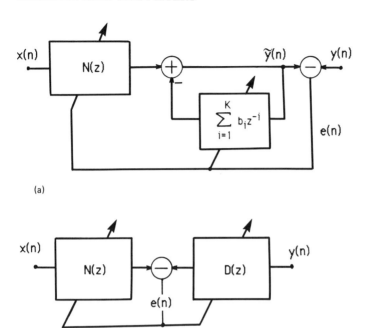

(a)

(b)

Figure 4.11 Simplified gradient IRR adaptive filters: (a) Parallel type; (b) Series-parallel type.

vector after convergence, B_∞, for the parallel case. Equations (4.126) and (4.127) yield

$$B_\infty = E[\tilde{Y}(n)\tilde{Y}^t(n)]^{-1}E\{\tilde{Y}(n)[y(n+1) - A^t(n)X(n+1)]\} \qquad (4.132)$$

The parallel-series type yields a similar equation, but with $E[Y(n)Y^t(n)]^{-1}$; if the output error is approximated by a white noise with power σ_e^2, then

$$E[Y(n)Y^t(n)] = \sigma_e^2 I_N + E[\tilde{Y}(n)\tilde{Y}^t(n)] \qquad (4.133)$$

and a bias is introduced on the recursive coefficients. The above equation clearly illustrates the stability hazards associated with using $\tilde{Y}(n)$, because the matrix can become singular. Therefore, the residual error is larger with the parallel-series approach, while the adaptation speed is not significantly modified, particularly for small step sizes, because the initial error sequences are about the same for both types.

Finally, several structures are available, and IIR gradient adaptive filters can be an attractive alternative to FIR filters in relevant applications.

4.12 NONLINEAR FILTERING

The digital filters considered up to now have been linear filters, which means
that the output is a linear function of the input data. We can have a nonlinear
scalar function of the input data vector:

$$\tilde{y}(n) = f[X(n)] \tag{4.134}$$

The Taylor series expansion of the function $f(X)$ about the vector zero is

$$f(X) = \sum_{k=0}^{\infty} \frac{1}{k!} \left[\sum_{i=1}^{N} x_i \frac{\partial}{\partial x_i} \right]^k f(X) \tag{4.135}$$

with differential operator notation. When limited to second order, the
expansion is

$$\tilde{y}(n) = y_0 + H^t X(n) + \text{trace}(M X(n) X^t(n)) \tag{4.136}$$

where y_0 is a constant, H is the vector of the linear coefficients, and M is the
square matrix of the quadratic coefficients, the filter length N being the
number of elements of the data vector $X(n)$. This nonlinear filter is called the
second-order Volterra filter (SVF) [21].

The quadratic coefficient matrix M is symmetric because the data matrix
$X(n)X^t(n)$ is symmetric. Also, if the input and reference signals are assumed to
have zero mean, $\tilde{y}(n)$ must also have zero mean, which implies

$$E[\tilde{y}(n)] = y_0 + \text{trace}(MR) \tag{4.137}$$

Therefore (4.136) can be rewritten as

$$\tilde{y}(n) = H^t X(n) + \text{trace}(M[X(n)X^t(n) - R]) \tag{4.138}$$

When this structure is used in an adaptive filter configuration, the coefficients
must be calculated to minimize the output MSE, $E\{(y(n) - \tilde{y}(n))^2\}$.
For Gaussian signals, the optimum coefficients are [22]

$$H_{\text{opt}} = R^{-1}E[y(n)X(n)]$$
$$M_{\text{opt}} = \tfrac{1}{2}R^{-1}E[y(n)X(n)X^t(n)]R^{-1} \tag{4.139}$$

It is worth pointing out that the linear operator of the optimum SVF, in these
conditions, is exactly the optimum linear filter. Thus, the nonlinear filter can
be constructed by adding a quadratic section in parallel to the linear filter.
The minimum output MSE is

$$E_{\min} = E[y^2(n)] - E[y(n)X(n)]^t R^{-1} E[y(n)X(n)]$$
$$- \tfrac{1}{2}\text{trace}(R^{-1}E[y(n)X(n)X^t(n)]R^{-1}E[y(n)X(n)X^t(n)]) \tag{4.140}$$

The gradient techniques can be implemented by calculating the derivatives
of the output error with respect to the coefficients. The gradient adaptive SVF

equations are

$$e(n + 1) = y(n + 1) - H^t(n)X(n + 1)$$
$$- \text{trace}(M(n)[X(n + 1)X^t(n + 1) - R])$$
$$H(n + 1) = H(n) + \delta_h X(n + 1)e(n + 1)$$
$$M(n + 1) = M(n) + \delta_m X(n + 1)X^t(n + 1)e(n + 1) \quad (4.141)$$

where δ_h and δ_m are the adaptation steps.

The zeroth-order term trace($M(n)R$) is not constant in the adaptive implementation. It can be replaced by an estimate of the mean value of the quadratic section output, for example, using the recursive estimator of Section 3.3.

The stability bounds for the adaptation steps can be obtained as in Section 4.2 by considering the a posteriori error $\varepsilon(n + 1)$:

$$\varepsilon(n + 1) = e(n + 1)[1 - \delta_h X^t(n + 1)X(n + 1)$$
$$- \delta_m \text{trace}(X(n + 1)X^t(n + 1)[X(n + 1)X^t(n + 1) - R])]$$
$$(4.142)$$

Assuming that the linear operator acts independently, we adopt condition (4.7) for δ_h. Now, the stability condition for δ_m is

$$|1 - \delta_m(\text{trace } E[X(n + 1)X^t(n + 1)X(n + 1)X^t(n + 1)] - \text{trace } R^2)| < 1$$

The following approximation can be made:

$$\text{trace } E[X(n + 1)X^t(n + 1)X(n + 1)X^t(n + 1)] \approx (N\sigma_x^2)^2 > \text{trace } R^2 \quad (4.143)$$

Hence, we have the stability condition

$$0 < \delta_m < \frac{2}{(N\sigma_x^2)^2} \quad (4.144)$$

The total output error is the sum of the minimum error $E_{\min}$ given by (4.140) and the excess MSEs of the linear and quadratic sections. Using developments as in Section 4.3, one can show the excess MSE of the quadratic section E_M can be approximated by

$$E_M \approx \frac{\delta_m}{8} E_{\min}[(N\sigma_x^2)^2 + 2 \text{ trace } R^2] \quad (4.145)$$

In practice, the quadratic section in general serves as a complement to the linear section. Indeed the improvement must be worth the price paid in additional computational complexity [23].

4.13 STRENGTHS AND WEAKNESSES OF GRADIENT FILTERS

The strong points of the gradient adaptive filters, illustrated throughout this chapter, are their ease of design, their simplicity of realization, their flexibility, and their robustness against signal characteristic evolution and computation errors.

The stability conditions have been derived, the residual error has been estimated, and the learning curves have been studied. Simple expressions have been given for the stability bound, the residual error, and the time constant in terms of the adaptation step size. Word-length limitation effects have been investigated, and estimates have been derived for the coefficient and internal data word lengths as a function of the specifications. Useful variations from the classical LMS algorithm have been discussed. In short, all the knowledge necessary for a smart and successful engineering application has been provided.

Although gradient adaptive filters are attractive, their performance is severely limited in some applications. Their main weakness is their dependence on signal statistics, which can lead to low speed or large residual errors. They give their best results with flat spectrum signals, but if the signals have a fine structure they can be inefficient and unable, for example, to perform simple analysis tasks. For these cases LS adaptive filters offer an attractive solution.

EXERCISES

1. A sinusoidal signal $x(n) = \sin(n\pi/2)$ is applied to a second-order linear predictor as in Figure 4.3. Calculate the theoretical ACF of the signal and the prediction coefficients. Verify that the zeros of the FIR prediction filter are on the unit circle at the right frequency.

    Using the LMS algorithm (4.3) with $\delta = 0.1$, show the evolution of the coefficients from time $n = 0$ to $n = 10$. How is that evolution modified if the MLAV algorithm (4.77) and the sign algorithm (4.87) are used instead.

2. A second-order adaptive FIR filter has the above $x(n)$ as input and

    $$y(n) = x(n) + x(n - 1) + 0.5x(n - 2)$$

    as reference signal. Calculate the coefficients, starting from zero initial values, from time $n = 0$ to $n = 10$. Calculate the theoretical residual error and the time constant and compare with the experimental results.

3. *Adaptive line enhancer.* Consider an adaptive third-order FIR predictor. The input signal is

   $$x(n) = \sin(n\omega_0) + b(n)$$

   where $b(n)$ is a white noise with power σ_b^2. Calculate the optimal coefficients $a_{i,opt}$, $1 \leqslant i \leqslant 3$. Give the noise power in the sequence

   $$s(n) = \sum_{i=1}^{3} a_{i,opt} x(n - i)$$

   as well as the signal power. Calculate the SNR enhancement.
   The predictor is now assumed to be adaptive with step $\delta = 0.1$. Give the SNR enhancement.

4. In a transmission system, an echo path is modeled as an FIR filter, and an adaptive echo canceler with 500 coefficients is used to remove the echo. At unity input signal power, the theoretical system gain, the echo attenuation, is 53 dB, and the time constant specification is 800 sampling periods. Calculate the range of the adaptation step size δ if the actual system gain specification is 50 dB.

   Assuming the echo path to be passive, estimate the coefficient and internal data word lengths, considering that the power of the signals can vary in a 40-dB range.

5. An adaptive notch filter is used to remove a sinusoid from an input signal. The filter transfer function is

   $$H(z) = \frac{1 + az^{-1} + z^{-2}}{1 + 0.9az^{-1} + 0.81z^{-2}}$$

   Give the block diagram of the adaptive filter. Calculate the error gradient. Simplify the error gradient and give the coefficient updating equation. The signal $x(n) = \sin(n\pi/4)$ is fed to the filter from time zero on. For an initial coefficient value of zero what are the trajectories, in the z-plane, of the zeros and poles of the notch filter. Verify experimentally with $\delta = 0.1$.

6. An order 4 FIR predictor is realized as a cascade of two second-order sections. Show that only one section is needed to compute the error gradient and give the block diagram. What happens for any input signal if the filter is made adaptive and the initial coefficient values are zero. Now the predictor transfer function is

   $$H(z) = (1 - az^{-1} + az^{-2})(1 + bz^{-1} + bz^{-2})$$

   and the coefficients a and b are updated. Give the trajectories, in the z-plane, of the predictor zeros.

Calculate the maximum prediction gain for the signal $x(2p + 1) = 1$, $x(2p) = 0$.

7. Give the block diagram of the gradient second-order Volterra adaptive filter according to equations (4.141). Evaluate the computational complexity in terms of number of multiplications and additions per sampling period and point out the cost of the quadratic section.

REFERENCES

1. R. W. Lucky, "Techniques for Adaptive Equalization of Digital Communication Systems," *Bell System Tech. J.* **45**, 255–286 (1966).

2. B. Widrow and S. D. Stearns, *Adaptive Signal Processing*, Prentice-Hall, Englewood Cliffs, N.J., 1985.

3. W. A. Gardner, "Learning Characteristics of Stochastic Gradient Descent Algorithms: A General Study, Analysis and Critique," in *Signal Processing*, No. 6, North-Holland, 1984, pp. 113–133.

4. O. Macchi and E. Eweda, "Convergence Analysis of Self-Adaptive Equalizers," *IEEE Trans.* **IT-30**, 161–176 (March 1984).

5. L. L. Horowitz and K. D. Senne, "Performance Advantage of Complex LMS for Controlling Narrow-Band and Adaptive Arrays," *IEEE Trans.* **CAS-28**,562–576 (June 1981).

6. C. N. Tate and C. C. Goodyear, "Note on the Convergence of Linear Predictive Filters, Adapted Using the LMS Algorithm," *IEE Proc.* **130**, 61–64 (April 1983).

7. M. S. Mueller and J. J. Werner, Adaptive Echo Cancellation with Dispersion and Delay in the Adjustment Loop," *IEEE Trans.* **ASSP-33**, 520–526 (June 1985).

8. C. Caraiscos and B. Liu, "A Round-Off Error Analysis of the LMS Adaptive Algorithm," *IEEE Trans.* **ASSP-32**, 34–41 (Feb. 1984).

9. A. Segalen and G. Demoment, "Constrained LMS Adaptive Algorithm," *Electronics Lett.* **18**, 226–227 (March 1982).

10. A. Gersho, "Adaptive Filtering with Binary Reinforcement," *IEEE Trans.* **IT-30**, 191–199 (March 1984).

11. T. Claasen and W. Mecklenbrauker, "Comparison of the Convergence of Two Algorithms for Adaptive FIR Digital Filters," *IEEE Trans.* **ASSP-29**, 670–678 (June 1981).

12. N. J. Bershad, "On the Optimum Data Non-Linearity in LMS Adaptation," *IEEE Trans.* **ASSP-34**, 69–76 (February 1986).

13. B. Widrow, J. McCool, M. Larimore, and R. Johnson, "Stationary and Nonstationary Learning Characteristics of the LMS Adaptive Filter," *Proc. IEEE* **64**, 1151–1162 (August 1976).

14. D. Mitra and B. Gotz, "An Adaptive PCM System Designed for Noisy Channels and Digital Implementation," *Bell System Tech. J.* **57**, 2727–2763 (September 1978).

15. S. Abu El Ata, "Asymptotic Behavior of an Adaptive Estimation Algorithm with Application to M-Dependent Data," *IEEE Trans.* **AC-27**, 1225–1257 (December 1982).

16. L. Xieting, S. Xiangyi, and J. Zhonghong, "The DLMS Algorithm Suitable for the Pipelined Realization of Adaptive Filters," *Proc. IEEE-ASSP Workshop*, Academia Sinica, Beijing, 1986, pp. 399–402.

17. L. B. Jackson and S. L. Wood, "Linear Prediction in Cascade Form," *IEEE Trans.* **ASSP-26**, 518–528 (December 1978).

18. J. R. Treichler, "Adaptive Algorithms for IIR Filters," in *Adaptive Filters*, Prentice-Hall, Englewood Cliffs, N.J., 1985, Chap. 4.

19. C. R. Johnson, "Adaptive IIR Filtering: Current Results and Open Issues," *IEEE Trans.* **IT-30**, 237–250 (March 1984).

20. I. D. Landau, "A Feedback System Approach to Adaptive Filtering," *IEEE Trans.* **IT-30**, 251–262 (March 1984).

21. E. Biglieri, "Theory of Volterra Processors and Some Applications," *Proc.* **ICASSP-82**, Paris, 1982, pp. 294–297.

22. T. Koh and E. Powers, "Second Order Volterra Filtering and Its Application to Non Linear System Identification," *IEEE Trans.* **ASSP-33**, 1445–1455 (December 1985).

23. G. L. Sicuranza, "Non Linear Digital Filter Realization by Distributed Arithmetic," *IEEE Trans.* **ASSP-33**, 939–945 (August 1985).

5

Linear Prediction Error Filters

Linear prediction error filters are included in adaptive filters based on FLS algorithms, and they represent a significant part of the processing. They crucially influence the operation and performance of the complete system. Therefore it is important to have a good knowledge of the theory behind these filters, of the relations between their coefficients and the signal parameters, and of their implementation structures. Moreover, they are needed as such in some application areas like signal compression or analysis [1].

5.1 DEFINITION AND PROPERTIES

Linear prediction error filters form a class of digital filters characterized by constraints on the coefficients, specific design methods, and some particular implementation structures.

In general terms, a linear prediction error filter is defined by its transfer function $H(z)$, such that

$$H(z) = 1 - \sum_{i=1}^{N} a_i z^{-i} \tag{5.1}$$

where the coefficients are computed so as to minimize a function of the output $e(n)$ according to a given criterion. If the output power is minimized, then the definition agrees with that given in Section 2.8 for linear prediction.

135

When the number of coefficients N is a finite integer, the filter is a FIR type. Otherwise the filter is IIR type, and its transfer function often takes the form of a rational fraction:

$$H(z) = \frac{1 - \sum_{i=1}^{L} a_i z^{-i}}{1 - \sum_{i=1}^{M} b_i z^{-i}} \tag{5.2}$$

For simplicity, the same number of coefficients $N = L = M$ is often assumed in the numerator and denominator of $II(z)$, implying that some may take on zero values.

The block diagram of the filter associated with equation (5.2) is shown in Figure 5.1, where the recursive and the nonrecursive sections are represented.

As seen in Section 2.5, linear prediction corresponds to the modeling of the signal as the output of a generating filter fed by a white noise, and the linear prediction error filter transfer function is the inverse of the generating filter transfer function. Therefore, the linear prediction error filter associated with $H(z)$ in (5.2) is sometimes designated by extension as an ARMA (L, M) predictor, which means that the AR section of the signal model has L coefficients and the MA section has M coefficients.

For a stationary signal, the linear prediction coefficients can be calculated by LS techniques. A direct application of the general method presented in Section 1.4 yields the set of N equations:

$$\frac{\partial}{\partial a_j} E[e^2(n)] = r(j) - \sum_{i=1}^{N} a_i r(j - i) = 0, \qquad 1 \leqslant j \leqslant N$$

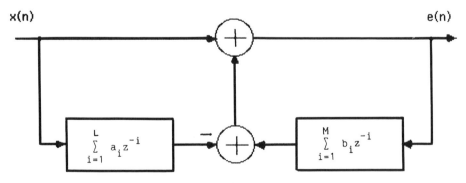

Figure 5.1   IIR linear prediction error filter.

which can be completed by the power relation (4.16)

$$E_{aN} = E[e^2(n)] = r(0) - \sum_{i=1}^{N} a_i r(i)$$

In concise form, the linear prediction matrix equation is

$$R_{N+1} \begin{bmatrix} 1 \\ -A_N \end{bmatrix} = \begin{bmatrix} E_{aN} \\ 0 \end{bmatrix} \tag{5.3}$$

where A_N is the N-element prediction coefficient vector and E_{aN} is the prediction error power. The $(N+1) \times (N+1)$ signal AC matrix, R_{N+1}, is related to R_N by

$$R_{N+1} = \begin{bmatrix} r(0) & r(1) & \cdots & r(N) \\ r(1) & & & \\ \vdots & & R_N & \\ r(N) & & & \end{bmatrix}, \qquad R_N = E[X(n)X^t(n)] \tag{5.4}$$

The linear prediction equation is also the AR modeling equation (2.63) given in Section 2.5.

The above coefficient design method is valid for any stationary signal. An alternative and illustrative approach can be derived, which is useful when the signal is made of determinist, or predictable, components in noise.

Let us assume that the input signal is

$$x(n) = s(n) + b(n) \tag{5.5}$$

where $s(n)$ is a useful signal with power spectral density $S(\omega)$ and $b(n)$ a zero mean white noise with power σ_b^2. The independence relation between the sequences $s(n)$ and $b(n)$ leads to

$$E_{aN} = E[e^2(n)] = \frac{1}{2\pi} \int_{-\pi}^{\pi} |H(\omega)|^2 S(\omega) \, d\omega + \sigma_b^2 (1 + A_N^t A_N) \tag{5.6}$$

The factor $|H(\omega)|^2$ is a function of the prediction coefficients which can be calculated to minimize E_{aN} by setting to zero the derivatives of (5.6) with respect to the coefficients. The two terms on the right side of (5.6) can be characterized as the residual prediction error and the amplified noise, respectively. Indeed their relative values reflect the predictor performance and the degradation caused by the noise added to the useful signal.

If $E_{aN} = 0$, then there is no noise, $\sigma_b^2 = 0$, and the useful signal is predictable; in other words, it is the sum of at most N cisoids. In that case, the zeros of the prediction error filter are on the unit circle, at the signal frequencies, like those of the minimal eigenvalue filter. These filters are identical, up to a constant factor, because the prediction equation

$$R_{N+1} \begin{bmatrix} 1 \\ -A_N \end{bmatrix} = 0 \tag{5.7}$$

is also an eigenvalue equation, corresponding to $\lambda_{min} = 0$.

A characteristic property of linear prediction error filters is that they are minimum phase, as shown in Section 2.8; all of their zeros are within or on the unit circle in the complex plane.

As an illustration, first- and second-order FIR predictors are studied next.

5.2 FIRST- AND SECOND-ORDER FIR PREDICTORS

The transfer function of the first-order FIR predictor is

$$H(z) = 1 - az^{-1} \tag{5.8}$$

Indeed its potential is very limited. It can be applied to a constant signal in white noise with power σ_b^2:

$$x(n) = 1 + b(n)$$

The prediction error power is

$$E[e^2(n)] = |H(1)|^2 + \sigma_b^2(1 + a^2) = (1 - a)^2 + \sigma_b^2(1 + a^2) \tag{5.9}$$

Setting to zero the derivative of $E[e^2(n)]$ with respect to the coefficient a yields

$$a = \frac{1}{1 + \sigma_b^2} \tag{5.10}$$

The zero of the filter is on the real axis in the z-plane when $\sigma_b^2 = 0$ and moves away from the unit circle toward the origin when the noise power is increased.

The ratio of residual prediction error to amplified noise power is maximal for $\sigma_b^2 = \sqrt{2}$, which corresponds to a SNR ratio of -1.5 dB. Its maximum value is about 0.2, which means that the residual prediction error power is much smaller than the amplified noise power.

The transfer function of the second-order FIR predictor is

$$H(z) = 1 - a_1 z^{-1} - a_2 z^{-2} \tag{5.11}$$

It can be applied to a sinusoid in noise:

$$x(n) = \sqrt{2} \sin(n\omega_0) + b(n)$$

The prediction error power is

$$E[e^2(n)] = |H(\omega_0)|^2 + \sigma_b^2(1 + a_1^2 + a_2^2)$$

Hence,

$$a_1 = 2 \cos \omega_0 \frac{\sin^2 \omega_0 + \dfrac{\sigma_b^2}{2}}{\sin^2 \omega_0 + \sigma_b^2(2 + \sigma_b^2)}$$

and

$$a_2 = -1\left[1 - \sigma_b^2 \frac{1 + \sigma_b^2 + 2 \cos^2 \omega_0}{(1 + \sigma_b^2)^2 - \cos^2 \omega_0}\right] \tag{5.13}$$

When the noise power vanishes, the filter zeros reach the unit circle in the complex plane and take on the values $e^{\pm j\omega_0}$. They are complex if $a_1^2 + 4a_2 < 0$, which is always verified as soon as $|\cos \omega_0| < \frac{\sqrt{2}}{2}$; that is, $\frac{\pi}{4} \leqslant \omega_0 \leqslant \frac{3\pi}{4}$. Otherwise the zeros are complex when the noise power is small enough. The noise power limit σ_{bL}^2 is the solution of the following third-degree equation in the variable $x = 1 + \sigma_b^2$:

$$x^3 + x^2 \frac{3 \cos^2 \omega_0}{8 \cos^2 \omega_0 - 4} - x \frac{3}{2} \cos^2 \omega_0 + \frac{4 \cos^6 \omega_0 + \cos^2 \omega_0}{8 \cos^2 \omega_0 - 4} = 0 \tag{5.14}$$

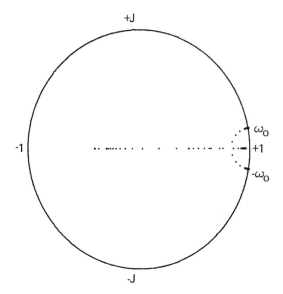

Figure 5.2 Location of the zeros of a second-order FIR predictor applied to a sinusoid in noise with varying power.

This equation has only one positive and real solution for the relevant values of the frequency ω_0. So, σ_{bL}^2 can be calculated; a simple approximation is [2]

$$\sigma_{bL}^2 \approx 1.33\omega_0^3 \qquad (\omega_0 \text{ in radians}) \tag{5.15}$$

The trajectory of the zeros in the complex plane when the additive noise power varies is shown in Figure 5.2. When the noise power increases from zero, the filter zeros move from the unit circle on a circle centered at $+1$ and with radius approximately ω_0; beyond σ_{bL}^2 they move on the real axis toward the origin.

The above results are useful for the detection of sinusoids in noise.

5.3 FORWARD AND BACKWARD PREDICTION EQUATIONS

The linear prediction error is also called the process innovation to illustrate the fact that new information has become available. However, when a limited fixed number of data is handled, as in FIR or transversal filters, the oldest data sample is discarded every time a new sample is acquired. Therefore, to fully analyze the system evolution, one must characterize the loss of the oldest data sample, which is achieved by backward linear prediction.

The forward linear prediction error $e_a(n)$ is

$$e_a(n) = x(n) - \sum_{i=1}^{N} a_i x(n - i)$$

or, in vector notation,

$$e_a(n) = x(n) - A_N^t X(n - 1) \tag{5.16}$$

The backward linear prediction error $e_b(n)$ is defined by

$$e_b(n) = x(n - N) - B_N^t X(n) \tag{5.17}$$

where B_N is the vector of the backward coefficients. The two filters are shown in Figure 5.3.

The minimization of $E[e_b^2(n)]$ with respect to the coefficients yields the backward linear prediction matrix equation

$$R_{N+1} \begin{bmatrix} -B_N \\ 1 \end{bmatrix} = \begin{bmatrix} 0 \\ E_{bN} \end{bmatrix} \tag{5.18}$$

Premultiplication by the co-identity matrix J_{N+1} gives

$$J_{N+1} R_{N+1} \begin{bmatrix} -B_N \\ 1 \end{bmatrix} = \begin{bmatrix} E_{bN} \\ 0 \end{bmatrix}$$

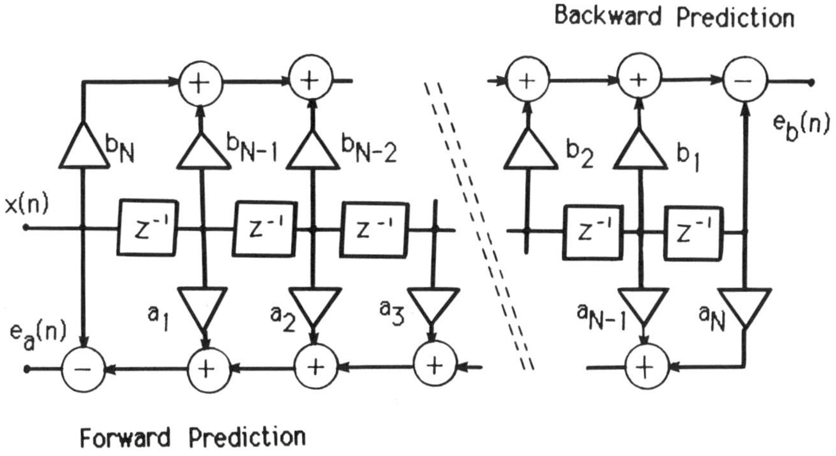

Backward Prediction

Forward Prediction

Figure 5.3 Forward and backward linear prediction error filters.

which, considering relation (3.57) in Chapter 3, yields

$$R_{N+1} \begin{bmatrix} 1 \\ -J_N B_N \end{bmatrix} = \begin{bmatrix} E_{bN} \\ 0 \end{bmatrix} \tag{5.19}$$

Hence

$$A_N = J_N B_N, \qquad E_{aN} = E_{bN} = E_N \tag{5.20}$$

For a stationary input signal, forward and backward prediction error powers are equal and the coefficients are the same, but in reverse order. Therefore, in theory, linear prediction analysis can be performed by the forward and backward approaches. However, it is in the transition phases that a difference appears, as seen in the next chapter. When the AC matrix is estimated, the best performance is achieved by combining both approaches, which gives the forward-backward linear prediction (FBLP) technique presented in Section 9.6.

Since the forward linear prediction error filter is minimum phase, the backward filter is maximum phase, due to (5.20).

An important property of backward linear prediction is that it provides a set of uncorrelated signals. The errors $e_{bi}(n)$ for successive orders $0 \leqslant i \leqslant N$ are not correlated. To show this useful result, let us express the vector of backward prediction errors in terms of the corresponding coefficients by repeatedly applying equation (5.18):

$$
\begin{bmatrix} e_{b0}(n) \\ e_{b1}(n) \\ e_{b2}(n) \\ \vdots \\ e_{b(N-1)}(n) \end{bmatrix}^t = X_N^t(n) \begin{bmatrix} 1 & -B_1 & & & \\ 0 & 1 & -B_2 & & \\ 0 & 0 & 1 & & -B_{N-1} \\ \vdots & \vdots & \vdots & \ddots & \\ 0 & 0 & 0 & \cdots & 1 \end{bmatrix} \tag{5.21}
$$

In more concise form, (5.21) is

$$
[e_b(n)]^t = X_N^t(n) M_B
$$

To check for the correlation, let us compute the backward error covariance matrix:

$$
E\{[e_b(n)][e_b(n)]^t\} = M_B^t R_N M_B \tag{5.22}
$$

By definition it is a symmetrical matrix. The product $R_N M_B$ is a lower triangular matrix, because of equation (5.18). The main diagonal consists of the successive prediction error powers $E_i (0 \leqslant i \leqslant N-1)$. But M_B^t is also a lower triangular matrix. Therefore, the product must have the same structure; since it must be symmetrical, it can only be a diagonal matrix. Hence

$$
E\{[e_b(n)][e_b(n)]^t\} = \mathrm{diag}(E_i) \tag{5.23}
$$

and the backward prediction error sequences are uncorrelated. It can be verified that the same reasoning cannot be applied to forward errors.

The AC matrix R_{N+1} used in the above prediction equations contains R_N, as shown in decomposition (5.4), and order iterative relations can be derived for linear prediction coefficients.

5.4 ORDER ITERATIVE RELATIONS

To simplify the equations, let

$$
r_N^a = \begin{bmatrix} r(1) \\ r(2) \\ \vdots \\ r(N) \end{bmatrix}, \qquad r_N^b = J_N r_N^a \tag{5.24}
$$

Now, the following equation is considered, in view of deriving relations between order N and order $N-1$ linear prediction equations:

$$
\begin{bmatrix} R_N & \vdots & r_N^b \\ \cdots & & \cdots \\ (r_N^b)^t & \vdots & r(0) \end{bmatrix} \begin{bmatrix} 1 \\ -A_{N-1} \\ 0 \end{bmatrix} = \begin{bmatrix} E_{N-1} \\ 0 \\ K_N \end{bmatrix} \tag{5.25}
$$

where

$$K_N = r(N) - \sum_{i=1}^{N-1} a_{i,N-1} r(N-i) \tag{5.26}$$

For backward linear prediction, using (5.20), we have

$$\begin{bmatrix} r(0) & \vdots & (r_N^a)^t \\ \hdashline r_N^a & \vdots & R_N \end{bmatrix} \begin{bmatrix} 0 \\ \hdashline -B_{N-1} \\ 1 \end{bmatrix} = \begin{bmatrix} K_N \\ 0 \\ E_{N-1} \end{bmatrix} \tag{5.27}$$

Multiplying both sides by the factor $k_N = K_N/E_{N-1}$ yields

$$R_{N+1} \left[\begin{bmatrix} 0 \\ -B_{N-1} \\ 1 \end{bmatrix} k_N \right] = \begin{bmatrix} k_N^2 E_{N-1} \\ 0 \\ K_N \end{bmatrix} \tag{5.28}$$

Subtracting (5.28) from (5.25) leads to the order N linear prediction equation, which for the coefficients implies the recursion

$$A_N = \begin{bmatrix} A_{N-1} \\ 0 \end{bmatrix} - k_N \begin{bmatrix} B_{N-1} \\ -1 \end{bmatrix} \tag{5.29}$$

and

$$E_N = E_{N-1}(1 - k_N^2) \tag{5.30}$$

for the prediction error power. The last row of recursion (5.29) gives the important relation

$$a_{NN} = k_N \tag{5.31}$$

Finally the order N linear prediction matrix equation (5.3) can be solved recursively by the procedure consisting of equations (5.28), (5.31), (5.29), and (5.30) and called the Levinson-Durbin algorithm. It is given in Figure 5.4, and the corresponding FORTRAN subroutine to solve a linear system is given in Annex 5.1. Solving a system of N linear equations when the matrix to be inverted is Toeplitz requires N divisions and $N(N+1)$ multiplications, instead of the $\frac{N^3}{3}$ multiplications mentioned in Section 3.4 for the triangular factorization.

An alternative approach to compute the scalars k_i is to use the cross-correlation variables h_{jN} defined by

$$h_{jN} = E[x(n)e_{aN}(n-j)] \tag{5.32}$$

where $e_{aN}(n)$ is the output of the forward prediction error filter having N

Initialization :

$$E_0 = r(0)$$

For $1 \leq j \leq N$:

$$k_j = \frac{1}{E_{j-1}} \left[r(j) - \sum_{i=1}^{j-1} a_{i\,j-1}\, r(j-i) \right]$$

$$a_{jj} = k_j$$

$$a_{ij} = a_{i\,j-1} - k_j\, a_{j-i,\,j-1} \quad ; \quad 1 \leq i \leq j-1$$

$$E_j = E_{j-1}\, (1 - k_j^2)$$

Figure 5.4 The Levinson-Durbin algorithm for solving the linear prediction equation.

coefficients [3]. As mentioned in Section 2.5, the sequence h_{jN} is the impulse response of the generating filter when $x(n)$ is an order N AR signal. From the definition (5.16) for $e_{aN}(n)$, the variables h_{jN} are expressed by

$$h_{jN} = r(j) - \sum_{i=1}^{N} a_{iN} r(i+j)$$

or, in vector notation,

$$h_{jN} = r(j) - (r_{jN})^t A_N \tag{5.33}$$

where

$$(r_{jN})^t = [r(j+1), r(j+2), \ldots, r(j+N)]$$

Clearly, the above definition leads to

$$h_{0N} = E_N \tag{5.34}$$

and

$$k_N = \frac{h_{(-N)(N-1)}}{E_{N-1}} = \frac{h_{(-N)(N-1)}}{h_{0(N-1)}} \tag{5.35}$$

A recursion can be derived from the prediction coefficient recursion (5.29) as follows:

$$h_{jN} = h_{j(N-1)} + k_N(r_{jN})^t \begin{bmatrix} B_{N-1} \\ -1 \end{bmatrix}$$ (5.36)

Developing the second term on the right gives

$$(r_{jN})^t \begin{bmatrix} B_{N-1} \\ -1 \end{bmatrix} = -h_{(-j-N)(N-1)}$$

Thus

$$h_{jN} = h_{j(N-1)} - k_N h_{(-j-N)(N-1)}$$ (5.38)

which yields, as a particular case if we take relation (5.35) into account,

$$h_{0N} = h_{0(N-1)}(1 - k_N^2) = E_N$$ (5.39)

Now a complete algorithm is available to compute the coefficients k_i. It is based entirely on the variables h_{ji} and consists of equations (5.35) and (5.38). The FORTRAN subroutine is given in Annex 5.2. The initial conditions are given by definition (5.33):

$$h_{j0} = r(j)$$ (5.40)

According to the h_{jN} definition (5.32) and the basic decorrelation property of linear prediction, the following equations hold:

$$h_{ji} = 0, \qquad -i \leqslant j \leqslant -1$$ (5.41)

If N coefficients k_i have to be computed, the indexes of the variables h_{ji} involved are in the range $(-N, N-1)$, as can be seen from equations (5.35) and (5.38). The multiplication count is about $N(N-1)$.

An additional property of the above algorithm is that the variables h_{ij} are bounded, which is useful for fixed-point implementation. Considering the definition (5.32), the cross-correlation inequality (3.10) of Chapter 3 yields

$$|h_{jN}| = |E[x(n)e(n-j)]| \leqslant \tfrac{1}{2}(r(0) + E_N)$$

Since $E_N \leqslant r(0)$ for all N,

$$|h_{jN}| \leqslant r(0)$$ (5.42)

The variables h_{jN} are bounded in magnitude by the signal power.

The order recursions can be associated with a particular structure, the lattice prediction filter.

5.5 THE LATTICE LINEAR PREDICTION FILTER

The coefficients k_i establish direct relations between forward and backward prediction errors for consecutive orders. From the definition of the order N forward prediction error $e_{aN}(n)$, we have

$$e_{aN}(n) = x(n) - A_N^t X(n-1) \tag{5.43}$$

and the coefficient recursion (5.29), we derive

$$e_{aN}(n) = e_{a(N-1)}(n) - k_N[-B_{N-1}^t, 1]X(n-1) \tag{5.44}$$

The order N backward prediction error $e_{bN}(n)$ is

$$e_{bN}(n) = x(n-N) - B_N^t X(n) \tag{5.45}$$

For order $N-1$,

$$e_{b(N-1)}(n) = x(n+1-N) - \sum_{i=1}^{N-1} b_{i(N-1)}x(n+1-i) = [-B_{N-1}^t, 1]X(n) \tag{5.46}$$

Therefore, the prediction errors can be rewritten as

$$e_{aN}(n) = e_{a(N-1)}(n) - k_N e_{b(N-1)}(n-1) \tag{5.47}$$

and

$$e_{bN}(n) = e_{b(N-1)}(n-1) - k_N e_{a(N-1)}(n) \tag{5.48}$$

The corresponding structure is shown in Figure 5.5; it is called a lattice filter section, and a complete FIR filter of order N is realized by cascading N such sections. Indeed, to start, $e_{b0}(n) = x(n)$. Now the lattice coefficients k_i can be further characterized. Consider the cross-correlation

$$E[e_{aN}(n)e_{bN}(n-1)] = r(N+1) - B_N^t r_N^a - A_N^t J_N r_N^a + A_N^t R_N B_N \tag{5.49}$$

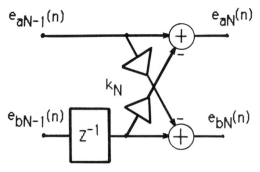

Figure 5.5 Lattice linear prediction filter section.

Because of the backward prediction equation

$$R_N B_N = r_N^b = J_N r_N^a \tag{5.50}$$

the sum of the last two terms in the above cross-correlation is zero. Hence

$$E[e_{aN}(n)e_{bN}(n-1)] = r(N+1) - B_N^t r_N^a = K_{N+1}$$

and

$$k_N = \frac{E[e_{a(N-1)}(n)e_{b(N-1)}(n-1)]}{E_{N-1}} \tag{5.51}$$

The lattice coefficients represent a normalized cross-correlation of forward and backward prediction errors. They are often called the PARCOR coefficients [4]. Due to wave propagation analogy, they are also called the reflection coefficients.

The lattice coefficient k_N is related to the N zeros z_i of the order N FIR prediction error filter, whose transfer function is

$$H_N(z) = 1 - \sum_{i=1}^{N} a_{iN} z^{-i} = \prod_{i=1}^{N} (1 - z_i z^{-1}) \tag{5.52}$$

Since $k_N = a_{NN}$, we have

$$k_N = (-1)^{N+1} \prod_{i=1}^{N} z_i \tag{5.53}$$

From the filter linear phase property, we know that $|z_i| \leqslant 1$, which yields

$$|k_N| \leqslant 1 \tag{5.54}$$

Conversely, using (5.29), it can be shown iteratively that, if the lattice coefficient absolute values are bounded by unity, then the prediction error filter has all its roots inside the unit circle and, thus, it is minimum phase. Therefore, it is very easy to check for the minimum phase property of a lattice FIR filter. Just check that the magnitude of the lattice coefficients does not exceed unity.

The correspondence between PARCOR and the transversal filter coefficients is provided by recursion (5.29). In order to get the set of $a_{iN}(1 \leqslant i \leqslant N)$ from the set of $k_i(1 \leqslant i \leqslant N)$, we need to iterate the recursion N times with increasing indexes. To get the k_i from the a_{iN}, we must proceed in the reverse order and calculate the intermediate coefficients $a_{ji}(N-1 \geqslant i \geqslant 1; j \leqslant i)$ by the following expression:

$$a_{j(i-1)} = \frac{1}{1-k_i^2} [a_{ji} + k_i a_{(i-j)i}], \qquad k_i = a_{ii} \tag{5.55}$$

The procedure is stopped if $k_i = 1$, which means that the signal consists of i sinusoids without additive noise.

A set of interesting properties of the transversal filter coefficients can be deduced from the magnitude limitation of the PARCOR coefficients [5]. For example,

$$|a_{iN}| < \frac{N!}{(N-i)!i!} \leqslant 2^{N-1} \tag{5.56}$$

which can be useful for coefficient scaling in fixed-point implementation and leads to

$$\sum_{i=1}^{N} |a_{iN}| < 2^N - 1 \tag{5.57}$$

and

$$\|A_N\|^2 = A_N^t A_N < \frac{(2n)!}{(n!)^2} - 1 \tag{5.58}$$

This bound is reached for the two theoretical extreme cases where $k_i = -1$ and $k_i = (-1)^{i-1}$ $(1 \leqslant i \leqslant N)$.

The results we have obtained in linear prediction now allow us to complete our discussion on AC matrices and, particularly, their inverses.

5.6 THE INVERSE AC MATRIX

When computing the inverse of a matrix, first compute the determinant. The linear prediction matrix equation is

$$\begin{bmatrix} 1 \\ -A_N \end{bmatrix} = \begin{bmatrix} r(0) & (r_N^a)^t \\ r_N^a & R_N \end{bmatrix}^{-1} \begin{bmatrix} E_N \\ 0 \end{bmatrix} \tag{5.59}$$

The first row yields

$$1 = \frac{\det R_N}{\det R_{N+1}} E_N \tag{5.60}$$

which, using the Levinson recursions, leads to

$$\det R_N = [r(0)]^N \prod_{i=1}^{N-1} (1 - k_i^2)^{N-i} \tag{5.61}$$

To exploit further equation (5.59), let us denote by V_i the column vectors of the inverse matrix R_{N+1}^{-1}.

Considering the forward and backward linear prediction equations, we can write the vectors V_1 and V_{N+1} as

$$V_1 = \frac{1}{E_N} \begin{bmatrix} 1 \\ -A_N \end{bmatrix}, \qquad V_{N+1} = \frac{1}{E_N} \begin{bmatrix} -B_N \\ 1 \end{bmatrix} \tag{5.62}$$

Thus, the prediction coefficients show up directly in the inverse AC matrix, which can be completely expressed in terms of these coefficients.

Let us consider the $2(N + 1) \times (N + 1)$ rectangular matrix M_A defined by

$$M_A^t = \begin{bmatrix} 1 & -a_{1N} & -a_{2N} & \cdots & -a_{NN} & 0 & \cdots & 0 & 0 \\ 0 & 1 & -a_{1N} & \cdots & -a_{(N-1)N} & -a_{NN} & \cdots & 0 & 0 \\ \vdots & \vdots & \vdots & & \vdots & \vdots & & \vdots & \vdots \\ 0 & 0 & 0 & \cdots & 1 & -a_{1N} & \cdots & -a_{NN} & 0 \end{bmatrix}$$

$$\underbrace{\phantom{1 \quad -a_{1N} \quad -a_{2N} \quad \cdots \quad -a_{NN}}}_{N+1} \qquad \underbrace{}_{N+1}$$

$$(5.63)$$

The prediction equation (5.3) and relations (2.64) and (2.72) for AR signals yield the equality

$$M_A^t R_{2(N+1)} M_A = E_N I_{N+1} \tag{5.64}$$

where $R_{2(N+1)}$ is the AC matrix of the order N AR signal. Pre- and postmultiplying by M_A and M_A^t; respectively, gives

$$(M_A M_A^t) R_{2(N+1)} (M_A M_A^t) = (M_A M_A^t) E_N \tag{5.65}$$

The expression of the matrix R_{N+1}^{-1} is obtained by partitioning M_A into two square $(N + 1) \times (N + 1)$ matrices M_{A1} and M_{A2},

$$M_A^t = [M_{A1}^t, M_{A2}^t] \tag{5.66}$$

and taking into account the special properties of the triangular matrices involved:

$$R_{N+1}^{-1} = \frac{1}{E_N} (M_{A1}^t M_{A1} - M_{A2} M_{A2}^t) \tag{5.67}$$

This expression shows that the inverse AC matrix is doubly symmetric. If the signal is AR with order less than N, then R_{N+1}^{-1} is Toeplitz in the center, but edge effects appear in the upper left and lower right corners. A simple example is given in Section 3.4.

Decomposition (5.67) can be extended to matrices which are not doubly symmetric. In that case, the matrices M_{B1} and M_{B2} of the backward prediction coefficients are involved, and the equation becomes

$$R_{N+1}^{-1} = \frac{1}{E_{aN}} (M_{A1}^t M_{B1} - M_{B2} M_{A2}^t) \tag{5.67a}$$

An alternative decomposition of R_{N+1}^{-1} can be derived from the cross-correlation properties of data and error sequences.

Since the error signal $e_N(n)$ is not correlated with the input data $x(n-1), \ldots, x(n-N)$, the sequences $e_{N-i}(n-i)$, $0 \leqslant i \leqslant N$, are not correlated. In vector form they are written

$$
\begin{bmatrix}
e_N(n) \\
e_{N-1}(n-1) \\
\vdots \\
e_0(n-N)
\end{bmatrix}
=
\begin{bmatrix}
1 & \cdots & -A_N & \cdots & \cdots \\
0 & 1 & \cdots & -A_{N-1} & \cdots \\
\vdots & \vdots & \cdot & \cdot & \\
0 & 0 & \cdots & & 1
\end{bmatrix}
\begin{bmatrix}
x(n) \\
x(n-1) \\
\vdots \\
x(n-N)
\end{bmatrix}
\tag{5.68}
$$

The covariance matrix is the diagonal matrix of the prediction errors. After algebraic manipulations we have

$$
R_{N+1}^{-1} =
\begin{bmatrix}
1 & 0 & \cdots & 0 \\
-a_{1N} & 1 & \cdots & 0 \\
-a_{2N} & -a_{2(N-1)} & \cdots & 0 \\
\vdots & \vdots & \ddots & \vdots \\
-a_{NN} & -a_{(N-1)(N-1)} & \cdots & 1
\end{bmatrix}
\begin{bmatrix}
E_N^{-1} & 0 & \cdots & 0 \\
0 & E_{N-1}^{-1} & \cdots & 0 \\
\vdots & \vdots & \ddots & \vdots \\
0 & 0 & \cdots & E_0^{-1}
\end{bmatrix}
$$

$$
\times
\begin{bmatrix}
1 & -a_{1N} & -a_{2N} & \cdots & -a_{NN} \\
0 & 1 & -a_{1(N-1)} & \cdots & -a_{(N-1)(N-1)} \\
\vdots & \vdots & \cdot & \cdots & \vdots \\
0 & 0 & \cdots & & 1
\end{bmatrix}
\tag{5.69}
$$

This is the triangular Cholesky decomposition of the inverse AC matrix. It can also be obtained by considering the backward prediction errors, which are also uncorrelated, as shown in Section 5.3.

The important point in this section is that the inverse AC matrix is completely represented by the forward prediction error power and the prediction coefficients. Therefore, LS algorithms which implement R_N^{-1} need not manipulate that matrix, but need only calculate the forward prediction error power and the forward and backward prediction coefficients. This is the essence of FLS algorithms.

5.7 THE NOTCH FILTER AND ITS APPROXIMATIONS

The ideal predictor is the filter which cancels the predictable components in the signal without amplifying the unpredictable ones. That favorable situation occurs with sinusoids in white noise, and the ideal filter is the notch filter with frequency response

$$
H_{NI}(\omega) = 1 - \sum_{i=1}^{M} \delta(\omega - \omega_i)
\tag{5.70}
$$

where $\delta(x)$ is the Dirac distribution and the ω_i, $1 \leqslant i \leqslant M$, are the frequencies of the sinusoids. Clearly, such a filter completely cancels the sinusoids and does not amplify the input white noise.

An arbitrarily close realization $H_N(\omega)$ of the ideal filter is achieved by

$$
H_N(z) = \frac{\displaystyle\prod_{i=1}^{M} (1 - e^{j\omega_i}z^{-1})}{\displaystyle\prod_{i=1}^{M} (1 - (1 - \varepsilon)e^{j\omega_i}z^{-1})}
\tag{5.71}
$$

where the positive scalar ε is made arbitrarily small [6]. The frequency response of a second-order notch filter is shown in Figure 5.6, with the location of poles and zeros in the z-plane.

The notch filter cannot be realized by an FIR predictor. However, it can be approximated by developing in series the factors in the denominator of $H_N(z)$, which yields

$$
\frac{1}{1 - P_i z^{-1}} = 1 + \sum_{n=1}^{\infty} (P_i z^{-1})^n
\tag{5.72}
$$

This approach is used to figure out the location in the z-plane of the zeros and poles of linear prediction filters.

5.8 ZEROS OF FIR PREDICTION ERROR FILTERS

The first-order notch filter $H_{N1}(z)$ is adequate to handle zero frequency signals:

$$
H_{N1}(z) = \frac{1 - z^{-1}}{1 - (1 - \varepsilon)z^{-1}}
\tag{5.73}
$$

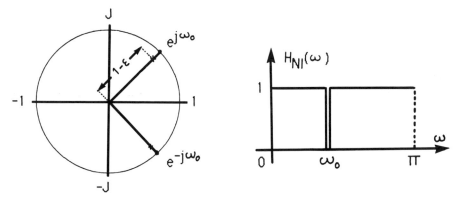

Figure 5.6 The notch filter response, zeros and poles.

A simple Tchebycheff FIR approximation is

$$H(z) = \frac{1 - z^{-1}}{1 - (1 - \varepsilon)z^{-1}} [1 - (bz^{-1})^N]$$

where b is a positive real constant. Now, a realizable filter is obtained for $b = 1 - \varepsilon$, because

$$H(z) = (1 - z^{-1})[1 + bz^{-1} + \cdots + b^{N-1}z^{-(N-1)}] \tag{5.74}$$

The constant b can be calculated to minimize the prediction error power. For a zero frequency signal $s(n)$ of unit power, a white input noise with power σ_b^2, the output power of the filter with transfer function $H(z)$ given by (5.74) is

$$E[e^2(n)] = 2\sigma_b^2 \frac{1 + b^{2N-1}}{1 + b} \tag{5.75}$$

The minimum is reached by setting to zero the derivative with respect to b; thus

$$b = \left[\frac{1}{2N - 1 + (2N - 2)b} \right]^{1/2(N-1)} \tag{5.76}$$

For b reasonably close to unity the following approximation is valid:

$$b \approx \left[\frac{1}{4N - 3} \right]^{1/2(N-1)} \tag{5.77}$$

According to (5.74) the zeros of the filter which approximates the prediction error filter are located at $+1$ and $be^{j2\pi l/N}$, $1 \leqslant i \leqslant N - 1$, in the complex plane. And the circle radius b does not depend on the noise power. For large N, b comes close to unity, and estimate (5.77) is all the better. Figure 5.7(a) shows true and estimated zeros for a 12-order prediction error filter.

A refinement in the above procedure is to replace $1 - z^{-1}$ by $1 - az^{-1}$ in $H(z)$ and optimize the scalar a because, in the prediction of noisy signals, the filter zeros are close to but not on the unit circle, as pointed out earlier, particularly in Section 5.2.

The above approach can be extended to estimate the prediction error filter zeros when the input signal consists of M real sinusoids of equal amplitude and uniformly distributed on the frequency axis. The approximating transfer function is

$$H(z) = \frac{1 - z^{-2M}}{1 - b^{2M}z^{-2M}} (1 - b^N z^{-N}) \tag{5.78}$$

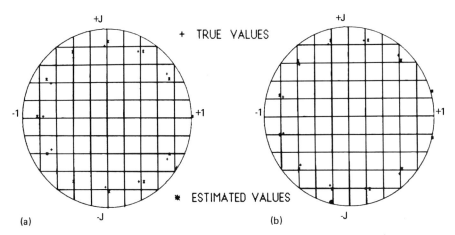

Figure 5.7   Zeros of a 12-order predictor applied to (a) a zero frequency signal, (b) a $\frac{\pi}{12}$ frequency sinusoid.

If $N = k\,2M$, for integer k, the output error power is

$$E[e^2(n)] = 2\sigma_b^2 \frac{1 + b^{2N-2M}}{1 + b^{2M}} \tag{5.79}$$

the minimization procedure leads to

$$b \approx \left[\frac{M}{2N - 3M} \right]^{1/2(N-2M)} \tag{5.80}$$

Equation (5.77) corresponds to the above expression when $M = \frac{1}{2}$. Note that the zero circle radius b depends on the number $N - 2M$, which can be viewed as the number of free or uncommitted zeros in the filter; the mission of these zeros is to bring down the amplification of the input noise power. If the noise is not flat, they are no longer on a circle within the unit circle.

The validity of the above derivation might look rather restricted, since the sinusoidal frequencies have to be uniformly distributed and the filter order N must be a multiple of the number of sinusoids M. Expression (5.80) remains a reasonably good approximation of the zero modulus as soon as $N > 2M$. For example, the true and estimated zeros of an order 12 linear prediction error filter, applied to the sinusoid with frequency $\frac{\pi}{12}$, are shown in Figure 5.7(b).

When the sinusoidal frequencies are arbitrarily distributed on the frequency, the output noise power is increased with respect to the uniform case and the zeros in excess of $2M$ come closer to the unit circle center. Therefore expression (5.80) may be regarded as an estimation of the upper bound of the distance of the zeros in excess of $2M$ to the center of the unit circle. That result is useful for the retrieval of sinusoids in noise [7].

The foregoing results provide useful additional information about the magnitude of the PARCOR coefficients.

When the PARCOR coefficients k_i are calculated iteratively, their magnitudes grow, monotonically or not, up to a maximum value which, because of equation (5.53), corresponds to the prediction filter order best fitted to the signal model. Beyond, the k_i decrease in magnitude, due to the presence of the zeros in excess.

If the signal consists of M real sinusoids, then

$$|k_N| \approx b^{N-2M}, \qquad N \geqslant 2M \tag{5.81}$$

Substituting (5.80) into (5.81) gives

$$k_N \approx \left(\frac{M}{2N - 3M}\right)^{1/2} \qquad N \geqslant 2M \tag{5.82}$$

Equation (5.82) is a decreasing law which can be extended to any signal and considered as an upper bound estimate for the lattice coefficient magnitudes for predictor orders exceeding the signal model order. In Figure 5.8 true lattice coefficients are compared with estimates for sinusoids at frequencies $\frac{\pi}{2}$ and $\frac{\pi}{12}$.

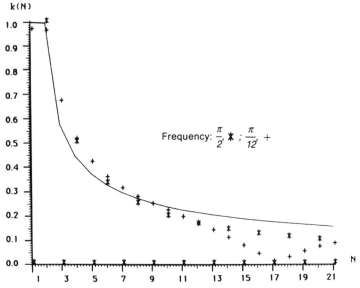

Figure 5.8 Lattice coefficients vs. predictor order for sinusoids.

The magnitude of the maximum PARCOR coefficient is related to the input SNR. The relation is simple for M sinusoids uniformly distributed on the frequency axis, because the order $2M$ prediction error filter is

$$H(z) = 1 - b^{2M}z^{-2M} \tag{5.83}$$

The optimum value of b is derived from the prediction error power as before, so

$$b^{2M} = |k_{2M}| = \frac{\text{SNR}}{1 + \text{SNR}} \tag{5.84}$$

The approach taken to locate the predictor zeros can also be applied to the poles of an IIR filter.

5.9 POLES OF IIR PREDICTION ERROR FILTERS

The transfer function of a purely recursive IIR filter of order N is

$$H(z) = \frac{1}{1 - \displaystyle\sum_{i=1}^{N} b_i z^{-i}} \tag{5.85}$$

Considering a zero frequency signal in noise, to begin with, we can obtain a Tchebycheff approximation of the prediction error filter $1 - az^{-1}$ by the expression

$$H(z) = \frac{1 - az^{-1}}{1 - a^{N+1}z^{-(N+1)}} = \frac{1}{1 + az^{-1} + \cdots + a^N z^{-N}} \tag{5.86}$$

where $0 \ll a < 1$. Now the prediction error power is

$$E[e^2(n)] = |H(1)|^2 + \sigma_b^2\left(\sum_{i=0}^{\infty} h_i^2\right)$$

where the h_i is the filter impulse response. A simple approximation is

$$E[e^2(n)] \approx |H(1)|^2 + \sigma_b^2 \frac{1 + a^2}{1 - a^{2(N+1)}} \tag{5.87}$$

The parameter a is obtained by setting to zero the derivative of the prediction error power. However, a simple expression is not easily obtained.

Two different situations must be considered separately, according to the noise power σ_b^2. For small noise power

$$\frac{\partial}{\partial a}\left(\frac{1 - a}{1 - a^{N+1}}\right)^2 \approx -\frac{1}{N+1}, \quad \frac{\partial}{\partial a}\left[\frac{1 + a^2}{1 - a^{2(N+1)}}\right] \approx \frac{1}{N+1}\frac{1}{(1 - a)^2}$$

and

$$a \approx 1 - \sigma_b \tag{5.88}$$

On the other hand, for large noise power, simple approximations are

$$\frac{\partial}{\partial a}\left(\frac{1-a}{1-a^{N+1}}\right)^2 \approx -2(1-a), \quad \frac{\partial}{\partial a}\left[\frac{1+a^2}{1-a^{2(N+1)}}\right] \approx 2a$$

which yield

$$a \approx \frac{1}{1+\sigma_b^2} \tag{5.89}$$

In any case, for a zero frequency signal the poles of the IIR filter are uniformly distributed in the complex plane on a circle whose radius depends on the SNR. We can rewrite $H(z)$ as

$$H(z) = \frac{1}{\displaystyle\prod_{n=1}^{N}(1-ae^{jn\omega_0}z^{-1})}, \quad \omega_0 = \frac{2\pi}{N+1} \tag{5.90}$$

There is no pole at the signal frequency and, in some sense, the IIR predictor operates by default.

The prediction gain is limited. Since $|a| < 1$ for stability reasons, we derive a simple bound $E_{\min}$ for the prediction error power from (5.87) and (5.86), neglecting the input noise:

$$E_{\min} = \frac{1}{(N+1)^2} \tag{5.91}$$

The above derivations can be extended to signals made of sinusoids in noise. The results show, as above, that the purely recursive IIR predictors are not as efficient as their FIR counterparts.

5.10 GRADIENT ADAPTIVE PREDICTORS

The gradient techniques described in the previous chapter can be applied to prediction filters. A second-order FIR filter is taken as an example in Section 4.4. The reference signal is the input signal itself, which simplifies some expressions, such as coefficient and internal data word-length estimations (4.61) and (4.65) in Chapter 4, which in linear prediction become

$$b_c \approx \log_2(\tau_e) + \log_2(G_S) + \log_2(a_{\max}) \tag{5.92}$$

and

$$b_i \approx 2 + \tfrac{1}{2}\log_2(\tau_e) + \log_2(G_p) \tag{5.93}$$

where G_p^2 is the prediction gain, defined, according to equation (4.9) in Chapter 4, as the input signal-to-prediction-error power ratio. The maximum magnitude of the coefficients, a_{max}, is bounded by 2^{N-1} according to inequality (5.56).

The purely recursive IIR prediction error filter in Figure 5.9 is a good illustration of adaptive IIR filters. Its equations are

$$e(n + 1) = x(n + 1) - B^t(n)E(n)$$
$$B(n + 1) = B(n) + \delta e(n + 1)E(n) \tag{5.94}$$

with

$$B^t(n) = [b_1(n), \ldots, b_N(n)], \qquad E^t(n) = [e(n), \ldots, e(n + 1 - N)]$$

The coefficient updating equation can be rewritten as

$$B(n + 1) = [I_N - \delta E(n)E^t(n)]B(n) + \delta x(n + 1)E(n) \tag{5.95}$$

The steady-state position is reached when the error $e(n + 1)$ is no longer correlated with the elements of the error vector; the filter tends to decorrelate the error sequence. The steady-state coefficient vector B_∞ is

$$B_\infty = (E[E(n)E^t(n)])^{-1}E[x(n + 1)E(n)] \tag{5.96}$$

and the error covariance matrix should be close to a diagonal matrix:

$$E[E(n)E^t(n)] \approx \sigma_e^2 I_N \tag{5.97}$$

The output power is

$$E[e^2(n + 1)] = E[x^2(n + 1)] - B_\infty^t E[E(n)E^t(n)]B_\infty \tag{5.98}$$

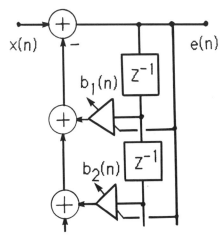

Figure 5.9 Purely recursive IIR prediction filter.

which yields the prediction gain

$$G_p^2 = \frac{\sigma_x^2}{\sigma_e^2} \approx 1 + B_\infty^t B_\infty \tag{5.99}$$

Therefore the coefficients should take as large values as possible.

Note that, in practice, a local instability phenomenon can occur with recursive gradient predictors [8]. As indicated in the previous section, the additive input noise keeps the poles inside the unit circle. If that noise is small enough, in a gradient scheme with given step δ, the poles jump over the unit circle. The filter becomes unstable, which can be interpreted as the addition to the filter input of a spurious sinusoidal component, exponentially growing in magnitude and at the frequency of the pole. The adaptation process takes that component into account, reacts exponentially as well, and the pole is pushed back in the unit circle, which eliminates the above spurious component. Hence the local instability, which can be prevented by the introduction of a leakage factor as in Section 4.6, which yields the coefficient updating equation

$$B(n + 1) = (1 - \gamma)B(n) + \delta e(n + 1)E(n) \tag{5.100}$$

The bound on the adaptation step size δ can be determined, as in Section 4.2, by considering the a posteriori error

$$\varepsilon(n + 1) = e(n + 1)[1 - \delta E^t(n)E(n)] \tag{5.101}$$

which leads to the bound

$$0 < \delta < \frac{2}{N\sigma_e^2} \tag{5.102}$$

Since the output error power is at most equal to the input signal power, the bound is the same as for the FIR structure. The initial time constant is also about the same, if the step size is small enough, due to the following approximation, which is valid for small coefficient magnitudes:

$$\frac{1}{1 + \sum\limits_{i=1}^{N} b_i z^{-i}} \approx 1 - \sum\limits_{i=1}^{N} b_i z^{-i} \tag{5.103}$$

As an illustration, the trajectories of the six poles of a purely recursive IIR prediction error filter applied to a sinusoid with frequency $\frac{2\pi}{3}$ are shown in Figure 5.10. After the initial phase, there are no poles at frequencies $\pm\frac{2\pi}{3}$.

The lattice structure presented in Section 5.5 can also be implemented in a gradient adaptive prediction error filter, as shown in Figure 5.11 for the FIR case. Several criteria can be used to update the coefficients k_i. A simple one is

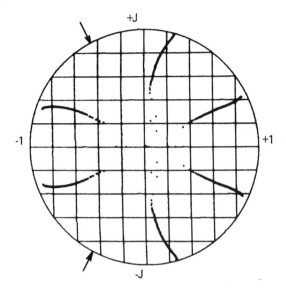

Figure 5.10 Pole trajectories of a gradient adaptive IIR predictor applied to a sinusoid at frequency $2\frac{\pi}{3}$.

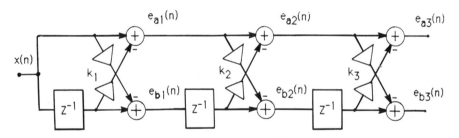

Figure 5.11 FIR lattice prediction error filter.

the minimization of the sum of forward and backward prediction error powers at each stage. The derivation of equations (5.47) and (5.48) with respect to the coefficients leads to the updating relations ($1 \leqslant i \leqslant N$)

$$k_i(n + 1) = k_i(n) + \frac{\delta}{2}[e_{ai}(n + 1)e_{b(i-1)}(n) + e_{bi}(n + 1)e_{a(i-1)}(n + 1)]$$

$$(5.104)$$

which, from (5.47) and (5.48), can be rewritten as

$$k_i(n + 1) = k_i(n) + \delta \left[e_{a(i-1)}(n + 1)e_{b(i-1)}(n + 1) \right.$$

$$\left. - k_i(n) \frac{e_{b(i-1)}^2(n) + e_{a(i-1)}^2(n + 1)}{2} \right] \qquad (5.105)$$

Clearly, the steady-state solution $k_{i\infty}$ agrees with the PARCOR coefficient definition (5.51).

The performance of the lattice gradient algorithm can be assessed through the methods developed in Chapter 4, and comparisons can be made with the transversal FIR structure, including computation accuracies [9, 10]. However, the lattice filter is made of sections which have to be analyzed in turn.

The coefficient updating for the first lattice section, according to Figure 5.11, is

$$k_1(n + 1) = k_1(n) + \delta \left[x(n + 1)x(n) - k_1(n) \frac{x^2(n + 1) + x^2(n)}{2} \right] \qquad (5.106)$$

For comparison, the updating equation of the coefficient of the first-order FIR filter can be written as

$$a(n + 1) = a(n) + \delta [x(n + 1)x(n) - a(n)x^2(n)] \qquad (5.107)$$

The only difference resides in the better power estimation performed by the last term on the right side of (5.106), and it can be assumed that the first lattice section performs like a first-order FIR prediction error filter, which leads to the residual error

$$E_{1R} = (1 - k_1^2)\sigma_x^2 \left(1 + \frac{\delta}{2}\sigma_x^2 \right) \qquad (5.108)$$

To assess the complete lattice prediction error filter, we now consider the subsequent sections. However, the adaptation step sizes are adjusted in these sections to reflect the decrease in signal powers. To make the time constant homogeneous, the adaptation step sizes in different sections are made inversely proportional to the input signal powers.

In such conditions, the first section is crucial for global performance and accuracy requirements. For example, the first section is the major contributor to the filter excess output noise power, and E_{1R} can be taken as the total lattice filter residual error.

Thus, transversal and lattice filters have the same excess output noise power if the following equality holds:

$$\sigma_x^2 \prod_{i=1}^{N} (1 - k_i^2) \frac{\delta}{2} N\sigma_x^2 = (1 - k_1^2)\sigma_x^2 \frac{\delta}{2} \sigma_x^2$$

Therefore, the lattice gradient filter is attractive, under the above hypotheses, if

$$\prod_{i=2}^{N} \frac{1}{1 - k_i^2} < N \qquad (5.109)$$

that is, when the system gain is small and when the first section is very efficient, which can be true in linear prediction of speech, for example. Combinations of lattice and transversal adaptive filters can be envisaged, and the above results suggest cascading a lattice section and a transversal filter [11].

As for computational accuracy, the coefficient magnitudes of lattice filters are bounded by unity. Therefore, the coefficient word length for the lattice prediction error filter can be estimated by

$$b_{cl} \approx \log_2(\tau_e) + \log_2(G_p) \qquad (5.110)$$

which can be appreciably smaller than estimate (5.92) for the transversal counterpart.

Naturally, simplified adaptive approaches, like LAV and sign algorithms, can also be used in linear prediction with any structure.

5.11 LINEAR PREDICTION AND HARMONIC DECOMPOSITION

Two different representations of a signal given by the first $N + 1$ terms $[r(0), r(1), \ldots, r(N)]$ of its ACF have been obtained. The harmonic decomposition presented in Section 2.11 corresponds to the modeling by a set of sinusoids and is also called composite sinusoidal modeling (CSM); it yields the following expression for the signal spectrum $S(\omega)$ according to relation (2.127) of Chapter 2:

$$S(\omega) = \sum_{k=1}^{N/2} |S_k|^2 [\delta(\omega - \omega_k) + \delta(\omega + \omega_k)] \qquad (5.111)$$

Linear prediction provides a representation of the signal spectrum by

$$S(\omega) = \frac{\sigma_e^2}{\left| 1 - \sum_{i=1}^{N} a_i e^{-ji\omega} \right|^2} \qquad (5.112)$$

Relations between these two approaches can be established by considering the decomposition of the z-transfer function of the prediction error filter into two parts with symmetric and antisymmetric coefficients, which is the line spectrum pair (LSP) representation [12].

The order recursion (5.29) is expressed in terms of z-polynomials by

$$1 - A_N(z) = 1 - A_{N-1}(z) - k_N z^{-N}[1 - A_{N-1}(z^{-1})] \tag{5.113}$$

where

$$1 - A_N(z) = 1 - \sum_{i=1}^{N} a_{iN} z^{-i} \tag{5.114}$$

Let us consider now the order $N + 1$ and denote by $P_N(z)$ the polynomial obtained when $k_{N+1} = 1$:

$$P_N(z) = 1 - A_N(z) - z^{-(N+1)}[1 - A_N(z^{-1})] \tag{5.115}$$

Let $Q_N(z)$ be the polynomial obtained when $k_{N+1} = -1$:

$$Q_N(z) = 1 - A_N(z) + z^{-(N+1)}[1 - A_N(z^{-1})] \tag{5.116}$$

Clearly, this is a decomposition of the polynomial (5.114):

$$1 - A_N(z) = \tfrac{1}{2}[P_N(z) + Q_N(z)] \tag{5.117}$$

and $\tfrac{1}{2}P_N(z)$ and $\tfrac{1}{2}Q_N(z)$ are polynomials with antisymmetric and symmetric coefficients, respectively.

Since $k_{N+1} = \pm 1$, due to the results in Section 5.5 and equation (5.53), $P_N(z)$ and $Q_N(z)$ have all their zeros on the unit circle. Furthermore, if N is even, it is readily verified that $P_N(1) = 0 = Q_N(-1)$. Therefore, the following factorization is obtained:

$$P_N(z) = (1 - z^{-1}) \prod_{i=1}^{N/2} (1 - 2\cos(\theta_i)z^{-1} + z^{-2})$$

$$Q_N(z) = (1 + z^{-1}) \prod_{i=1}^{N/2} (1 - 2\cos(\omega_i)z^{-1} + z^{-2}) \tag{5.118}$$

The two sets of parameters θ_i and ω_i ($1 \leqslant i \leqslant N$) are called the LSP parameters.

If $z_0 = e^{j\omega_0}$ is a zero of the polynomial $1 - A(z)$ on the unit circle, it is also a zero of $P_N(z)$ and $Q_N(z)$. Now if this zero moves inside the unit circle, the corresponding zeros of $P_N(z)$ and $Q_N(z)$ move on the unit circle in opposite directions from ω_0. A necessary and sufficient condition for the polynomial $1 - A(z)$ to be minimum phase is that the zeros of $P_N(z)$ and $Q_N(z)$ be simple and alternate on the unit circle [13].

The above approach provides a realization structure for the prediction error filter in Fig. 5.12. The z-transfer functions $F(z)$ and $G(z)$ are the linear

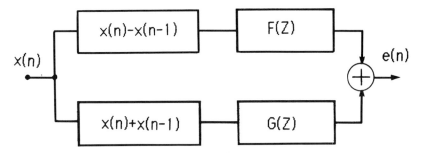

Figure 5.12 Line pair spectrum predictor.

phase factors in (5.118). This structure is amenable to implementation as a cascade of second-order sections, and the overall minimum phase property is checked by observing the alternation of the z^{-1} coefficients. It can be used for predictors with poles and zeros [14].

Equations (5.118) show that the LSP parameters θ_i and ω_i are obtained by harmonic decomposition of the sequences $x(n) - x(n - 1)$ and $x(n) + x(n - 1)$. This is an interesting link between harmonic decomposition, or CSM, and linear prediction.

5.12 CONCLUSION

Linear prediction error filters have been studied. Properties and coefficient design techniques have been presented. The analysis of first- and second-order filters yields simple results which are useful in signal analysis, particularly for the detection of sinusoidal components in a spectrum. Backward linear prediction provides a set of uncorrelated sequences. Combined with forward prediction, it leads to order iterative relations which correspond to a particular structure, the lattice filter. The lattice or PARCOR coefficients enjoy a number of interesting properties, and they can be calculated from the signal ACF by efficient algorithms.

The inverse AC matrix, which is involved in LS algorithms, can be expressed in terms of forward and backward prediction coefficients and prediction error power. To manipulate prediction filters and fast algorithms, it is important that we be able to locate the zeros in the unit circle; the analysis based on the notch filter and carried out for sinusoids in noise provides an insight useful for more general signals.

The gradient adaptive techniques apply to linear prediction filters with a number of simplifications, and the lattice structure is an appealing alternative to the transversal structure. An additional realization option is offered by the LSP approach, which provides an interesting link between linear prediction and harmonic decomposition.

EXERCISES

1. Calculate the impulse responses h_{ji} ($1 \leqslant j \leqslant 3; 0 \leqslant i \leqslant 6$) corresponding to the following z-transfer functions:

$$H_1(z) = (1 + z^{-1} + 0.5z^{-2})^2$$
$$H_2(z) = \tfrac{1}{2}(1 + z^{-1} + 0.5z^{-2})(1 + 2z^{-1} + 2z^{-2})$$
$$H_3(z) = \tfrac{1}{4}(1 + 2z^{-1} + 2z^{-2})^2$$

Calculate the functions

$$E_j(n) = \sum_{i=0}^{n} h_{ji}^2, \qquad 0 \leqslant n \leqslant 6, \ 1 \leqslant j \leqslant 3$$

and draw the curves $E_j(n)$ versus n.

Explain the differences between minimum phase, linear phase, and maximum phase.

2. Calculate the first four terms of the ACF of the signal

$$x(n) = \sqrt{2} \sin\left(n\frac{\pi}{4}\right)$$

Using the normal equations, calculate the coefficients of the predictor of order $N = 3$. Locate the zeros of the prediction error filter in the complex z-plane. Perform the same calculations when a white noise with power $\sigma_b^2 = 0.1$ is added to the signal and compare with the above results.

3. Consider the signal

$$x(n) = \sin(n\omega_1) + \sin(n\omega_2)$$

Differentiating (5.6) with respect to the coefficients and setting these derivatives to zero, calculate the coefficients of the predictor of order $N = 2$. Show the equivalence with solving linear prediction equations. Locate the zeros of the prediction error filter in the complex z-plane and comment on the results.

4. Calculate the coefficients a_1 and a_2 of the notch filter with transfer function

$$H(z) = \frac{1 + a_1 z^{-1} + a_2 z^{-2}}{1 + (1 - \varepsilon)a_1 z^{-1} + (1 - \varepsilon)^2 a_2 z^{-2}}, \qquad \varepsilon = 0.1$$

which cancels the signal $x(n) = \sin(0.7n)$.

Locate the poles and zeros in the complex plane. Give the frequencies which satisfy $|H(e^{j\omega})| = 1$ and calculate $H(1)$ and $H(-1)$. Draw the function $|H(\omega)|$.

Express the white noise amplification factor of the filter as a function of the parameter ε.

5. Use the Levinson-Durbin algorithm to compute the PARCOR coefficients associated with the correlation sequence

$$r(0) = 1, \qquad r(n) = \alpha 0.9^n \qquad 0 < \alpha \leqslant 1$$

Give the diagram of the lattice filter with three sections. Comment on the case $\alpha = 1$.

6. Calculate the inverse of the 3×3 AC matrix R_3. Express the prediction coefficients a_1 and a_2 and the prediction error E_2. Compute R_3^{-1} using relation (5.67) and compare with the direct calculation result.

7. Consider the ARMA signal

$$x(n) = e(n) - 0.5e(n - 1) - 0.9x(n - 1)$$

where $e(n)$ is a unit power white noise. Express the coefficients of the FIR predictor of infinite order.

Using the results of Section 2.6 on ARMA signals, calculate the AC function $r(n)$ for $0 \leqslant n \leqslant 3$. Give the coefficients of the prediction filters of orders 1, 2, and 3 and compare with the first coefficients of the infinite predictor. Locate the zeros in the complex plane.

8. The continuous signal $x(n) = 1$ is applied from time zero on to the adaptive IIR prediction error filter, whose equations are

$$e(n + 1) = x(n + 1) - b(n)e(n)$$
$$b(n + 1) = b(n) + \delta e(n + 1)e(n)$$

For $\delta = 0.2$ and zero initial conditions, calculate the coefficient sequence $b(n)$, $1 \leqslant n \leqslant 20$. How does the corresponding pole move in the complex z-plane?

A noise with power σ_b^2 is added to the input signal. Calculate the optimum value of the first-order IIR predictor. Give a lower bound for σ_b^2 which prevents the pole from crossing the unit circle. When there is no noise, what value of the leakage factor has the same effect.

9. Give the LSP decomposition of the prediction filter

$$1 - A_N(z) = (1 - 1.6z^{-1} + 0.9z^{-2})(1 - z^{-1} + z^{-2})$$

Locate the zeros of the polynomials obtained. Give the diagram of the adaptive realization, implemented as a cascade of second-order filter sections.

ANNEX 5.1 LEVINSON ALGORITHM

```
      SUBROUTINE LEV(N,Q,X,B)
C
C        SOLVES THE SYSTEM : [R]X=B  WITH [R] TOEPLITZ MATRIX
C        N = SYSTEM ORDER ( 2 < N < 17 )
C        Q = N+1 ELEMENT AUTOCORRELATION VECTOR : r(0,....,N)
C        X = SOLUTION VECTOR
C        B = RIGHT SIDE VECTOR
C
      DIMENSION Q(1),X(1),B(1),A(16),Y(16)
      A(1)=-Q(2)/Q(1)
      X(1)=B(1)/Q(1)
      RE=Q(1)+A(1)*Q(2)
      DO60I=2,N
      T=Q(I+1)
      DO10J=1,I-1
   10 T=T+Q(I-J+1)*A(J)
      A(I)=-T/RE
      DO20J=1,I-1
   20 Y(J)=A(J)
      DO30J=1,I-1
   30 A(J)=Y(J)+A(I)*Y(I-J)
      S=B(I)
      DO40J=1,I-1
   40 S=S-Q(I-J+1)*X(J)
      X(I)=S/RE
      DO50J=1,I-1
   50 X(J)=X(J)+X(I)*Y(I-J)
      RE=RE+A(I)*T
   60 CONTINUE
      RETURN
      END
```

ANNEX 5.2 LEROUX-GUEGUEN ALGORITHM

```
      SUBROUTINE LGPC(N,R,RK)
C
C        LEROUX-GUEGUEN  Algorithm for computing the PARCOR
C        coeff. from AC-function.
C        N =Number of coefficients
C        R =Correlation coefficients (INPUT)
C        RK=Reflexion coefficients (OUTPUT)
C
      DIMENSION R(20),RK(20),RE(20),RH(20)
      RK(1)=R(2)/R(1)
      RE(1)=R(2)
      RE(2)=R(1)-RK(1)*R(2)
      DO10I=2,N
      X=R(I+1)
      RH(1)=X
```

```
       I1=I-1
       DO20J=1,I1
       RH(J+1)=RE(J)-RK(J)*X
       X=X-RK(J)*RE(J)
  20   RE(J)=RH(J)
       RK(I)=X/RE(I)
       RE(I+1)=RE(I)-RK(I)*X
       RE(I)=RH(I)
  10   CONTINUE
       RETURN
       END
```

REFERENCES

1. J. Makhoul, "Linear Prediction: A Tutorial Review," *Proc. IEEE* **63**, 561–580 (April 1975).

2. J. L. Lacoume, M. Gharbi, C. Latombe, and J. L. Nicolas, "Close Frequency Resolution by Maximal Entropy Spectral Estimators," *IEEE Trans.* **ASSP-32**, 977–983 (October 1984).

3. J. Leroux and C. Gueguen, "A Fixed Point Computation of Partial Correlation Coefficients," *IEEE Trans.* **ASSP-25**, 257–259 (June 1977).

4. J. D. Markel and A. H. Gray, *Linear Prediction of Speech*, Springer-Verlag, New York, 1976.

5. B. Picinbono and M. Benidir, "Some Properties of Lattice Autoregressive Filters," *IEEE Trans.* **ASSP-34**, 342–349 (April 1986).

6. D. V. B. Rao and S. Y. Kung, "Adaptive Notch Filtering for the Retrieval of Sinusoids in Noise," *IEEE Trans.* **ASSP-32**, 791–802 (August 1984).

7. J. M. Travassos-Romano and M. Bellanger, "Zeros and Poles of Linear Prediction Digital Filters," *Proc. EUSIPCO-86*, North-Holland, The Hague, 1986, pp. 123–126.

8. M. Jaidane-Saidane and O. Macchi, "Self Stabilization of IIR Adaptive Predictors, with Application to Speech Coding," *Proc. EUSIPCO-86*, North-Holland, The Hague, 1986, pp. 427–430.

9. M. Honig and D. Messerschmitt, *Adaptive Filters, Structures, Algorithms and Applications*, Kluwer Academic, Boston, 1985, Chaps. 5–7.

10. G. Sohie and L. Sibul, "Stochastic Convergence Properties of the Adaptive Gradient Lattice," *IEEE Trans.* **ASSP-32**, 102–107 (February 1984).

11. P. M. Grant and M. J. Rutter, "Application of Gradient Adaptive Lattice Filters to Channel Equalization," *Proc. IEEE* **131F**, 473–479 (August 1984).

12. S. Sagayama and F. Itakura, "Duality Theory of Composite Sinusoidal Modeling and Linear Prediction," *Proc. of ICASSP-86*, Tokyo, 1986, pp. 1261–1265.
13. H. W. Schussler, "A Stability Theorem for Discrete Systems," *IEEE Trans.* **ASSP-24**, 87–89 (February 1976).
14. K. Hosoda and A. Fukasawa, "ADPCM Codec Composed by the Prediction Filter Including Poles and Zeros," *Proc. EUSIPCO-83*, Elsevier, 1983, pp. 391–394.

6

Fast Least Squares Transversal Adaptive Filters

Least squares techniques require the inversion of the input signal AC matrix. In adaptive filtering, which implies real-time operations, recursive methods provide means to update the inverse AC matrix whenever new information becomes available. However, the inverse AC matrix is completely determined by the prediction coefficients and error power. The same applies to the real-time estimation of the inverse AC matrix, which is determined by FBLP coefficients and prediction error power estimations. In these conditions, all the information necessary for recursive LS techniques is contained in these parameters, which can be calculated and updated. Fast transversal algorithms perform that function efficiently for FIR filters in direct form.

The first-order LS adaptive filter is an interesting case, not only because it provides a gradual introduction to the recursive mechanisms, the initial conditions, and the algorithm performance, but also because it is implemented in several approaches and applications.

6.1 THE FIRST-ORDER LS ADAPTIVE FILTER

The first-order filter, whose diagram is shown in Figure 6.1, has a single coefficient $h_0(n)$ which is computed to minimize at time n a cost function, which is the error energy

$$E_1(n) = \sum_{p=1}^{n} [y(p) - h_0(n)x(p)]^2 \qquad (6.1)$$

169

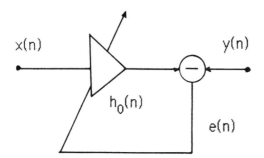

Figure 6.1 Adaptive filter with a single coefficient.

The solution, obtained by setting to zero the derivative of $E_1(n)$ with respect to $h_0(n)$ is

$$h_0(n) = \frac{\sum_{p=1}^{n} y(p)x(p)}{\sum_{p=1}^{n} x^2(p)} = \frac{r_{yx}(n)}{r_{xx}(n)} \tag{6.2}$$

In order to derive a recursive procedure, let us consider

$$h_0(n + 1) = r_{xx}^{-1}(n + 1)[r_{yx}(n) + y(n + 1)x(n + 1)] \tag{6.3}$$

From expression (6.2), we have

$$[r_{xx}(n + 1) - x^2(n + 1)]h_0(n) = r_{yx}(n) \tag{6.4}$$

Hence

$$h_0(n + 1) = h_0(n) + r_{xx}^{-1}(n + 1)x(n + 1)[y(n + 1) - h_0(n)x(n + 1)] \tag{6.5}$$

The filter coefficient is updated using the new data and the a priori error, defined previously by

$$e(n + 1) = y(n + 1) - h_0(n)x(n + 1) \tag{6.6}$$

Recall that this error is named "a priori" because it uses the preceding coefficient value.

The scalar $r_{xx}(n + 1)$ is the input signal power estimate; it is updated by

$$r_{xx}(n + 1) = r_{xx}(n) + x^2(n + 1) \tag{6.7}$$

Together expressions (6.5) and (6.7) make a recursive procedure for the first-order LS adaptive filter. However, in practice, the recursive approach cannot be exactly equivalent to the theoretical LS algorithm, because of the initial conditions.

At time $n = 1$, a coefficient initial value $h_0(0)$ is needed by equation (6.5). If it is taken as zero, relation (6.5) yields

$$h_0(1) = \frac{y(1)}{x(1)} \tag{6.8}$$

which is the solution. However, in the second equation (6.7) it is not possible to take $r_{xx}(0) = 0$ because there is a division in (6.5) and $r_{xx}(1)$ has to be greater than zero. Thus, the algorithm is started with a positive value, $r_{xx}(0) = r_0$, and the actual coefficient updating equation is

$$h_0(n + 1) = h(n) + \frac{x(n + 1)}{r_0 + \sum_{p=1}^{n+1} x^2(p)} [y(n + 1) - h_0(n)x(n + 1)], \qquad n \geqslant 0 \tag{6.9}$$

This equation still is a LS equation, but the criterion is different from (6.1). Instead, it can be verified that it is

$$E_1'(n) = \sum_{p=1}^{n} [y(p) - h_0(n)x(p)]^2 + r_0 h_0^2(n) \tag{6.10}$$

The consequence is the introduction of a time constant, which can be evaluated by considering the simplified case $y(n) = x(n) = 1$. With these signals, the coefficient evolution equation is

$$h_0(n + 1) = h_0(n) + \frac{1}{r_0 + (n + 1)} [1 - h_0(n)], \qquad n \geqslant 0$$

or

$$h_0(n + 1) = \left(1 - \frac{1}{r_0 + n + 1}\right) h_0(n) + \frac{1}{r_0 + n + 1}, \qquad n \geqslant 0 \tag{6.11}$$

which, assuming $h_0(0) = 0$, leads to

$$h_0(n) = \frac{n}{r_0 + n} \tag{6.12}$$

The evolution of the coefficient is shown in Figure 6.2 for different values of the initial constant r_0. Note that negative values can also be taken for r_0.

Definition (4.10) in Chapter 4 yields the coefficient time constant $\tau_c \approx r_0$. Clearly, the initial constant r_0 should be kept as small as possible; the lower limit is determined by the computational accuracy in the realization.

Adaptive filters, in general, are designed with the capability of handling nonstationary signals, which is achieved through the introduction of a limited memory. An efficient approach consists of introducing a memory-limiting or

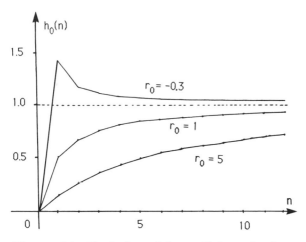

Figure 6.2   Evolution of the coefficient of a first-order LS adaptive filter.

-forgetting factor $W(0 \ll W < 1)$, which corresponds to an exponential weighting operation in the cost function:

$$E_{W1}(n) = \sum_{p=1}^{n} W^{n-p}[y(p) - h_0(n)x(p)]^2 \tag{6.13}$$

Taking into account the initial constant r_0, we obtain the actual cost function

$$E'_{W1}(n) = \sum_{p=1}^{n} W^{n-p}[y(p) - h_0(n)x(p)]^2 + W^n r_0 h_0^2(n) \tag{6.14}$$

The updating equation for the coefficient becomes

$$h_0(n+1) = h_0(n) + \frac{x(n+1)}{r_0 W^{n+1} + \sum_{p=1}^{n+1} W^{n+1-p}x^2(p)}$$
$$\times [y(n+1) - h_0(n)x(n+1)], \qquad n \geqslant 0 \tag{6.15}$$

In the simplified case $x(n) = y(n) = 1$, if we assume $h_0(0) = 0$, we get

$$h_0(n+1) = h_0(n) + \frac{1}{r_0 W^{n+1} + (1 - W^{n+1})/(1 - W)}[1 - h_0(n)], \qquad n \geqslant 0 \tag{6.16}$$

Now the coefficient time constant is $\tau_{cW} \approx W r_0$. But for n sufficiently large, the updating equation approaches

$$h_0(n+1) = W h_0(n) + 1 - W \tag{6.17}$$

which corresponds to the long-term time constant

$$\tau \approx \frac{1}{1 - W}$$

(6.18)

The curves $1 - h_0(n)$ versus time are shown in Figure 6.3 for $r_0 = 1$ and $W = 0.95$ and $W = 1$. Clearly, the weighting factor W can accelerate the convergence of $h_0(n)$ toward its limit.

For the LMS algorithm with step size δ under the same conditions, one gets

$$h_0(n) = 1 - (1 - \delta)^n$$

(6.17a)

The corresponding curve in Figure 6.3 illustrates the advantage of LS techniques in the initial phase.

In the recursive procedure, only the input signal power estimate is affected by the weighting operation, and equation (6.7) becomes

$$r_{xx}(n + 1) = Wr_{xx}(n) + x^2(n + 1)$$

(6.19)

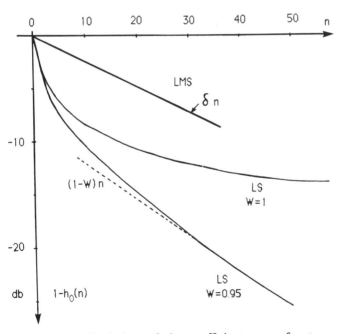

Figure 6.3 Evolution of the coefficient error for two weighting factor values.

In transversal filters with several coefficients, the above scalar operations become matrix operations and a recursive procedure can be worked out to avoid matrix inversion.

6.2 RECURSIVE EQUATIONS FOR THE ORDER N FILTER

The adaptive filter of order N is defined in matrix equations by

$$e(n+1) = y(n+1) - H^t(n)X(n+1) \tag{6.20}$$

where the vectors $H(n)$ and $X(n)$ have N elements. The cost function, which is the error energy

$$E_N(n) = \sum_{p=1}^{n} W^{n-p}[y(p) - H^t(n)X(p)]^2 \tag{6.21}$$

leads, as shown in Section 1.4, to the least squares solution

$$H(n) = R_N^{-1}(n)r_{yx}(n) \tag{6.22}$$

with

$$R_N(n) = \sum_{p=1}^{n} W^{n-p}X(p)X^t(p), \qquad r_{yx}(n) = \sum_{p=1}^{n} W^{n-p}y(p)X(p) \tag{6.23}$$

The corresponding recurrence equation for the coefficients is

$$H(n+1) = H(n) - R_N^{-1}(n+1)X(n+1)[y(n+1) - X^t(n+1)H(n)] \tag{6.24}$$

The matrix $R_N^{-1}(n+1)$ in that expression can be updated recursively with the help of a matrix identity called the matrix inversion lemma [1]. Given matrices A, B, C, and D satisfying the equation

$$A = B + CDC^t$$

the inverse of matrix A is

$$A^{-1} = B^{-1} - B^{-1}C[C^tB^{-1}C + D^{-1}]^{-1}C^tB^{-1} \tag{6.25}$$

The matrix A^{-1} can appear in various forms, which can be derived from the identity

$$(B - UDV)^{-1} = [I_N - B^{-1}UDV]^{-1}B^{-1}$$

where B is assumed nonsingular, through the generic power series expansion

$$(I_N - B^{-1}UDV)^{-1}B^{-1} = [I_N + B^{-1}UDV + (B^{-1}UDV)^2 + \cdots]B^{-1} \tag{6.25a}$$

The convergence of the series is obtained if the eigenvalues of $(B^{-1}UDV)$ are less than unity. Expression (6.25a) is a generalized matrix inversion lemma [2]. Consider, for example, regrouping and summing all terms but the first in (6.25a) to obtain

$$(B - UDV)^{-1} = I_N + B^{-1}U[I_N - DVB^{-1}U]^{-1}DVB^{-1} \qquad (6.25b)$$

which is another form of (6.25).

This lemma can be applied to the calculation of $R_N^{-1}(n + 1)$ in such a way that no matrix inversion is needed, just division by a scalar. Since

$$R_N(n + 1) = WR_N(n) + X(n + 1)X^t(n + 1) \qquad (6.26)$$

let us choose

$$B = WR_N(n), \quad C = X(n + 1), \quad D = 1$$

then, lemma (6.25) yields

$$R_N^{-1}(n + 1) = \frac{1}{W}\left[R_N^{-1}(n) - \frac{R_N^{-1}(n)X(n + 1)X^t(n + 1)R_N^{-1}(n)}{W + X^t(n + 1)R_N^{-1}(n)X(n + 1)} \right] \qquad (6.27)$$

It is convenient to define the adaptation gain $G(n)$ by

$$G(n) = R_N^{-1}(n)X(n) \qquad (6.28)$$

which, using (6.27) and after adequate simplifications, leads to

$$G(n + 1) = \frac{1}{W + X^t(n + 1)R_N^{-1}(n)X(n + 1)} R_N^{-1}(n)X(n + 1) \qquad (6.29)$$

Now, expression (6.27) and recursion (6.24) can be rewritten as

$$R_N^{-1}(n + 1) = \frac{1}{W}[R_N^{-1}(n) - G(n + 1)X^t(n + 1)R_N^{-1}(n)] \qquad (6.30)$$

and

$$H(n + 1) = H(n) + G(n + 1)[y(n + 1) - X^t(n + 1)H(n)] \qquad (6.31)$$

Relations (6.29)–(6.31) provide a recursive procedure to perform the filter coefficient updating without matrix inversion. Clearly, a nonzero initial value $R_N^{-1}(0)$ is necessary for the procedure to start; that point is discussed in a later section.

The number of arithmetic operations represented by the above procedure is proportional to N^2, because of the matrix multiplications involved. Matrix manipulations can be completely avoided, and the computational complexity made proportional to N only by considering that $R_N(n)$ is a real-time estimate of the input signal AC matrix and that, as shown in Chapter 5, its inverse can be represented by prediction parameters.

Before introducing the corresponding fast algorithms, several useful relations between LS variables are derived.

6.3 RELATIONSHIPS BETWEEN LS VARIABLES

In deriving the recursive least squares (RLS) procedure, the matrix inversion is avoided by the introduction of an appropriate scalar. Let

$$\varphi(n + 1) = \frac{W}{W + X^t(n + 1)R_N^{-1}(n)X(n + 1)} \tag{6.32}$$

It is readily verified, using (6.27), that

$$\varphi(n + 1) = 1 - X^t(n + 1)R_N^{-1}(n + 1)X(n + 1) \tag{6.33}$$

The scalar $\theta(n)$, defined by

$$\theta(n) = X^t(n)R_N^{-1}(n)X(n) \tag{6.34}$$

has a special interpretation in signal processing. First, it is clear from

$$\theta(n + 1) = X^t(n + 1)[WR_N(n) + X(n + 1)X^t(n + 1)]^{-1}X(n + 1)$$

that, assuming the existence of the inverse matrix

$$\theta(n + 1) \leqslant X^t(n + 1)[X(n + 1)X^t(n + 1)]^{-1}X(n + 1)$$

Since

$$[X(n + 1)X^t(n + 1)]X(n + 1) = \|X(n + 1)\|X(n + 1) \tag{6.35}$$

where $\|X\|$ the Euclidean norm of the vector X, the inverse matrix $[X(n + 1) \quad X(n + 1)]^{-1}$ by definition satisfies

$$[X(n + 1) \quad X^t(n + 1)]^{-1}X(n + 1) = \|X(n + 1)\|^{-1}X(n + 1) \tag{6.35a}$$

and the variable $\theta(n)$ is bounded by

$$0 \leqslant \theta(n) \leqslant 1 \tag{6.36}$$

Now, from Section 2.12, it appears that the term in the exponent of the joint density of N zero mean Gaussian variables, has a form similar to $\theta(n)$, which can be interpreted as its sample estimate—hence the name of likelihood variable given to $\theta(n)$ in estimation theory [3]. Thus, $\theta(n)$ is a measure of the likelihood that the N most recent input data samples come from a Gaussian process with AC matrix $R_N(n)$ determined from all the available past observations. A small value of $\theta(n)$ indicates that the recent input data are likely samples of a Gaussian signal, and a value close to unity indicates that the observations are unexpected; in the latter case, $X(n + 1)$ is out of the current estimated signal space, which can be due to the time-varying nature of the signal statistics. As a consequence, $\theta(n)$ can be used to detect changes in the signal statistics. If the adaptation gain $G(n)$ is available, as in the fast algorithms presented below, $\theta(n)$ can be readily calculated by

$$\theta(n) = X^t(n)G(n) \tag{6.37}$$

From the definitions, $\varphi(n)$ and $\theta(n)$ have similar properties. Those relevant to LS techniques are presented next.

Postmultiplying both sides of recurrence relation (6.26) by $R_N^{-1}(n)$ yields

$$R_N(n + 1)R_N^{-1}(n) = WI_N + X(n + 1)X^t(n + 1)R_N^{-1}(n) \tag{6.38}$$

Using the identity

$$\det[I_N + V_1 V_2^t] = 1 + V_1^t V_2 \tag{6.39}$$

where V_1 and V_2 are N-element vectors, and the definition of $\varphi(n)$, one gets

$$\varphi(n + 1) = W^N \frac{\det R_N(n)}{\det R_N(n + 1)} \tag{6.40}$$

Because of the definition of $R_N(n)$ and its positiveness and recurrence relation (6.26), the variable $\varphi(n)$ is bounded by

$$0 \leqslant \varphi(n) \leqslant 1 \tag{6.41}$$

which, through a different approach, confirms (6.36). This is a crucial property, which can be used to check that the LS conditions are satisfied in realizations of fast algorithms.

Now, we show that the variable $\varphi(n)$ has a straightforward physical meaning. The RLS procedure applied to forward linear prediction is based on a cost function, which is the prediction error energy

$$E_a(n) = \sum_{p=1}^{n} W^{n-p}[x(p) - A^t(n)X(p - 1)]^2 \tag{6.42}$$

The coefficient vector is

$$A(n) = R_N^{-1}(n - 1)r_N^a(n) \tag{6.43}$$

with

$$r_N^a(n) = \sum_{p=1}^{n} W^{n-p}x(p)X(p - 1) \tag{6.44}$$

The index $n - 1$ in (6.43) is typical of forward linear prediction, and the RLS coefficient updating equation is

$$A(n + 1) = A(n) + G(n)e_a(n + 1) \tag{6.45}$$

where

$$e_a(n + 1) = x(n + 1) - A'(n)X(n) \tag{6.46}$$

is the a priori forward prediction error.

The updated coefficients $A(n + 1)$ are used to calculate the a posteriori prediction error

$$\varepsilon_a(n + 1) = x(n + 1) - A'(n + 1)X(n) \tag{6.47}$$

or

$$\varepsilon_a(n + 1) = e_a(n + 1)[1 - G'(n)X(n)] \tag{6.48}$$

From definition (6.33) we have

$$\varphi(n) = \frac{\varepsilon_a(n + 1)}{e_a(n + 1)} \tag{6.49}$$

and $\varphi(n)$ is the ratio of the forward prediction errors at the next time. This result can lead to another direct proof of inequality (6.41).

A similar result can also be obtained for backward linear prediction. The cost function used for the RLS procedure is the backward prediction error energy

$$E_b(n) = \sum_{p=1}^{n} W^{n-p}[x(p - N) - B'(n)X(p)]^2 \tag{6.50}$$

The backward coefficient vector is

$$B(n) = R_N^{-1}(n)r_N^b(n) \tag{6.51}$$

with

$$r_N^b(n) = \sum_{p=1}^{n} W^{n-p}x(p - N)X(p) \tag{6.52}$$

The coefficient updating equation is now

$$B(n + 1) = B(n) + G(n + 1)e_b(n + 1) \tag{6.53}$$

with

$$e_b(n + 1) = x(n + 1 - N) - B'(n)X(n + 1) \tag{6.54}$$

The backward a posteriori prediction error is

$$\varepsilon_b(n + 1) = x(n + 1 - N) - B'(n + 1)X(n + 1) \tag{6.55}$$

Substituting (6.53) into (6.55) gives

$$\varphi(n + 1) = \frac{\varepsilon_b(n + 1)}{e_b(n + 1)} \tag{6.56}$$

which shows that $\varphi(n)$ is the ratio of the backward prediction errors at the same time index.

The forward prediction error energy can be computed recursively. Substituting equation (6.43) into the expression of $E_a(n + 1)$ yields

$$E_a(n + 1) = \sum_{p=1}^{n+1} W^{n+1-p}x^2(p) - A^t(n + 1)r_N^a(n + 1) \tag{6.57}$$

The recurrence relations for $A(n + 1)$ and $r_N^a(n + 1)$, in connection with the definitions for the adaptation gain and the prediction coefficients, yield after simplification

$$E_a(n + 1) = WE_a(n) + e_a(n + 1)\varepsilon_a(n + 1) \tag{6.58}$$

Similarly, the backward prediction error energy can be calculated by

$$E_b(n + 1) = WE_b(n) + e_b(n + 1)\varepsilon_b(n + 1) \tag{6.59}$$

These are fundamental recursive computations which are used in the fast algorithms.

6.4 FAST ALGORITHM BASED ON A PRIORI ERRORS

In the RLS procedure, the adaptation gain $G(n)$ used to update the coefficients is itself updated with the help of the inverse input signal AC matrix. In fast algorithms, prediction parameters are used instead [4].

Let us consider the $(N + 1) \times (N + 1)$ AC matrix $R_{N+1}(n + 1)$; as pointed out in Chapter 5, it can be partitioned in two different manners, exploited in forward and backward prediction equations:

$$R_{N+1}(n + 1) = \begin{bmatrix} \sum_{p=1}^{n+1} W^{n+1-p}x^2(p) & [r_N^a(n + 1)]^t \\ r_N^a(n + 1) & R_N(n) \end{bmatrix} \tag{6.60}$$

and

$$R_{N+1}(n + 1) = \begin{bmatrix} R_N(n + 1) & r_N^b(n + 1) \\ [r_N^b(n + 1)]^t & \sum_{p=1}^{n+1} W^{n+1-p}x^2(p - N) \end{bmatrix} \tag{6.61}$$

The objective is to find $G(n + 1)$ satisfying

$$R_N(n + 1)G(n + 1) = X(n + 1) \tag{6.62}$$

Since $R_N(n)$ is present in (6.60), let us calculate

$$R_{N+1}(n + 1)\begin{bmatrix} 0 \\ G(n) \end{bmatrix} = \begin{bmatrix} [r_N^a(n + 1)]^t G(n) \\ X(n) \end{bmatrix} \tag{6.63}$$

From definitions (6.28) for the adaptation gain and (6.43) for the optimal forward prediction coefficients, we have

$$[r_N^a(n + 1)]^t G(n) = A^t(n + 1)X(n) \tag{6.64}$$

Introducing the a posteriori prediction error, we get

$$R_{N+1}(n + 1)\begin{bmatrix} 0 \\ G(n) \end{bmatrix} = X_1(n + 1) - \begin{bmatrix} \varepsilon_a(n + 1) \\ 0 \end{bmatrix} \tag{6.65}$$

where $X_1(n)$ is the vector of the $N + 1$ most recent input data. Similarly, partitioning (6.61) leads to

$$R_{N+1}(n + 1)\begin{bmatrix} G(n + 1) \\ 0 \end{bmatrix} = \begin{bmatrix} X(n + 1) \\ [r_N^b(n + 1)]^t G(n + 1) \end{bmatrix} \tag{6.66}$$

From definitions (6.62) and (6.51), we have

$$[r_N^b(n + 1)]^t G(n + 1) = B^t(n + 1)X(n + 1) \tag{6.67}$$

and

$$R_N(n + 1)\begin{bmatrix} G(n + 1) \\ 0 \end{bmatrix} = X_1(n + 1) - \begin{bmatrix} 0 \\ \varepsilon_b(n + 1) \end{bmatrix} \tag{6.68}$$

Now, the adaptation gain at dimension $N + 1$, denoted $G_1(n + 1)$ with the above notation, is defined by

$$R_{N+1}(n + 1)G_1(n + 1) = X_1(n + 1) \tag{6.69}$$

Then, equation (6.65) can be rewritten as

$$R_{N+1}(n + 1)\begin{bmatrix} G_1(n + 1) - \begin{bmatrix} 0 \\ G(n) \end{bmatrix} \end{bmatrix} = \begin{bmatrix} \varepsilon_a(n + 1) \\ 0 \end{bmatrix} \tag{6.70}$$

Equation (6.68) becomes

$$R_{N+1}(n + 1)\begin{bmatrix} G_1(n + 1) - \begin{bmatrix} G(n + 1) \\ 0 \end{bmatrix} \end{bmatrix} = \begin{bmatrix} 0 \\ \varepsilon_b(n + 1) \end{bmatrix} \tag{6.71}$$

Now, linear prediction matrix equations will be used to compute $G_1(n + 1)$ from $G(n)$, and then $G(n + 1)$ from $G_1(n + 1)$. The forward linear prediction matrix equation, combining (6.43) and (6.57), is

$$R_{N+1}(n + 1)\begin{bmatrix} 1 \\ -A(n + 1) \end{bmatrix} = \begin{bmatrix} E_a(n + 1) \\ 0 \end{bmatrix} \tag{6.72}$$

Identifying factors in (6.72) and (6.70) yields

$$G_1(n + 1) = \begin{bmatrix} 0 \\ G(n) \end{bmatrix} + \frac{\varepsilon_a(n + 1)}{E_a(n + 1)}\begin{bmatrix} 1 \\ -A(n + 1) \end{bmatrix} \tag{6.73}$$

The backward linear prediction matrix equation is

$$R_{N+1}(n+1)\begin{bmatrix} -B(n+1) \\ 1 \end{bmatrix} = \begin{bmatrix} 0 \\ E_b(n+1) \end{bmatrix} \tag{6.74}$$

Identifying factors in (6.74) and (6.71) yields

$$G_1(n+1) - \begin{bmatrix} G(n+1) \\ 0 \end{bmatrix} = \frac{\varepsilon_b(n+1)}{E_b(n+1)} \begin{bmatrix} -B(n+1) \\ 1 \end{bmatrix} \tag{6.75}$$

The scalar factor on the right side need not be calculated; it is already available. Let us partition the adaptation gain vector

$$G_1(n+1) = \begin{bmatrix} M(n+1) \\ m(n+1) \end{bmatrix} \tag{6.76}$$

with $M(n+1)$ having N elements; the scalar $m(n+1)$ is given by the last line of (6.75):

$$m(n+1) = \frac{\varepsilon_b(n+1)}{E_b(n+1)} \tag{6.77}$$

The N-element adaptation gain is updated by

$$G(n+1) = M(n+1) + m(n+1)B(n+1) \tag{6.78}$$

But the updated adaptation gain is needed to get $B(n+1)$. Substituting (6.53) into (6.78) provides an expression of the gain as a function of available quantities:

$$G(n+1) = \frac{1}{1 - m(n+1)e_b(n+1)} [M(n+1) + m(n+1)B(n)] \tag{6.79}$$

Note that, instead, (6.78) can be substituted into the coefficient updating equation, allowing the computation of $B(n+1)$ first:

$$B(n+1) = \frac{1}{1 - m(n+1)e_b(n+1)} [B(n) + M(n+1)e_b(n+1)] \tag{6.80}$$

In these equations, a new scalar is showing up. Since one must always be careful with dividers, it is interesting to investigate its physical interpretation and appreciate its magnitude range. Combining (6.77) and the energy updating equation (6.59) yields

$$1 - m(n+1)e_b(n+1) = 1 - \frac{\varepsilon_b(n+1)e_b(n+1)}{E_b(n+1)} = \frac{WE_b(n)}{E_b(n+1)} \tag{6.81}$$

ALGORITHM F.L.S.1

AVAILABLE AT TIME n:

$$
\begin{aligned}
\text{COEFFICIENTS OF ADAPTIVE FILTER} &: H(n) \\
\text{\quad\quad\quad\quad\quad\quad FORWARD PREDICTOR} &: A(n) \\
\text{\quad\quad\quad\quad\quad\quad BACKWARD PREDICTOR} &: B(n) \\
\text{DATA VECTOR} &: X(n) \\
\text{ADAPTATION GAIN} &: G(n) \\
\text{FORWARD PREDICTION ERROR ENERGY} &: E_a(n)
\end{aligned}
$$

NEW DATA AT TIME n:

input Signal : $x(n+1)$; Reference : $y(n+1)$

ADAPTATION GAIN UPDATING :

$$e_a(n+1) = x(n+1) - A^t(n) X(n)$$

$$A(n+1) = A(n) + G(n) e_a(n+1)$$

$$\epsilon_a(n+1) = x(n+1) - A^t(n+1) X(n)$$

$$E_a(n+1) = W E_a(n) + e_a(n+1) \epsilon_a(n+1)$$

$$G(n+1) = \begin{bmatrix} 0 \\ G(n) \end{bmatrix} + \frac{\epsilon_a(n+1)}{E_a(n+1)} \begin{bmatrix} 1 \\ -A(n+1) \end{bmatrix} = \begin{bmatrix} M(n+1) \\ m(n+1) \end{bmatrix}$$

$$e_b(n+1) = x(n+1-N) - B^t(n) X(n+1)$$

$$G(n+1) = \frac{1}{1 - m(n+1) e_b(n+1)} (M(n+1) + m(n+1) B(n))$$

$$B(n+1) = B(n) + G(n+1) e_b(n+1)$$

ADAPTIVE FILTER :

$$e(n+1) = y(n+1) - H^t(n) X(n+1)$$

$$H(n+1) = H(n) + G(n+1) e(n+1)$$

Figure 6.4 Computational organization of the fast algorithm based on a priori errors.

Thus, the divider $1 - m(n + 1)e_b(n + 1)$ is the ratio of two consecutive values of the backward prediction error energy, and its theoretical range is

$$0 < 1 - m(n + 1)e_b(n + 1) \leqslant 1 \tag{6.82}$$

Clearly, as time goes on, its value approaches unity, more so than when the prediction error is small. Incidentally, equation (6.81) is an alternative to

(6.59) to update the backward prediction error energy. Overall a fast algorithm is available and the sequence of operations is given in Figure 6.4. The corresponding FORTRAN subroutine is given in Annex 6.1.

It is sometimes called the fast Kalman algorithm [4]. The LS initialization is obtained by taking $A(n) = B(n) = G(n) = 0$ and $E_a(0) = E_0$, a small positive constant, as discussed in a later section.

The adaptation gain updating requires $8N + 4$ multiplications and two divisions in the form of inverse calculations; in the filtering, $2N$ multiplications are involved. Approximately $6N$ memories are needed to store the coefficients and variables. The progress with respect to RLS algorithms is impressive; however, it is still possible to improve these figures.

The above algorithm is mainly based on the a priori errors; for example, the backward a posteriori prediction error is not calculated. If all the prediction errors are exploited, a better balanced and more efficient algorithm is derived [5, 6].

6.5 ALGORITHM BASED ON ALL PREDICTION ERRORS

Let us define an alternative adaptation gain vector with N elements, $G'(n)$, by

$$R_N(n)G'(n + 1) = X(n + 1) \tag{6.83}$$

Because of the term $R(n)$ in $G'(n + 1)$, it is also called the a priori adaptation gain, in contrast with the a posteriori gain $G(n + 1)$.

Similarly at order $N + 1$

$$R_{N+1}(n)G_1'(n + 1) = X_1(n + 1) \tag{6.84}$$

Exploiting, as in the previous section, the two different partitionings, (6.60) and (6.61), of the AC matrix estimation $R_{N+1}(n)$, one gets

$$R_{N+1}(n)\begin{bmatrix} G'(n + 1) \\ 0 \end{bmatrix} = X_1(n + 1) - \begin{bmatrix} 0 \\ e_b(n + 1) \end{bmatrix} \tag{6.85}$$

and

$$R_{N+1}(n)\begin{bmatrix} 0 \\ G'(n) \end{bmatrix} = X_1(n + 1) - \begin{bmatrix} e_a(n + 1) \\ 0 \end{bmatrix} \tag{6.86}$$

Now, substituting definition (6.84) into (6.85) yields

$$R_{N+1}(n)\begin{bmatrix} G_1'(n + 1) - \begin{bmatrix} G'(n + 1) \\ 0 \end{bmatrix} \end{bmatrix} = \begin{bmatrix} 0 \\ e_b(n + 1) \end{bmatrix} \tag{6.87}$$

Identifying with the backward prediction matrix equation (6.74) gives a first expression for the order $N + 1$ adaptation gain:

$$G_1'(n + 1) = \begin{bmatrix} G'(n + 1) \\ 0 \end{bmatrix} + \frac{e_b(n + 1)}{E_b(n)} \begin{bmatrix} -B(n) \\ 1 \end{bmatrix} \qquad (6.88)$$

Similarly (6.86) and (6.84) lead to

$$R_{N+1}(n) \left[G_1'(n + 1) - \begin{bmatrix} 0 \\ G'(n) \end{bmatrix} \right] = \begin{bmatrix} e_a(n + 1) \\ 0 \end{bmatrix} \qquad (6.89)$$

Identifying with the forward prediction matrix equation (6.72) provides another expression for the gain:

$$G_1'(n + 1) = \begin{bmatrix} 0 \\ G'(n) \end{bmatrix} + \frac{e_a(n + 1)}{E_a(n)} \begin{bmatrix} 1 \\ -A(n) \end{bmatrix} \qquad (6.90)$$

The procedure for calculating $G'(n + 1)$ consists of calculating $G_1'(n + 1)$ from the forward prediction parameters by (6.90) and then using (6.88).

Once the alternative gain $G'(n)$ is updated, it can be used in the filter coefficient recursion, provided it is adequately modified. It is necessary to replace $R_N^{-1}(n + 1)$ by $R_N^{-1}(n)$ in equation (6.24). At time $n + 1$ the optimal coefficient definition (6.22) is

$$[WR_N(n) + X(n + 1)X^t(n + 1)]H(n + 1)$$
$$= Wr_{yx}(n) + y(n + 1)X(n + 1)$$

which, after some manipulation, leads to

$$H(n + 1) = H(n) + W^{-1}R_N^{-1}(n)X(n + 1)[y(n + 1) - X^t(n + 1)H(n + 1)] \qquad (6.91)$$

The a posteriori error

$$\varepsilon(n + 1) = y(n + 1) - X^t(n + 1)H(n + 1) \qquad (6.92)$$

has to be calculated from available data; this is achieved with the help of the variable $\varphi(n)$ defined by (6.32), which is the ratio of a posteriori to a priori errors. From (6.32) we have

$$W + X^t(n + 1)G'(n + 1) = \frac{W}{\varphi(n + 1)} = \alpha(n + 1) \qquad (6.93)$$

The variable $\alpha(n + 1)$ is actually calculated in the algorithm.

Substituting $H(n + 1)$ from (6.91) into (6.92) yields the kind of relationship already obtained for prediction:

$$\varepsilon(n + 1) = \varphi(n + 1)e(n + 1) \qquad (6.94)$$

Now the coefficient updating equation is

$$H(n + 1) = H(n) + \frac{e(n + 1)}{\alpha(n + 1)} G'(n + 1) \tag{6.95}$$

Note that, from the above derivations, the two adaptation gains are related by the scalar $\alpha(n + 1)$ and an alternative definition of $G'(n + 1)$ is

$$G'(n + 1) = [W + X^t(n + 1)R_N^{-1}(n)X(n + 1)]G(n + 1)$$
$$= \alpha(n + 1)G(n + 1) \tag{6.96}$$

The variable $\alpha(n + 1)$ can be calculated from its definition (6.93). However, a recursive procedure, similar to the one worked out for the adaptation gain, can be obtained. The variable corresponding to the order $N + 1$ is $\alpha_1(n + 1)$, defined by

$$\alpha_1(n + 1) = W + X_1^t(n + 1)G_1'(n + 1) \tag{6.93a}$$

The two different expressions for $G_1'(n + 1)$, (6.88) and (6.90), yield

$$\alpha_1(n + 1) = \alpha(n) + \frac{e_a^2(n + 1)}{E_a(n)} = \alpha(n + 1) + \frac{e_b^2(n + 1)}{E_b(n)} \tag{6.97}$$

which provides the recursion for $\alpha(n + 1)$ and $\varphi(n + 1)$.

Since $\varphi(n + 1)$ is available, it can be used to derive the a posteriori prediction errors $\varepsilon_a(n + 1)$ and $\varepsilon_b(n + 1)$, with only one multiplication instead of the N multiplications and additions required by the definitions.

The backward a priori prediction error can be obtained directly. If the $N + 1$ dimension vector gain is partitioned,

$$G_1'(n + 1) = \begin{bmatrix} M'(n + 1) \\ m'(n + 1) \end{bmatrix} \tag{6.98}$$

the last line of matrix equation (6.88) is

$$m'(n + 1) = \frac{e_b(n + 1)}{E_b(n)} \tag{6.99}$$

which provides $e_b(n + 1)$ through just a single multiplication. However, due to roundoff problems discussed in a later section, this simplification is not recommended. The overall algorithm is given in Figure 6.5.

The LS initialization corresponds to

$$A(n) = B(n) = G'(n) = 0, \quad E_a(0) = E_0, \quad E_b(0) = W^{-N}E_0 \tag{6.100}$$

where E_0 is a small positive constant. Definition (6.93) also yields $\alpha(0) = W$.

The adaptation gain updating section requires $6N + 9$ multiplications and three divisions in the form of inverse calculations. The filtering section has

ALGORITHM F.L.S.2
_____

AVAILABLE AT TIME n:
    COEFFICIENTS OF ADAPTIVE FILTER   : $H(n)$
        "   "   FORWARD PREDICTOR : $A(n)$
        "   "   BACKWARD PREDICTOR : $B(n)$
                        DATA VECTOR   : $X(n)$
                    ADAPTATION GAIN   : $G'(n)$
    PREDICTION ERROR ENERGIES : $E_a(n)$, $E_b(n)$
        PREDICTION ERROR RATIO : $\alpha(n)$
            WEIGHTING FACTOR : W

NEW DATA AT TIME n:
    Input signal : $x(n+1)$; Reference : $y(n+1)$

ADAPTATION GAIN UPDATING:

$$e_a(n+1) = x(n+1) - A^t(n)\, X(n)$$

$$A(n+1) = A(n) + G'(n)\, e_a(n+1)\, /\alpha(n)$$

$$E_a(n+1) = (\, E_a(n) + e_a(n+1)\, e_a(n+1)\, /\alpha(n)\,)\, W$$

$$G'_1(n+1) = \begin{bmatrix} 0 \\ G'(n) \end{bmatrix} + \frac{e_a(n+1)}{E_a(n)} \begin{bmatrix} 1 \\ -A(n) \end{bmatrix} = \begin{bmatrix} M(n+1) \\ m(n+1) \end{bmatrix}$$

$$e_b(n+1) = x(n+1-N) - B^t(n)\, X(n+1)$$

$$G'(n+1) = M(n+1) + m(n+1)\, B(n)$$

$$\alpha_1(n+1) = \alpha(n) + e_a(n+1)\, e_a(n+1)\, /\, E_a(n)$$

$$\alpha(n+1) = \alpha_1(n+1) - m(n+1)\, e_b(n+1)$$

$$E_b(n+1) = (E_b(n) + e_b(n+1)\, e_b(n+1)\, /\alpha(n+1))\, W$$

$$B(n+1) = B(n) + G'(n+1)\, e_b(n+1)\, /\alpha(n+1)$$

ADAPTIVE FILTER:

$$e(n+1) = y(n+1) - H^t(n)\, X(n+1)$$

$$H(n+1) = H(n) + G'(n+1)\, e(n+1)\, /\alpha(n+1)$$

Figure 6.5 Computational organization of the fast algorithm based on all prediction errors.

$2N + 1$ multiplications. Approximately $6N + 7$ memories are needed. Overall this second algorithm can bring an appreciable improvement in computational complexity over the first one, particularly for large order N.

6.6 STABILITY CONDITIONS FOR LS RECURSIVE ALGORITHMS

For a nonzero set of signal samples, the LS calculations provide a unique set of prediction coefficients. Recursive algorithms correspond to exact calculations at any time, and, therefore, their stability is guaranteed in theory for any weighting factor W. Since fast algorithms are mathematically equivalent to RLS, they enjoy the same property. Their stability is even guaranteed for a zero signal sequence, provided the initial prediction error energies are greater than zero. This is a very important and attractive theoretical property, which, unfortunately, is lost in realizations because of finite precision effects in implementations [7–10].

Fast algorithms draw their efficiency from a representation of LS parameters, the inverse input signal AC matrix, and cross-correlation estimations, which is reduced to a minimal number of variables. With the finite accuracy of arithmetic operations, that representation can only be approximate. So, the inverse AC matrix estimation $R_N^{-1}(n)$ appears in FLS algorithms through its product by the data vector $X(n)$, which is the adaptation gain $G(n)$. Since the data vector is by definition an exact quantity, the roundoff errors generated in the gain calculation procedure correspond to deviations of the actual inverse AC matrix estimation from its theoretical infinite accuracy value.

In Section 3.11, we showed that random errors on the AC matrix elements do not significantly affect the eigenvalues, but they alter the eigenvector directions. Conversely, a bias in estimating the ACF causes variations of eigenvalues.

When the data vector $X(n)$ is multiplied by the theoretical matrix $R_N^{-1}(n)$, the resulting vector has a limited range because $X(n)$ belongs to the signal space of the matrix.

However, if an approximation of $R_N^{-1}(n)$ is used, the data vector can have a significant projection outside of the matrix signal space; in that case, the norm of the resulting vector is no longer controlled, which can make variables exceed the limits of their magnitude range. Also, the eigenvalues can become negative because of long-term roundoff error accumulation.

Several variables have a limited range in FLS algorithms. A major step in the sequence of operations is the computation of a posteriori errors, from coefficients which have been updated with the adaptation gain and a priori errors. Therefore the accuracy of the representation of $R_N^{-1}(n)X(n)$ by $G(n)$

can be directly controlled by the ratio $\varphi(n)$ of a posteriori to a priori prediction errors. In realizations the variable $\varphi(n)$, introduced in Section 6.3, corresponds to

$$\varphi(n) = 1 - X^t(n)[R_N^q(n)]^{-1}X(n) \qquad (6.101)$$

where $R_N^q(n)$ is the matrix used instead of the theoretical $R_N(n)$. The variable $\varphi(n)$ can exceed unity if eigenvalues of $R_N^q(n)$ become negative; $\varphi(n)$ can become negative if the scalar $X^t(n)[R_N^q(n)]^{-1}X(n)$ exceeds unity.

Roundoff error accumulation, if present, takes place in the long run. The first precaution in implementing fast algorithms is to make sure that the scalar $X^t(n)[R_N^q(n)]^{-1}X(n)$ does not exceed unity.

To begin with, let us assume that the input signal is a white zero mean Gaussian noise with power σ_x^2. As seen in Section 3.11, for sufficiently large n one has

$$R_N(n) \approx \frac{\sigma_x^2}{1 - W} I_N \qquad (6.102)$$

Near the time origin, the actual matrix $R_N^q(n)$ is assumed to differ from $R_N(n)$ only by addition of random errors, which introduces a decoupling between $R_N^q(n)$ and $X(n)$. Hence the following approximation can be justified:

$$X^t(n)[R_N^q(n)]^{-1}X(n) \approx \frac{1 - W}{\sigma_x^2} X^t(n)X(n) \qquad (6.103)$$

The variable $X^t(n)X(n)$ is Gaussian with mean $N\sigma_x^2$ and variance $2N\sigma_x^4$. If a peak factor of 4 is assumed, a condition for keeping the prediction error ratio above zero is

$$(1 - W)(N + 4\sqrt{2N}) < 1 \qquad (6.104)$$

This inequality shows that a lower bound is imposed on W. For example, if $N = 10$, then $W > 0.95$.

Now, for a more general input signal, the extreme situation occurs when the data vector $X(n)$ has the direction of the eigenvector associated with the smallest eigenvalue $\lambda_{min}^q(n)$ of $R_N^q(n)$. Under the hypotheses of zero mean random error addition, neglecting long-term accumulation processes if any, the following approximation can be made:

$$\lambda_{min}^q(n) \approx \frac{\lambda_{min}}{1 - W} \qquad (6.105)$$

where λ_{min} is the smallest eigenvalue of the input signal AC matrix. If we further approximate $X^t(n)X(n)$ by $N\sigma_x^2$, the condition on $\varphi(n)$ becomes

$$(1 - W)\frac{N\sigma_x^2}{\lambda_{min}} < 1 \qquad (6.106)$$

This condition may appear extremely restrictive, since the ratio σ_x^2/λ_{min} can take on large values. For example, if $x(n)$ is a determinist signal with additive noise and the predictor order N is large enough, σ_x^2/λ_{min} is the SNR. Inequalities (6.104) and (6.106) have been derived under restrictive hypotheses on the effects of roundoff errors, and they must be used with care. Nevertheless, they show that the weighting factor W cannot be chosen arbitrarily small.

6.7 INITIAL VALUES OF THE PREDICTION ERROR ENERGIES

The recursive implementations of the weighted LS algorithms require the initialization of the state variables. If the signal is not known before time $n = 0$, it is reasonable to assume that it is zero and the prediction coefficients are zero. However, the forward prediction error energy must be set to a positive value, say E_0. For the algorithm to start on the right track, the initial conditions must correspond to a LS situation.

A positive forward prediction error energy, when the prediction coefficients are zero, can be interpreted as corresponding to a signal whose previous samples are all zero except for one. Moreover, if the gain $G(0)$ is also zero, then the input sequence is

$$x(-N) = (W^{-N}E_0)^{1/2}$$
$$x(n) = 0, \qquad n \leqslant 0, n \neq -N \tag{6.107}$$

The corresponding value for the backward prediction error energy is $E_b(0) = x^2(-N) = W^{-N}E_0$—hence the initialization (6.100).

In these conditions the initial value of the AC matrix estimation is

$$R_N(0) = \begin{bmatrix} 1 & 0 & \cdots & 0 \\ 0 & W^{-1} & \cdots & 0 \\ \vdots & \vdots & & \vdots \\ 0 & 0 & \cdots & W^{-(N-1)} \end{bmatrix} E_0 \tag{6.108}$$

and the matrix actually used to estimate the input AC matrix is $R_N^*(n)$, given by

$$R_N^*(n) = R_N(n) + W^n R_N(0) \tag{6.109}$$

The smallest eigenvalue of the expectation of $R_N^*(n)$, denoted $\lambda_{min}^*(n)$, is obtained, using (6.23), by

$$\lambda_{min}^*(n) = \frac{1 - W^n}{1 - W} \lambda_{min} + W^n E_0 \tag{6.110}$$

The first term on the right side is growing with n while the second is decaying. The transient phase and the steady state are put in the same situation as concerns stability if a lower bound is set on E_0. Equation (6.110) can be rewritten as

$$\lambda^*_{\min}(n) = \frac{\lambda_{\min}}{1 - W} = W^n \left(E_0 - \frac{\lambda_{\min}}{1 - W} \right) \tag{6.111}$$

Now, $\lambda^*_{\min}(n)$ is at least equal to $\lambda_{\min}/1 - W$ if E_0 itself is greater or equal to that quantity. From condition (6.106), we obtain

$$E_0 \geqslant N\sigma_x^2 \tag{6.112}$$

This condition has been derived under extremely restrictive hypotheses; it is, in general, overly pessimistic, and smaller values of the initial prediction error energy can work in practice. The representation of the matrix $R_N(n)$ in the system can stay accurate during a period of time longer than the transient phase as soon as the machine word length is sufficiently large. For example, extensive experiments carried out with a 16-bit microprocessor and fixed-point arithmetic have shown that a lower bound for E_0 is about $0.01\sigma_x^2$ [11]. If the word length is smaller, then E_0 must be larger. As an illustration, a unit

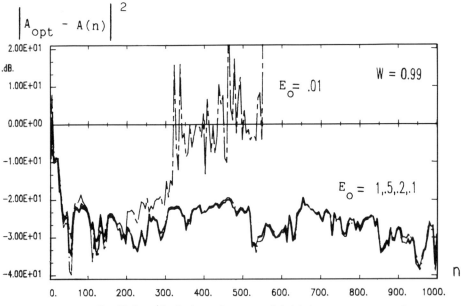

Figure 6.6 Coefficient deviations for several initial error energy values.

power AR signal is fed to a predictor with order $N = 4$, and the quadratic deviation of the coefficients from their ideal values is given in Figure 6.6 for several values of E_0. The weighting factor is $W = 0.99$ and a word length of 12 bits in fixed-point arithmetic is simulated. Satisfactory operation of the algorithm is obtained for $E_0 \geqslant 0.1$.

Finally, the above derivations show that the initial error energies cannot be taken arbitrarily small.

6.8 A STABILIZATION CONSTANT

The robustness of LS algorithms to roundoff errors can be improved by adding a noise sequence to the input signal. The smallest eigenvalue of the input AC matrix is increased by the additive noise power with that method, which can help satisfy inequality (6.106). However, as mentioned in Chapter 5, a bias is introduced on the prediction coefficients, and it is more desirable to use an approach bearing only on the algorithm operations.

When one considers condition (6.106), one can observe that, for W and N fixed, the only factor which can be manipulated is $\lambda_{\min}$, the minimal eigenvalue of the $N \times N$ input signal AC matrix. That factor is not available in the algorithm. However, it can be related to the prediction error energies, which are available.

From a different point of view, if the input signal is predictable, as seen in Section 2.9, the steady-state prediction error is zero for an order N sufficiently large. Consequently, the variables $E_a(n)$ and $E_b(n)$ can become arbitrarily small, and the rounding process eventually sets them to zero, which is unacceptable since they are used as divisors. Therefore a lower bound has to be imposed on error energies when the FLS algorithm is implemented in finite precision hardware. A simple method is to introduce a positive constant C in the updating equation

$$E_a(n + 1) = WE_a(n) + e_a(n + 1)\varepsilon_a(n + 1) + C \tag{6.113}$$

If σ_e^2 denotes the prediction error power associated with a stationary input signal, the expectation of $E_a(n)$ in the steady state is

$$E[E_a(n)] = \frac{\sigma_e^2 + C}{1 - W} \tag{6.114}$$

The same value would have been obtained with the weighting factor W' satisfying

$$\frac{\sigma_e^2 + C}{1 - W} = \frac{\sigma_e^2}{1 - W'} \tag{6.115}$$

and a first global assessment of the effect of introducing the constant C is that it increases the weighting factor from W to W', which helps satisfy condition (6.106).

As concerns the selection of a value for C, it can be related to the initial error energy E_0 and a reasonable choice can be:

$$C = (1 - W)E_0 \qquad (6.116)$$

In fact both E_0 and C depend on the performance objectives and the information available on the input signal characteristics.

Now, another aspect of introducing the constant C in equation (6.113), and a justification for its denomination, is that it counters the roundoff error accumulation which, as indicated in the next section, takes place in the backward prediction coefficient updating equation.

Adding a small constant C to $E_a(n + 1)$ leads to the adaptation gain

$$G_1^*(n + 1) \approx G_1(n + 1) - \frac{\varepsilon_a(n + 1)}{E_a^2(n + 1)}\left[\begin{array}{c} 1 \\ -A(n + 1) \end{array}\right]C \qquad (6.117)$$

The last element is

$$m^*(n + 1) \approx m(n + 1) + \frac{\varepsilon_a(n + 1)}{E_a^2(n + 1)} a_N(n + 1)C \qquad (6.118)$$

and the backward prediction updating equation in these conditions takes the form

$$B(n + 1) \approx (1 - \gamma_b)B(n) + G(n + 1)e_b(n + 1) \qquad (6.119)$$

with

$$E[\gamma_b] \approx C(1 - W)E[a_N{}^2(n + 1)]/E[E_a(n + 1)] \qquad (6.120)$$

However, it must be pointed out that, with the constant C, the algorithm is no longer in conformity with the LS theory and the theoretical stability is not guaranteed for any signals. The detailed analysis further reveals that the constant C increases the prediction error ratio $\varphi(n)$. Due to the range limitations for $\varphi(n)$ that can lead to the algorithm divergence for some signals. For example, with sinusoids as input signals, it can be seen, using the results given in Section 3.7 of Chapter 3, that $\varphi(n)$ can take on values very close to unity for sinusoids with close frequencies. In those cases the value of the constant C has to be very small and, consequently, a large machine wordlength is needed.

The roundoff error accumulation process is investigated next.

6.9 ROUNDOFF ERROR ACCUMULATION AND ITS CONTROL

Roundoff errors are generated by the quantization operations which generally take place after the multiplications and divisions. They are thought to come from independent sources, their spectrum is assumed flat, and their variance is $q^2/12$, where q is the quantization step size related to the internal word length of the machine used. The particularity of the FLS algorithms, presented in the previous sections, is that accumulation can take place [6–9]. Basically, the algorithm given in Figure 6.4, for example, consists of three overlapping recursions. The adaptation gain updating recursion makes the connection between forward and backward prediction coefficient recursions, and these recursions can produce roundoff noise accumulation [12].

Let us assume, for example, that an error vector $\Delta B(n)$ is added to the backward prediction coefficient vector $B(n)$ at time n. Then if we neglect the scalar term in (6.79) and consider the algorithm in Figure 6.4, the deviation at time $n + 1$ is

$$\Delta B(n + 1) = [I_N[1 + m(n + 1)e_b(n + 1)] - G(n + 1)X'(n + 1)]\Delta B(n)$$
$$- \Delta B(n)\Delta B'(n)m(n + 1)X(n + 1) \qquad (6.121)$$

If $\Delta B(n)$ is a random vector with zero mean, which is the case for a rounding operation, the mean of $\Delta B(n + 1)$ is not zero because of the matrix $\Delta B(n)\Delta B'(n)$ in (6.121) and because $m(n + 1)$ is related to the input signal, the expectation of the product $m(n + 1)X(n + 1)$ is, in general, not zero. The factor of $\Delta B(n)$ is close to a unity matrix—it can even have eigenvalues greater than 1—thus the introduction of error vectors $\Delta B(n)$ at each time n produces a drift in the coefficients. The effect is a shift of the coefficients from their optimal values, which degrades performance. However, if the minimum eigenvalue $\lambda_{1\min}$ of the $(N + 1) \times (N + 1)$ input AC matrix is close to the signal power σ_x^2, the prediction error power, also close to σ_x^2 because of (5.6), is an almost flat function of the coefficients and the drift can continue to the point where the resulting deviation of the eigenvalues and eigenvectors of the represented matrix $R_N^q(n)$ makes $\varphi(n)$ exceed its limits (6.41). Then, the algorithm is out of the LS situation and generally becomes unstable.

It is important to note that the long-term roundoff error accumulation affects the backward prediction coefficients but, except for the case $N = 1$, has much less effect on the forward coefficients. This is mainly due to the shift in the elements of the gain vector, which is performed by equation (6.73).

If the FLS algorithm has to be used with signals which can lead to instability and if the best performance is looked for, the roundoff error accumulation must be controlled. An efficient technique consists of finding a representative control variable and using it to prevent the coefficient drift [13].

The variable $\xi(n)$, defined by

$$\xi(n + 1) = [x(n + 1 - N) - B^t(n + 1)X(n + 1)] - m(n + 1)E_b(n + 1)$$

$$(6.122)$$

is representative of the long-term roundoff error accumulation because the two quantities of which it is a difference are updated in distinct computation loops. The coefficient vector $B(n + 1)$ is involved in calculating $\xi(n + 1)$, and a LS procedure can be used to correct the backward prediction coefficients. Such a procedure can be implemented in various ways. However, it is simplest with the algorithm based on all prediction errors, because all the necessary quantities are available and it is possible to implement it with no additional multiplications. In the equations given in Figure 6.5, it is sufficient to replace the backward coefficient updating recursion by

$$B(n + 1) = B(n) + \frac{G'(n + 1)[e_b(n + 1) + e_b(n + 1) - m'(n + 1)E_b(n)]}{\alpha(n + 1)} \qquad (6.123)$$

The FORTRAN program of the corresponding algorithm, including roundoff error accumulation control in the simplest version, is given in Annex 6.2.

An alternative way of escaping roundoff error accumulation is to avoid using backward prediction coefficients altogether.

6.10 A SIMPLIFIED ALGORITHM

When the input signal is stationary, the steady-state backward prediction coefficients are equal to the forward coefficients, as shown in Chapter 5, and the following equalities hold:

$$B(n) = J_N A(n), \qquad E_a(n) = E_b(n) \qquad (6.124)$$

This suggests replacing backward coefficients by forward coefficients in FLS algorithms. However, the property of theoretical stability of the LS principle is lost. Therefore it is necessary to have means to detect out-of-range values of LS variables. The variable $\alpha(n) = W/\varphi(n)$ can be used in combination with the gain vector $G'(n)$. The simplified algorithm obtained is given in Figure 6.7. It requires $7N + 5$ multiplications and two divisions (inverse calculations). The stability in the initial phase, starting from the idle state, can be critical. Therefore, the magnitude of $\alpha(n)$ is monitored, and if it falls below W the system is reinitialized.

In some cases, particularly with AR input signals when the prediction order exceeds the model order, the simplified algorithm turns out to provide

---

ALGORITHM S. F.L.S.

<u>AVAILABLE AT TIME n:</u>
COEFFICIENTS OF ADAPTIVE FILTER	: $H(n)$
" " FORWARD PREDICTOR	: $A(n)$
DATA VECTOR	: $X(n)$
ADAPTATION GAIN	: $G'(n)$
PREDICTION ERROR ENERGY	: $E_a(n)$
PREDICTION ERROR RATIO	: $\alpha(n)$
WEIGHTING FACTOR	: W

<u>NEW DATA AT TIME n:</u>
  Input signal : $x(n+1)$; Reference : $y(n+1)$

<u>ADAPTATION GAIN UPDATING:</u>
$$e_a(n+1) = x(n+1) - A^t(n) X(n)$$
$$A(n+1) = A(n) + G'(n) e_a(n+1) / \alpha(n)$$
$$E_a(n+1) = (E_a(n) + e_a(n+1) e_a(n+1) / \alpha(n)) W$$

$$G'_1(n+1) = \begin{bmatrix} 0 \\ \\ G'(n) \end{bmatrix} + \frac{e_a(n+1)}{E_a(n)} \begin{bmatrix} 1 \\ \\ -A(n) \end{bmatrix} = \begin{bmatrix} M(n+1) \\ \\ m(n+1) \end{bmatrix}$$

$$G'(n+1) = M(n+1) + m(n+1) J_N A(n)$$
$$\alpha(n+1) = W + X^t(n+1) G'(n+1)$$

<u>ADAPTIVE FILTER:</u>
$$e(n+1) = y(n+1) - H^t(n) X(n+1)$$
$$H(n+1) = H(n) + G'(n+1) e(n+1) / \alpha(n+1)$$

Figure 6.7   Computational organization of a simplified LS-type algorithm.

faster convergence than the standard FLS algorithms with the same parameters because the backward coefficients start with a value which is not zero but can be close to the final one.

6.11 PERFORMANCE OF LS ADAPTIVE FILTERS

The main specifications for adaptive filters concern, as in Section 4.2, the time constant and the system gain. Before investigating the initial transient phase, let us consider the filter operation after the first data have become available.

The set of output errors from time 1 to n constitute the vector

$$
\begin{bmatrix} e(1) \\ e(2) \\ e(3) \\ \vdots \\ e(N) \\ \vdots \\ e(n) \end{bmatrix} = \begin{bmatrix} y(1) \\ y(2) \\ y(3) \\ \vdots \\ y(N) \\ \vdots \\ y(n) \end{bmatrix} - \begin{bmatrix} x(1) & 0 & 0 & \cdots & 0 \\ x(2) & x(1) & 0 & \cdots & 0 \\ x(3) & x(2) & x(1) & \cdots & 0 \\ \vdots & & \vdots & & \vdots \\ x(N) & x(N-1) & & \cdots & x(1) \\ \vdots & & \vdots & & \vdots \\ x(n) & x(n-1) & & \cdots & x(n+1-N) \end{bmatrix}
$$

$$
\times \begin{bmatrix} h_0(n) \\ h_1(n) \\ \vdots \\ h_{N-1}(n) \end{bmatrix} \tag{6.125}
$$

Recall that the coefficients at time n are calculated to minimize the sum of the squares of the output errors. Clearly, for $n = 1$ the solution is

$$
h_0(1) = \frac{y(1)}{x(1)}, \qquad h_i(1) = 0, \quad 2 \leqslant i \leqslant N - 1 \tag{6.126}
$$

For $n = 2$,

$$
h_0(2) = \frac{y(1)}{x(1)}
$$

$$
h_1(2) = \frac{y(2)}{x(1)} - h_0(2)\frac{x(2)}{x(1)}
$$

$$
h_i(2) = 0, \qquad 3 \leqslant i \leqslant N - 1 \tag{6.127}
$$

The output of the adaptive LS filter is zero from time 1 to N, and the coefficients correspond to an exact solution of the minimization problem. In fact, the system of equations becomes overdetermined, and the LS procedure starts only at time $N + 1$.

In order to get simple expressions for the transient phase, we first analyze the system identification, shown in Figure 6.8. The reference signal is

$$
y(n) = X^t(n)H_{\text{opt}} + b(n) \tag{6.128}
$$

where $b(n)$ is a zero mean white observation noise with power $E_{\min}$, uncorrelated with the input $x(n)$. H_{opt} is the vector of coefficients which the adaptive filter has to find.

The coefficient vector of the LS adaptive filter at time n is

$$
H(n) = R_N^{-1}(n) \sum_{p=1}^{n} W^{n-p}[X(p)X^t(p)H_{\text{opt}} + X(p)b(p)] \tag{6.129}
$$

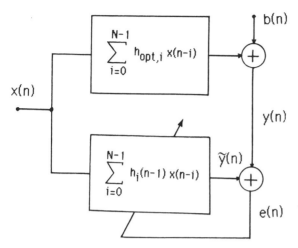

Figure 6.8 Adaptive system identification.

or, in concise form,

$$H(n) = H_{\text{opt}} + R_N^{-1}(n) \sum_{p=1}^{n} W^{n-p} X(p) b(p) \tag{6.130}$$

Denoting by $\Delta H(n)$ the coefficient deviation

$$\Delta H(n) = H(n) - H_{\text{opt}} \tag{6.131}$$

and assuming that, for a given sequence $x(p)$, $b(p)$ is the only random variable, we obtain the covariance matrix

$$E[\Delta H(n)\Delta H^t(n)] = E_{\min} R_N^{-1}(n) \left[\sum_{p=1}^{n} W^{2(n-p)} X(p) X^t(p) \right] R_N^{-1}(n) \tag{6.132}$$

For $W = 1$,

$$E[\Delta H(n)\Delta H^t(n)] = E_{\min} R_N^{-1}(n) \tag{6.133}$$

At the output of the adaptive filter the error signal at time n is

$$e(n) = y(n) - X^t(n)H(n-1) = b(n) - X^t(n)\Delta H(n-1) \tag{6.134}$$

The variance is

$$E[e^2(n)] = E_{\min} + X^t(n)E[\Delta H(n-1)\Delta H^t(n-1)]X(n) \tag{6.135}$$

and, for $W = 1$,

$$E[e^2(n)] = E_{\min}[1 + X^t(n)E_N^{-1}(n-1)X(n)] \tag{6.136}$$

Now, the mean residual error power $E_R(n)$ is obtained by averaging over all input signal sequences. If the signal $x(n)$ is a realization of a stationary process with AC matrix R_{xx}, for large n one has

$$R_N(n) \approx n R_{xx} \tag{6.137}$$

Using the matrix equality

$$X^t(n)R_N^{-1}(n)X(n) = \text{trace}[R_N^{-1}(n)X(n)X^t(n)] \tag{6.138}$$

and (6.137), we have

$$E_R(n) = E_{\min}\left(1 + \frac{N}{n-1}\right) \tag{6.139}$$

If the first datum received is $x(1)$, then, since the LS process starts at time $N + 1$, the mean residual error power at time n is:

$$E_R(n) = E_{\min}\left(1 + \frac{N}{n-N}\right), \qquad n \geq N + 1 \tag{6.140}$$

Thus, at time $n = 2N$, the mean residual error power is twice, or 3 dB above, the minimal value. This result can be compared with that obtained for the LMS algorithm, which, for an input signal close to being a white noise and a step size corresponding to the fastest start, is

$$E(n) - E_{\min} = E(0)\left(1 - \frac{1}{N}\right)^{2n} \tag{6.141}$$

which was derived by applying results obtained in Section 4.4.

The corresponding curves in Figure 6.9 show the advantage of the theoretical LS approach over the gradient technique when the system starts from the idle state [14].

Now, when a weighting factor is used, the error variance has to be computed from (6.132). If the matrix $R_N(n)$ is approximated by its expectation as in (6.137), one has

$$R_N(n) \approx \frac{1 - W^n}{1 - W} R_{xx}$$

$$\sum_{p=1}^{n} W^{2(n-p)}X(p)X^t(p) \approx \frac{1 - W^{2n}}{1 - W^2} R_{xx} \tag{6.142}$$

which, using identity (6.138) again, gives

$$E_R(n) \approx E_{\min}\left[1 + N\frac{1-W}{1+W}\frac{1+W^n}{1-W^n}\right] \tag{6.143}$$

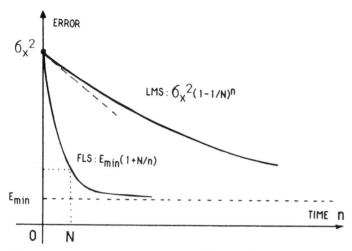

Figure 6.9 Learning curves for LS and LMS algorithms.

For $n \to \infty$,

$$E_R(\infty) = E_{min}\left[1 + N\frac{1 - W}{1 + W}\right] \tag{6.144}$$

This expression can be compared to the corresponding relation (4.35) in Chapter 4 for the gradient algorithm. The weighting factor introduces an excess MSE proportional to $1 - W$.

The coefficient learning curve is derived from recursion (6.24), which yields

$$\Delta H(n + 1) = [I_N - R_N^{-1}(n + 1)X(n + 1)X'(n + 1)]\Delta H(n)$$
$$+ R_N^{-1}(n + 1)X(n + 1)b(n + 1) \tag{6.145}$$

Assuming that $\Delta H(n)$ is independent of the input signal, which is true for large n, and using approximation (6.142), one gets

$$E[\Delta H(n + 1)] = \left(1 - \frac{1 - W}{1 - W^n}\right)E[\Delta H(n)] \tag{6.146}$$

Therefore, the learning curve of the filter of order N is similar to that of the first-order filter analyzed in Section 6.1, and for large n the time constant is $\tau = 1/(1 - W)$. It is that long-term time constant which has to be considered when a nonstationary reference signal is applied to the LS adaptive filter. In fact, $1/(1 - W)$ can be viewed as the observation time window of the filter, and, as in Section 4.8, its value is chosen to be compatible with the time period over which the signals can be considered as stationary; it is a trade-off between lag error and excess MSE.

6.12 SELECTING FLS PARAMETER VALUES

The performance of adaptive filters based on FLS algorithms differs from that of the theoretical LS filters because of the impact of the additional parameters they require. The value of the initial forward prediction error power E_0 affects the learning curve of the filter.

The matrix $R_N^*(n)$, introduced in Section 6.7, can be expressed by

$$R_N^*(n) = [I_N + W^n R_N(0)R_N^{-1}(n)]R_N(n) \qquad (6.147)$$

As soon as n is large enough, we can use (6.25a), to obtain its inverse:

$$[R_N^*(n)]^{-1} \approx R_N^{-1}(n)[I_N - W^n R_N(0)R_N^{-1}(n)] \qquad (6.148)$$

In these conditions, the deviation $\Delta A(n)$ of the prediction coefficients due to E_0 is

$$\Delta A(n) = W^n R_N^{-1}(n)R_N(0)A(n) \qquad (6.149)$$

and the corresponding excess MSE is

$$\Delta E(n) = [\Delta A(n)]^t R_{xx} \Delta A(n) \qquad (6.150)$$

Approximating $R_N(n)$ by its expectation and the initial matrix $R_N(0)$ by $E_0 I_N$ gives

$$\Delta E(n) \approx W^{2n} E_0^2 (1 - W)^2 A^t(n) R_{xx}^{-1} A(n) \qquad (6.151)$$

for W close to 1,

$$\ln[\Delta E(n)] \approx 2 \ln[E_0(1 - W)] + \ln[A^t(n)R_{xx}^{-1}A(n)] - 2n(1 - W) \qquad (6.152)$$

For example, the curves $\|\Delta A(n)\|^2$ as a function of n are given in Figure 6.10 for $N = 2$, $x(n) = \sin(n\frac{\pi}{4})$, $W = 0.95$, and three different values of the parameter E_0.

The impact of the initial parameter E_0 on the filter performance is clearly apparent from expression (6.152) and the above example. Smaller values of E_0 can be taken if the constant C of Section 6.8 is introduced.

The constant C in (6.113) increases the filter long-term time constant according to (6.115).

The ratio $(1 - W')/(1 - W)$ is shown in Figure 6.11 as a function of the prediction error σ_e^2. It appears that the starting value $\sigma_x^2/(\sigma_x^2 + C)$ should be made as close to unity as possible. So, C should be smaller than the input signal power σ_x^2, which in turn, through (6.106), means that W approaches unity.

If C is significantly smaller than σ_x^2, the algorithm can react quickly to large changes in input signal characteristics, and slowly to small changes. In other words, it has an adjustable time window.

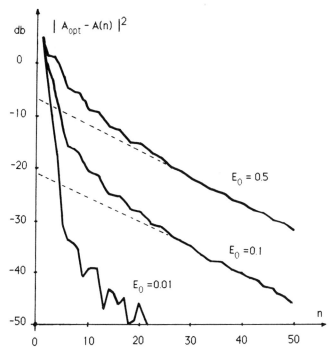

Figure 6.10 Coefficient deviations for several prediction error energy values with sinusoidal input.

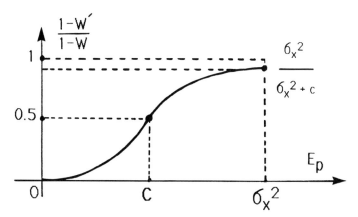

Figure 6.11 Weighting factor vs. prediction error power with constant C.

Another effect of C is to modify the excess misadjustment error power, according to equation (6.144), in which W' replaces W.

Nonstationary signals deserve particular attention. The range of values for C depends on E_0 and thus on the signal power. Thus, if the input signal is nonstationary, it can be interesting to use, instead of C, a function of the signal power. For example, the following equation can replace (6.113):

$$E_a(n + 1) = E_a(n) + e_a(n + 1)\varepsilon_a(n + 1) + (1 - W)N[C_1 + C_2 x^2(n + 1)]$$
$$(6.153)$$

where C_1 and C_2 are positive real constants, chosen in accordance with the characteristics of the input signal.

For example, an adequate choice for a speech sentence of unity long-term power has been found to be $C_1 = 1.5$ and $C_2 = 0.5$. The prediction gain obtained is shown in Figure 6.12 for several weighting factor values. As a comparison, the corresponding curve for the normalized LMS algorithm is also shown.

An additional parameter, the coefficient leakage factor, can be useful in FLS algorithms.

From the sequence of operations given in Figures 6.4 and 6.5, it appears that, if the signal $x(n)$ becomes zero, the prediction errors and adaptation gain decay to zero while the coefficients keep a fixed value. The system may be in the initial state considered in the previous sections, when the signal reappears, if a leakage factor is introduced in coefficient updating equations.

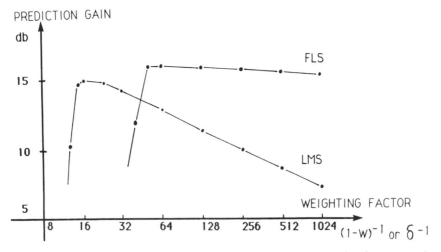

Figure 6.12 Prediction gain vs. weighting factor or step size for a speech sentence.

Furthermore, such a parameter offers the advantages already mentioned in Section 4.6—namely, it makes the filter more robust to roundoff errors and implementation constraints.

However, the corresponding arithmetic operations have to be introduced with care in FLS algorithms. They have to be performed outside the gain updating loop to preserve the ratio of a posteriori to a priori prediction errors. For example, in Figure 6.4 the two leakage operations

$$
\begin{aligned}
A(n + 1) &= (1 - \gamma)A(n + 1), \\
B(n + 1) &= (1 - \gamma)B(n + 1),
\end{aligned} \qquad 0 < \gamma \ll 1 \tag{6.154}
$$

can be placed at the end of the list of equations for the adaptation gain updating. Recall that the leakage factor introduces a bias given by expression (4.69) on the filter coefficients. Note also that, with the leakage factor, the algorithm is no longer complying with the LS theory and theoretical stability cannot be guaranteed for any signals.

6.13 WORD-LENGTH LIMITATIONS AND IMPLEMENTATION

The implementation of transversal FLS adaptive filters can follow the schemes used for gradient filters presented in Chapter 4. The operations in Figure 6.4, for example, correspond roughly to a set of five gradient filters adequately interconnected. However, an important point with FLS is the need for two divisions per iteration, generally implemented as inverse calculations.

The divider $E_a(n)$ is bounded by

$$
\min\left\{E_0, \frac{C}{1 - W}\right\} \leqslant E_a(n) \leqslant \frac{\sigma_x^2}{1 - W} \tag{6.155}
$$

and the constant C controls the magnitude range of its inverse. Recall that the other dividers are in the interval $]0, 1]$.

Overall, the estimations of word lengths for FLS filters can be derived using an approach similar to that which is used for LMS filters in Section 4.5. For example, let us consider the prediction coefficients.

In two extreme situations, the FLS algorithm is equivalent to an LMS algorithm with adaptation step sizes:

$$
\delta_{\max} = \frac{1}{\lambda_{\min}(n)}, \qquad \delta_{\min} = \frac{1}{\lambda_{\max}(n)} \tag{6.156}
$$

Now, taking $\lambda_{\max}(n) \approx \lambda_{\max}/(1 - W)$ and recalling that $\lambda_{\max} \leqslant N\sigma_x^2$, we obtain an estimation of the prediction coefficient word length b_c from equation (4.61) in Chapter 4:

$$b_c \approx \log_2 \left(\frac{N}{1 - W} \right) + \log_2(G_p) + \log_2(a_{max}) \tag{6.157}$$

where G_p is the prediction gain and a_{max} is the magnitude of the largest prediction coefficient, as in Section 4.5. Thus, it can be stated that FLS algorithms require larger word lengths than LMS algorithms, and the difference is about $\log_2 N$.

The implementation is guided by the basic constraint on updating operations, which have to be performed in a sample period. As shown in previous sections, there are different ways of organizing the computations, and that flexibility can be exploited to satisfy given realization conditions. In software, one can be interested in saving on the number of instructions or on the internal data memory capacity. In hardware, it may be important, particularly in high-speed applications using multiprocessor realizations, to rearrange the sequence of operations to introduce delays between internal filter sections and reach some level of pipelining [15]. For example, the algorithm based on a priori errors can be implemented by the following

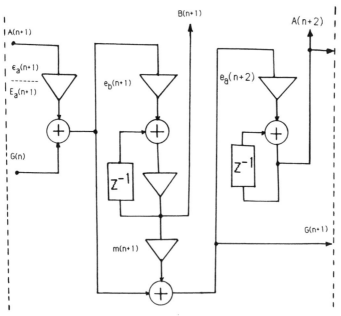

Figure 6.13 Adaptation section for a prediction coefficient in an FLS algorithm.

sequence at time $n + 1$:

$$e_a(n + 1) \to e_b(n) \to E_a(n) \to G_1(n) \to B(n) \to G(n) \to A(n + 1) \to \varepsilon_a(n + 1)$$

The corresponding diagram is shown in Figure 6.13 for a prediction coefficient adaptation section. With a single multiplier, the minimum multiply speed is five multiplications per sample period.

6.14 COMPARISON OF FLS AND LMS APPROACHES—SUMMARY

A geometrical illustration of the LS and gradient calculations is given in Fig. 6.14. It shows how the inverse input signal AC matrix R_{xx}^{-1} rotates the cost function gradient vector **Grad J** and adjusts its magnitude to reach the optimum coefficient values.

In FLS algorithms, real-time estimations of signal statistics are computed and the maximum convergence speed and accuracy can be expected. However, several parameters have to be introduced in realizations, which limit the performance; they are the weighting factor, initial prediction error energies, stabilization constant, and coefficient leakage factor. But if the values of these parameters are properly chosen, the performance can stay reasonably close to the theoretical optimum.

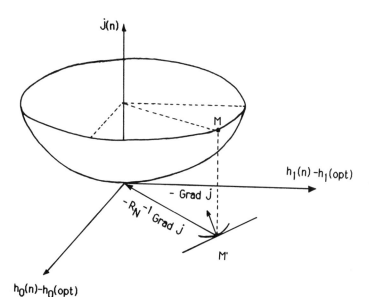

Figure 6.14 Geometrical illustration of LS and gradient calculations.

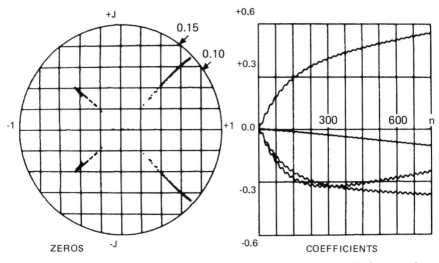

Figure 6.15 LMS adaptive prediction of two sinusoids with frequencies 0.1 and 0.15.

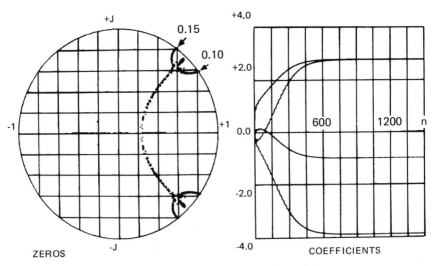

Figure 6.16 FLS adaptive prediction of two sinusoids.

In summary, the advantages of FLS adaptive filters are as follows:

Independence of the spread of the eigenvalues of the input signal AC matrix
Fast start from idle state
High steady-state accuracy

FLS adaptive filters can upgrade the adaptive filter overall performance in various existing applications. However, and perhaps more importantly, they can open up new areas. Consider, for example, spectral analysis, and let us assume that two sinusoids in noise have to be resolved with an order $N = 4$ adaptive predictor. The results obtained with the LMS algorithm are shown in Figure 6.15. Clearly, the prediction coefficient values cannot be used because they indicate the presence of a single sinusoid. Now, the same curves for the FLS algorithm, given in Figure 6.16, allow the correct detection after a few hundred iterations. That simple example shows that FLS algorithms can open new possibilities for adaptive filters in real-time spectral analysis.

EXERCISES

1. Verify, through matrix manipulations, the matrix inversion lemma (6.25). Use this lemma to find the inverse M^{-1} of the matrix

   $$M = \delta I_N + XX^t$$

   where X is an N-element nonzero vector. Give the limit of M^{-1} when $\delta \to 0$. Compare with (6.35).
2. Calculate the matrix $R_2(5)$ for the signal $x(n) = \sin(n\frac{\pi}{4})$ and $W = 0.9$. Compare the results with the signal AC matrix. Calculate the likelihood variable $\theta(5)$. Give bounds for $\theta(n)$ as $n \to \infty$.
3. Use the recurrence relationships for the backward prediction coefficient vector and the correlation vector to demonstrate the backward prediction error energy updating equation (6.59).
4. The signal

   $$x(n) = \sin\left(n\frac{\pi}{2}\right), \qquad n \geqslant 0$$

   $$x(n) = 0, \qquad\qquad n < 0$$

   is fed to an order $N = 4$ FLS adaptive predictor. Assuming initial conditions $A(0) = B(0) = G(0) = 0$, calculate the variables of the algorithm in Figure 6.4 for time $n = 1$ to 5 when $W = 1$ and for initial error energies $E_0 = 0$ and $E_0 = 1$. Compare the coefficient values to optimal values. Comment on the results.

5. In an FLS adaptive filter, the input signal $x(n)$ is set to zero at time N_0 and after. Analyze the evolution of the vectors $A(n)$, $B(n)$, $G(n)$ and the scalars $E_a(n)$ and $\alpha(n)$ for $n \geqslant N_0$.
6. Modify the algorithm of Figure 6.4 to introduce the scalar $\alpha(n)$ with the minimum number of multiplications. Give the computational organization, and count the multiplications, additions, and memories.
7. Study the hardware realization of the algorithm given in Figure 6.5. Find a reordering of the equations which leads to the introduction of sample period delays on the data paths interconnecting separate filter sections. Give the diagram of the coefficient adaptation section. Assuming a single multiplier per coefficient, what is the minimum multiply speed per sample period.

ANNEX 6.1 FLS ALGORITHM BASED ON A PRIORI ERRORS

```
      SUBROUTINE FLS1(N,X,VX,A,B,EA,G,W,IND)
C
C     COMPUTES THE ADAPTATION GAIN (FAST LEAST SQUARES)
C     N  = FILTER ORDER
C     X  = INPUT SIGNAL : x(n+1)
C     VX = N-ELEMENT DATA VECTOR : X(n)
C     A  = FORWARD PREDICTION COEFFICIENTS
C     B  = BACKWARD PREDICTION COEFFICIENTS
C     EA = PREDICTION ERROR ENERGY
C     G  = ADAPTATION GAIN
C     W  = WEIGHTING FACTOR
C     IND= TIME INDEX
C
      DIMENSION VX(15),A(15),B(15),G(15),G1(16)
      IF(IND.GT.1)GOTO30
C
C     INITIALIZATION
C
      DO20I=1,15
      A(I)=0.
      B(I)=0.
      G(I)=0.
      VX(I)=0.
   20 CONTINUE
      EA=1.
   30 CONTINUE
C
C     ADAPTATION GAIN CALCULATION
C
      EAV=X
      EPSA=X
      DO40I=1,N
   40 EAV=EAV-A(I)*VX(I)
      DO50I=1,N
      A(I)=A(I)+G(I)*EAV
      EPSA=EPSA-A(I)*VX(I)
```

```
      50   CONTINUE
           EA=W*EA+EAV*EPSA
           G1(1)=EPSA/EA
           DO60I=1,N
      60   G1(I+1)=G(I)-A(I)*G1(1)
           EAB=VX(N)
           DO70I=2,N
           J=N+1-I
      70   VX(J+1)=VX(J)
           VX(1)=X
           DO80I=1,N
      80   EAB=EAB-B(I)*VX(I)
           GG=1.0-EAB*G1(N+1)
           DO90I=1,N
           G(I)=G1(I)+G1(N+1)*B(I)
      90   G(I)=G(I)/GG
           DO100I=1,N
     100   B(I)=B(I)+G(I)*EAB
           RETURN
           END
```

ANNEX 6.2 FLS ALGORITHM BASED ON ALL THE PREDICTION ERRORS AND WITH ROUNDOFF ERROR CONTROL (SIMPLEST VERSION)

```
           SUBROUTINE FLS2(N,X,VX,A,B,EA,EB,GP,ALF,W,IND)
      C
      C    COMPUTES THE ADAPTATION GAIN (FAST LEAST SQUARES)
      C    N  = FILTER ORDER
      C    X  = INPUT SIGNAL : x(n+1)
      C    VX = N-ELEMENT DATA VECTOR : X(n)
      C    A  = FORWARD PREDICTION COEFFICIENTS
      C    B  = BACKWARD PREDICTION COEFFICIENTS
      C    EA = PREDICTION ERROR ENERGY  - EB
      C    GP = "A PRIORI" ADAPTATION GAIN
      C    ALF= PREDICTION ERROR RATIO
      C    W  = WEIGHTING FACTOR
      C    IND= TIME INDEX
      C
           DIMENSION VX(15),A(15),B(15),G(15),G1(16),GP(15)
           IF(IND.GT.1)GOTO30
      C
      C    INITIALIZATION
      C
           DO20I=1,15
           A(I)=0.
           B(I)=0.
           GP(I)=0.
           VX(I)=0.
      20   CONTINUE
           EA=1.
           EB=1./W**N
           ALF=W
      30   CONTINUE
      C
      C    ADAPTATION GAIN CALCULATION
      C
```

```
      EAV=X
      DO40I=1,N
40    EAV=EAV-A(I)*VX(I)
      EPSA=EAV/ALF
      G1(1)=EAV/EA
      EA=(EA+EAV*EPSA)*W
      DO50I=1,N
50    G1(I+1)=GP(I)-A(I)*G1(1)
      DO60I=1,N
60    A(I)=A(I)+GP(I)*EPSA
      EAB1=G1(N+1)*EB
      EAB=VX(N)-B(1)*X
      DO65I=2,N
      EAB=EAB-B(I)*VX(I-1)
65    CONTINUE
      DO70I=1,N
70    GP(I)=G1(I)+B(I)*G1(N+1)
      ALF1=ALF+G1(1)*EAV
      ALF=ALF1-G1(N+1)*EAB
      EPSB=(EAB+EAB-EAB1)/ALF
      EB=(EB+EAB*EPSB)*W
      DO80I=1,N
80    B(I)=B(I)+GP(I)*EPSB
      DO90I=2,N
      J=N+1-I
90    VX(J+1)=VX(J)
      VX(1)=X
      RETURN
      END
```

REFERENCES

1. A. A. Giordano and F. M. Hsu, *Least Squares Estimation With Applications to Digital Signal Processing*, Wiley, New York, 1985.
2. D. J. Tylavsky and G. R. Sohie, "Generalization of the Matrix Inversion Lemma," *Proc. IEEE*, **74**, 1050–1052 (July 1986).
3. J. M. Turner, "Recursive Least Squares Estimation and Lattice Filters," *Adaptive Filters*, Prentice-Hall, Englewood Cliffs, N.J., 1985, Chap. 5.
4. D. Falconer and L. Ljung, "Application of Fast Kalman Estimation to Adaptive Equalization," *IEEE Trans* **COM-26**, 1439–1446 (October 1978).
5. G. Carayannis, D. Manolakis, and N. Kalouptsidis, "A Fast Sequential Algorithm for LS Filtering and Prediction," *IEEE Trans.* **ASSP-31**, 1394–1402 (December 1983).
6. J. Cioffi and T. Kailath, "Fast Recursive Least Squares Transversal Filters for Adaptive Filtering," *IEEE Trans.* **ASSP-32** 304–337 (April 1984).
7. D. Lin, "On Digital Implementation of the Fast Kalman Algorithms," *IEEE Trans.* **ASSP-32**, 998–1005 (October 1984).

8. S. Ljung and L. Ljung, "Error Propagation Properties of Recursive Least Squares Adaptation Algorithms," *Automatica* **21**, 157–167 (1985).

9. J. M. Cioffi, "Limited Precision Effects in Adaptive Filtering," *IEEE Trans.* **CAS-** (1987).

10. S. H. Ardalan, "Floating Point Error Analysis of RLS and LMS Adaptive Filters," *Proc. IEEE/ICASSP-86*, Tokyo, 1986, pp. 513–516.

11. R. Alcantara, J. Prado, and C. Gueguen, "Fixed Point Implementation of the Fast Kalman Algorithm Using a TMS 32010 Microprocessor," *Proc. EUSIPCO-86*, North-Holland, The Hague, 1986, pp. 1335–1338.

12. P. Fabre and C. Gueguen, "Fast Recursive Least Squares: Preventing Divergence," *Proc. IEEE ICASSP-85*, Tampa, Fla., 1985, pp. 1149–1152.

13. J. L. Botto, "Stabilization of Fast Recursive Least Squares Transversal Filters," *Proc. IEEE/ICASSP-87*, Dallas, Texas, 1987, pp. 403–406.

14. M. L. Honig, "Echo Cancellation of Voice-Band Data Signals Using RLS and Gradient Algorithms," *IEEE Trans.* **COM-33**, 65–73 (January 1985).

15. V. B. Lawrence and S. K. Tewksbury, "Multiprocessor Implementation of Adaptive Digital Filters," *IEEE Trans.* **COM-31**, 826–835 (June 1983).

7

Other Adaptive Filter Algorithms

The derivation of FLS algorithms for transversal adaptive filters with N coefficients exploits the shifting property of the vector $X(n)$ of the N most recent input data, which is transferred to AC matrix estimations. Therefore, fast algorithms can be worked out whenever the shifting property exists. It means that variations of the basic algorithms can cope with different situations such as nonzero initial state variables and special observation time windows, and also that extensions to complex and multidimensional signals can be obtained.

A large family of algorithms can be constituted and, in this chapter, a selection is presented of those which may be of particular interest in different technical application fields.

If a set of N data $X(1)$ is already available at time $n = 1$, then when the filter is ready to start it may be advantageous to use that information in the algorithm rather than discard it. The so-called covariance algorithm is obtained [1].

7.1 COVARIANCE ALGORITHMS

The essential link in the derivation of the fast algorithms given in the previous chapter is provided by the $(N + 1) \times (N + 1)$ matrix $R_{N+1}(n + 1)$, which relates the adaptation gains $G(n + 1)$ and $G(n)$ at two consecutive instants. Here, a slightly different definition of that matrix has to be taken, because the first $(N + 1)$-element data vector which is available is $X_1(2)$:

$$[X_1(2)]^t = [x(2), X^t(1)]$$

Thus

$$R_{N+1}(n) = \sum_{p=2}^{n} W^{n-p} X_1(p) X_1^t(p) \tag{7.1}$$

The LS procedure for the prediction filters, because of the definitions, can only start at time $n = 2$, and the correlation vectors are

$$r_N^a(n) = \sum_{p=2}^{n} W^{n-p} x(p) X(p - 1)$$

$$r_N^b(n) = \sum_{p=2}^{n} W^{n-p} x(p - N) X(p) \tag{7.2}$$

The matrix $R_{N+1}(n + 1)$ can be partitioned in two ways:

$$R_{N+1}(n + 1) = \left[\begin{array}{c:c} \sum_{p=2}^{n+1} W^{n+1-p} x^2(p) & [r_N^a(n + 1)]^t \\ \hdashline r_N^a(n + 1) & R_N(n) \end{array}\right] \tag{7.3}$$

and

$$R_{N+1}(n + 1) = \left[\begin{array}{c:c} R_N(n + 1) - W^n X(1) X^t(1) & r_N^b(n + 1) \\ \hdashline [r_N^b(n + 1)]^t & \sum_{p=2}^{n+1} W^{n+1-p} x^2(p - N) \end{array}\right] \tag{7.4}$$

Now the procedure given in Section 6.4 can be applied again. However, several modifications have to be made because of the initial term $W^n X(1) X^t(1)$ in (7.4).

The $(N + 1)$-element adaptation gain vector $G_1(n + 1)$ can be calculated by equation (6.73) in Chapter 6, which yields $M(n + 1)$ and $m(n + 1)$. Equation (7.4) leads to

$$[R_N(n + 1) - W^n X(1) X^t(1)] M(n + 1) + m(n + 1) r_N^b(n + 1)$$
$$= X(n + 1) \tag{7.5}$$

Similarly the backward prediction matrix equation (6.74) in Chapter 6 combined with partitioning (7.4) leads to

$$[R_N(n + 1) - W^n X(1) X^t(1)] B(n + 1) = r_N^b(n + 1) \tag{7.6}$$

Now the definition of $G(n + 1)$ yields

$$G(n + 1) = R_N^{-1}(n + 1) X(n + 1) = [I_N - W^n R_N^{-1}(n + 1) X(1) X^t(1)]$$
$$\times [M(n + 1) + m(n + 1) B(n + 1)] \tag{7.7}$$

The difference with equation (6.78) of Chapter 6 is the initial term, which decays to zero as time elapses. The covariance algorithm, therefore, requires the same computations as the regular FLS algorithm with, in addition, the recursive computation of an initial transient variable. Let us consider the vector

$$D(n + 1) = W^n R_N^{-1}(n + 1)X(1) \tag{7.8}$$

A recursion is readily obtained by

$$R_N(n + 1)D(n + 1) = W^n X(1) \tag{7.9}$$

which at time n corresponds to

$$R_N(n)D(n) = W^{n-1}X(1)$$

Taking into account relationship (6.26) in Chapter 6 between $R_N(n)$ and $R_N(n + 1)$, one gets

$$D(n + 1) = [I_N - R_N^{-1}(n + 1)X(n + 1)X^t(n + 1)]D(n) \tag{7.10}$$

which with (7.7) and some algebraic manipulations yields

$$D(n + 1) = \frac{1}{1 - X^t(1)F(n + 1)X^t(n + 1)D(n)}[I_N - F(n + 1)X^t(n + 1)]D(n) \tag{7.11}$$

where

$$F(n) = M(n) + m(n)B(n) \tag{7.12}$$

The adaptation gain is obtained by rewriting (7.7) as

$$G(n + 1) = [I_N - D(n + 1)X^t(1)]F(n + 1) \tag{7.13}$$

Finally, the covariance version of the fast algorithm in Section 6.4 is obtained by incorporating equations (7.11) and (7.13) in the sequence of operations. The additional cost in computational complexity amounts to $4N$ multiplications and one division.

Some care has to be exercised in the initialization. If the prediction coefficients are zero, $A(1) = B(1) = 0$, since the initial data vector is nonzero, an initially constrained LS procedure has to be used, which, as mentioned in Section 6.7, corresponds to the following cost function for the filter [1]:

$$J_c(n) = \sum_{p=1}^{n} W^{n-p}[y(p) - X^t(p)H(n)]^2 + E_0 H^t(n)W(n)H(n) \tag{7.14}$$

where

$$W(n) = \text{diag}(W^n, W^{n-1}, \ldots, W^{n+1-N})$$

and E_0 is the initial prediction error energy.

In these conditions, the actual AC matrix estimate is

$$R_N^*(n) = \sum_{p=1}^{n} W^{n-p} X(p) X^t(p) + E_0 W(n) \qquad (7.15)$$

The value $R_N^{-1}(1)$ is needed because

$$D(1) = R_N^{-1}(1) X(1) = G(1)$$

It can be calculated with the help of the matrix inversion lemma. Finally,

$$D(1) = G(1) = \frac{1}{E_0 + X^t(1) W^{-1}(1) X(1)} W^{-1}(1) X(1) \qquad (7.16)$$

and for the prediction error energy $E_a(1) = W E_0$.

The weighting factor W introduces an exponential time observation window on the signal. Instead, it can be advantageous in some applications—for example, when the signal statistics can change abruptly—to use a constant time-limited window. The FLS algorithms can cope with that situation.

7.2 A SLIDING WINDOW ALGORITHM

The sliding window algorithms are characterized by the fact that the cost function $J_{SW}(n)$ to be minimized bears on the N_0 most recent output error samples:

$$J_{SW}(n) = \sum_{p=n+1-N_0}^{n} [y(p) - X^t(p) H(n)]^2 \qquad (7.17)$$

where N_0 is a fixed number representing the length of the observation time window, which slides on the time axis. In general, no weighting factor is used in that case, $W = 1$. Clearly, the AC matrix and cross-correlation vector estimations are

$$R_N(n) = \sum_{p=n+1-N_0}^{n} X(p) X^t(p), \qquad r_{yx} = \sum_{p=n+1-N_0}^{n} y(p) X(p) \qquad (7.18)$$

Again the matrix $R_{N+1}(n + 1)$ can be partitioned as

$$R_{N+1}(n + 1) = \sum_{p=n+2-N_0}^{n+1} \begin{bmatrix} x(p) \\ X(p-1) \end{bmatrix} [x(p), X^t(p - 1)]$$

$$= \begin{bmatrix} \sum_{p=n+2-N_0}^{n+1} x^2(n + 1) & \vdots & [r_N^a(n + 1)]^t \\ \hline r_N^a(n + 1) & \vdots & R_N(n) \end{bmatrix} \qquad (7.19)$$

and

$$R_{N+1}(n + 1) = \sum_{p=n+2-N_0}^{n} \begin{bmatrix} X(p) \\ x(p - N) \end{bmatrix} [X^t(p), x(p - N)]$$

$$= \begin{bmatrix} R_N(n + 1) & r_N^b(n + 1) \\ [r_N^b(n + 1)]^t & \sum_{p=n+2-N_0}^{n+1} x^2(p - N) \end{bmatrix} \qquad (7.20)$$

However, the recurrence relations become more complicated. For the AC matrix estimate, one has

$$R_N(n + 1) = R_N(n) + X(n + 1)X^t(n + 1)$$
$$- X(n + 1 - N_0)X^t(n + 1 - N_0) \qquad (7.21)$$

For the cross-correlation vector,

$$r_{yx}(n + 1) = r_{yx}(n) + y(n + 1)X(n + 1) - y(n + 1 - N_0)X(n + 1 - N_0)$$
$$(7.22)$$

The coefficient updating equation is obtained, as before, from

$$R_N(n + 1)H(n + 1) = r_{yx}(n + 1)$$

by substituting (7.22) and then, replacing $R_N(n)$ by its equivalent given by (7.21):

$$H(n + 1) = H(n) + R_N^{-1}(n + 1)X(n + 1)[y(n + 1) - X^t(n + 1)H(n)]$$
$$- R_N^{-1}(n + 1)X(n + 1 - N_0)$$
$$\times [y(n + 1 - N_0) - X^t(n + 1 - N_0)H(n)] \qquad (7.23)$$

Backward variables are showing up: the backward innovation error is

$$e_0(n + 1) = y(n + 1 - N_0) - X^t(n + 1 - N_0)H(n) \qquad (7.24)$$

and the backward adaptation gain is

$$G_0(n + 1) = R_N^{-1}(n + 1)X(n + 1 - N_0) \qquad (7.25)$$

In concise form, equation (7.23) is rewritten as

$$H(n + 1) = H(n) + G(n + 1)e(n + 1) - G_0(n + 1)e_0(n + 1)$$

These variables have to be computed and updated in the sliding window algorithms.

Partitioning (7.19) yields

$$R_{N+1}(n + 1) \begin{bmatrix} 0 \\ G_0(n) \end{bmatrix} = \begin{bmatrix} x(n + 1 - N_0) \\ X(n - N_0) \end{bmatrix} - \begin{bmatrix} \varepsilon_{0a}(n + 1) \\ 0 \end{bmatrix} \qquad (7.26)$$

with

$$\varepsilon_{0a}(n + 1) = x(n + 1 - N_0) - A^t(n + 1)X(n - N_0) \tag{7.27}$$

where the forward prediction coefficient vector is

$$A(n + 1) = R_N^{-1}(n) \sum_{p=n+2-N_0}^{n+1} x(p)X(p - 1) \tag{7.28}$$

Similarly, the second partitioning (7.20) yields

$$R_{N+1}(n + 1) \begin{bmatrix} G_0(n + 1) \\ 0 \end{bmatrix} = X_1(n + 1 - N_0) - \begin{bmatrix} 0 \\ \varepsilon_{0b}(n + 1) \end{bmatrix} \tag{7.29}$$

with

$$\varepsilon_{0b}(n + 1) = x(n + 1 - N_0 - N) - B^t(n + 1)X(n + 1 - N_0) \tag{7.30}$$

and

$$B(n + 1) = R_N^{-1}(n + 1)r_N^b(n + 1) \tag{7.31}$$

Now, combining the above equations with matrix prediction equations, as in Section 6.4, leads to

$$G_{01}(n + 1) = \begin{bmatrix} 0 \\ G(n) \end{bmatrix} - \frac{\varepsilon_{0a}(n + 1)}{E_a(n + 1)} \begin{bmatrix} 1 \\ -A(n + 1) \end{bmatrix} = \begin{bmatrix} M_0(n + 1) \\ m_0(n + 1) \end{bmatrix} \tag{7.32}$$

and

$$G_0(n + 1) = M_0(n + 1) + \frac{\varepsilon_{0b}(n + 1)}{E_b(n + 1)} B(n + 1)$$

$$m_0(n + 1) = \frac{\varepsilon_{0b}(n + 1)}{E_b(n + 1)} \tag{7.33}$$

Clearly, the updating technique is the same for both adaptation gains $G(n)$ and $G_0(n)$. The adequate prediction errors have to be employed.

The method used to derive the coefficient recursion (7.23) applies to linear prediction as well; hence

$$A(n + 1) = A(n) + R_N^{-1}(n)X(n)[x(n + 1) - X^t(n)A(n)]$$
$$- R_N^{-1}(n)X(n - N_0)[x(n + 1 - N_0) - X^t(n - N_0)A(n)] \tag{7.34}$$

or, in more concise form,

$$A(n + 1) = A(n) + G(n)e_a(n + 1) - G_0(n)e_{0a}(n + 1)$$

Now, the prediction error energy $E_a(n + 1)$, which appears in the matrix prediction equations, is

$$E_a(n + 1) = \sum_{p=n+2-N_0}^{n+1} x^2(p) - A^t(n + 1)r_N^a(n + 1) \tag{7.35}$$

Substituting (7.34) and the recursion for $r_N^a(n + 1)$ into the above expression, as in Section 6.3, leads to

$$E_a(n + 1) = E_a(n) + e_a(n + 1)\varepsilon_a(n + 1) - e_{0a}(n + 1)\varepsilon_{0a}(n + 1) \qquad (7.36)$$

The variables needed to perform the calculations in (7.32), and in the same equation for $G_1(n + 1)$, are available and the results can be used to get the updated gains.

The backward prediction coefficient vector is updated by

$$B(n + 1) = B(n) + G(n + 1)e_b(n + 1) - G_0(n + 1)e_{0b}(n + 1) \qquad (7.37)$$

which leads to the set of equations:

$$G(n + 1)[1 - m(n + 1)e_b(n + 1)]$$
$$= M(n + 1) + m(n + 1)B(n) - G_0(n + 1)e_{0b}(n + 1)m(n + 1)$$

$$G_0(n + 1)[1 + m_0(n + 1)e_{0b}(n + 1)]$$
$$= M_0(n + 1) + m_0(n + 1)B(n) + G(n + 1)e_b(n + 1)m_0(n + 1) \qquad (7.38)$$

Finally, letting

$$k = \frac{m(n + 1)}{1 + m_0(n + 1)e_{0b}(n + 1)}, \qquad k_0 = \frac{m_0(n + 1)}{1 - m(n + 1)e_b(n + 1)} \qquad (7.39)$$

we obtain the adaptation gains

$$G(n + 1) = \frac{1}{1 - ke_b(n + 1)} [M(n + 1) + kB(n) - ke_{0b}(n + 1)M_0(n + 1)]$$

$$G_0(n + 1) = \frac{1}{1 + k_0 e_{0b}(n + 1)} M_0(n + 1) + k_0 B(n) + k_0 e_b(n + 1)M(n + 1)]$$

$$(7.40)$$

The algorithm is then completed by the backward coefficient updating equation (7.37).

The initial conditions are those of the algorithm in Section 6.4, the extra operations being carried out only when the time index n exceeds the window length N_0.

Overall the sliding window algorithm based on a priori errors has a computational organization which closely follows that of the exponential window algorithm, but it performs the operations twice to update and use its two adaptation gains.

More efficient sliding window algorithms, but with a less regular structure, can be worked out by decomposing in two different steps the sequence of operations for each new input signal sample [2].

7.3 THE CASE OF COMPLEX SIGNALS

Complex signals take the form of sequences of complex numbers and are encountered in many applications, particularly in communications. Adaptive filtering techniques can be applied to complex signals in a straightforward manner, the main peculiarity being that the cost functions used in the optimization process must remain real and therefore moduli are involved.

For reasons of compatibility with the subsequent study of the multidimensional case, the cost function is taken as

$$J_{cX}(n) = \sum_{p=1}^{n} W^{n-p} |y(p) - \bar{H}^t(n)X(p)|^2 \tag{7.41}$$

or

$$J_{cX}(n) = \sum_{p=1}^{n} W^{n-p} e(n)\bar{e}(n)$$

where $\bar{e}(n)$ denotes the complex conjugate of $e(n)$, and the weighting factor W is assumed real.

Based on that cost function, FLS algorithms can be derived through the procedures presented previously [3].

The minimization of the cost function leads to

$$H(n) = R_N^{-1}(n) r_{yx}(n) \tag{7.42}$$

where

$$R_N(n) = \sum_{p=1}^{n} W^{n-p} X(p)\bar{X}^t(p)$$

$$r_{yx}(n) = \sum_{p=1}^{n} W^{n-p} \bar{y}(p)X(p) \tag{7.43}$$

Note that $[\bar{R}_N(n)]^t = R_N(n)$, which is the definition of a Hermitian matrix.

The connecting matrix $R_{N+1}(n+1)$ can be partitioned as

$$R_{N+1}(n+1) = \sum_{p=1}^{n+1} W^{n+1-p} \begin{bmatrix} x(p) \\ X(p-1) \end{bmatrix} [\bar{x}(p), \bar{X}^t(p-1)]$$

$$= \begin{bmatrix} \sum_{p=1}^{n+1} W^{n+1-p} |x(p)|^2 & [\bar{r}_N^a(n+1)]^t \\ r_N^a(n+1) & R_N(n) \end{bmatrix} \tag{7.44}$$

and

$$R_{N+1}(n+1) = \sum_{p=1}^{n+1} W^{n+1-p} \begin{bmatrix} X(p) \\ x(p-N) \end{bmatrix} [\bar{X}^t(p), \bar{x}(p-N)]$$

$$= \begin{bmatrix} R_N(n+1) & r_N^b(n+1) \\ [\bar{r}_N^b(n+1)]^t & \sum_{p=1}^{n+1} W^{n+1-p}|x(p-N)|^2 \end{bmatrix} \tag{7.45}$$

Following the definitions (7.42) and (7.43), the forward prediction coefficient vector is updated by

$$A(n+1) = R_N^{-1}(n)r_N^a(n+1) = A(n) + R_N^{-1}(n)X(n)[\bar{x}(n+1) - \bar{X}^t(n)A(n)]$$

or (7.46)

$$A(n+1) = A(n) + G(n)\bar{e}_a(n+1) \tag{7.47}$$

where the adaptation gain has the conventional definition and

$$e_a(n+1) = x(n+1) - \bar{A}^t(n)X(n)$$

Now, using the partitioning (7.44) as before, one gets

$$R_{N+1}(n+1)\begin{bmatrix} 0 \\ G(n) \end{bmatrix} = X_1(n+1) - \begin{bmatrix} \varepsilon_a(n+1) \\ 0 \end{bmatrix} \tag{7.48}$$

which, taking into account the prediction matrix equations, leads to the same equations as for real signals:

$$G_1(n+1) = \begin{bmatrix} 0 \\ G(n) \end{bmatrix} + \frac{\varepsilon_a(n+1)}{E_a(n+1)}\begin{bmatrix} 1 \\ -A(n+1) \end{bmatrix} = \begin{bmatrix} M(n+1) \\ m(n+1) \end{bmatrix}$$

The prediction error energy $E_a(n+1)$ can be updated by the following recursion, which is obtained through the method given in Section 6.3, for $R_N(n)$ Hermitian:

$$E_a(n+1) = WE_a(n) + e_a(n+1)\bar{e}_a(n+1) \tag{7.49}$$

The end of the procedure uses the partitioning of $R_{N+1}(n+1)$ given in equation (7.45) to express the order $N+1$ adaptation gain in terms of backward prediction variables. It can be verified that the conjugate of the backward prediction error

$$e_b(n+1) = x(n+1-N) - \bar{B}^t(n)X(n+1)$$

appears in the updated gain

$$G(n+1) = \frac{1}{1 - \bar{e}_b(n+1)m(n+1)}[M(n+1) + B(n)m(n+1)] \tag{7.50}$$

The backward prediction coefficients are updated by

$$B(n + 1) = B(n) + G(n + 1)\bar{e}_b(n + 1) \tag{7.51}$$

Finally the FLS algorithm for complex signals based on a priori errors is similar to the one given in Figure 6.4 for real data.

There is an identity between the complex signals and the two-dimensional signals which are considered in the next section. Algorithms for complex signals are directly obtained from those given in Figure 7.2 and Figure 7.3 by adding complex conjugation to transposition.

The prediction error ratio

$$\varphi(n) = \frac{\varepsilon_a(n + 1)}{e_a(n + 1)} = 1 - \bar{X}^t(n)R_N^{-1}(n)X(n) \tag{7.52}$$

is a real number, due to the Hermitian property of the AC matrix estimation $R_N(n)$. It is still limited to the interval $[0, 1]$ and can be used as a reliable checking variable.

7.4 MULTIDIMENSIONAL INPUT SIGNALS

The input and reference signals in adaptive filters can be vectors. To begin with, the case of an input signal consisting of K elements $x_i(n)(1 \leqslant i \leqslant k)$ and a scalar reference is considered. It is illustrated in Figure 7.1. The programmable filter, whose output $\tilde{y}(n)$ is a scalar like the reference $y(n)$, consists

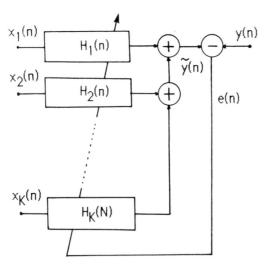

Figure 7.1 Adaptive filter with multidimensional input and scalar reference.

of a set of k different filters with coefficient vectors $H_i(n)(1 \leqslant i \leqslant k)$. These coefficients can be calculated to minimize a cost function in real time, through FLS algorithms.

Let $\chi(n)$ denote the k-element input vector

$$\chi^t(n) = [x_1(n), x_2(n), \ldots, x_k(n)]$$

Assuming that each filter coefficient vector $H_i(n)$ has N elements, let $X(n)$ denote the following input vector with KN elements:

$$X^t(n) = [\chi^t(n), \chi^t(n-1), \ldots, \chi^t(n+1-N)]$$

and let $H(n)$ denote the KN element coefficient vector

$$\overset{\longleftarrow \quad K \quad \longrightarrow \quad \longleftarrow \quad K \quad \longrightarrow \quad \quad \longleftarrow \quad K \quad \longrightarrow}{H^t(n) = [h_{11}(n), \ldots, h_{K1}(n), h_{12}(n), \ldots, h_{K2}(n), \ldots, h_{1N}(n), \ldots, h_{KN}(n)]}$$

The output error signal $e(n)$ is

$$e(n) = y(n) - H^t(n)X(n) \tag{7.53}$$

The minimization of the cost function $J(n)$ associated with an exponential time window,

$$J(n) = \sum_{p=1}^{n} W^{n-p} e^2(p)$$

leads to the set of equations

$$\frac{\partial J(n)}{\partial h_{ij}(n)} = 2 \sum_{p=1}^{n} W^{n-p}[y(p) - H^t(n)X(p)]x_i(p-j) = 0 \tag{7.54}$$

with $1 \leqslant i \leqslant K, 0 \leqslant j \leqslant N - 1$. Hence the optimum coefficient vector at time n is

$$H(n) = R_{KN}^{-1}(n)r_{KN}(n) \tag{7.55}$$

with

$$R_{KN}(n) = \sum_{p=1}^{n} W^{n-p}X(p)X^t(p)$$

$$r_{KN}(n) = \sum_{p=1}^{n} W^{n-p}y(p)X(p)$$

The matrix $R_{KN}(n)$ is a cross-correlation matrix estimation. The updating recursion for the coefficient vector takes the form

$$H(n+1) = H(n) + R_{KN}^{-1}(n+1)X(n+1)e(n+1) \tag{7.56}$$

and the adaptation gain

$$G_K(n) = R_{KN}^{-1}(n)X(n) \tag{7.57}$$

is a KN-element vector, which can be updated through a procedure similar to that of Section 6.4.

The connecting matrix $R_{KN1}(n + 1)$ is defined by

$$R_{KN1}(n + 1) = \sum_{p=1}^{n+1} W^{n+1-p} \begin{bmatrix} \chi(p) \\ X(p - 1) \end{bmatrix} [\chi^t(p), X^t(p - 1)] \tag{7.58}$$

and can be partitioned as

$$R_{KN1}(n + 1) = \begin{bmatrix} \sum_{p=1}^{n+1} W^{n+1-p}\chi(p)\chi^t(p) & [r_{KN}^a(n + 1)]^t \\ r_{KN}^a(n + 1) & R_{KN}(n) \end{bmatrix} \tag{7.59}$$

where $r_{KN}^a(n + 1)$ is the $KN \times K$ cross-correlation matrix

$$r_{KN}^a(n + 1) = \sum_{p=1}^{n+1} W^{n+1-p}X(p - 1)\chi^t(p) \tag{7.60}$$

From the alternative definition

$$R_{KN1}(n + 1) = \sum_{p=1}^{n+1} W^{n+1-p} \begin{bmatrix} X(p) \\ \chi(p - N) \end{bmatrix} [X^t(p), \chi^t(p - N)] \tag{7.61}$$

a second partitioning is obtained:

$$R_{KN1}(n + 1) = \begin{bmatrix} R_{KN}(n + 1) & r_{KN}^b(n + 1) \\ [r_{KN}^b(n + 1)]^t & \sum_{p=1}^{n+1} W^{n+1-p}\chi(n + 1 - N)\chi^t(n + 1 - N) \end{bmatrix} \tag{7.62}$$

where $r_{KN}^b(n + 1)$ is the $KN \times K$ matrix

$$r_{KN}^b(n + 1) = \sum_{p=1}^{n+1} W^{n+1-p}X(p)\chi^t(p - N) \tag{7.63}$$

The fast algorithms use the prediction equations. The forward prediction error takes the form of a K-element vector

$$e_{Ka}(n + 1) = \chi(n + 1) - A_K^t(n)X(n) \tag{7.64}$$

where the prediction coefficients form a $KN \times K$ matrix, which is computed to minimize the prediction error energy, defined by

$$E_a(n) = \sum_{p=1}^{n} W^{n-p}e_{Ka}^t(p)e_{Ka}(p) = \text{trace}[E_{Ka}(n)] \tag{7.65}$$

with the quadratic error energy matrix defined by

$$E_{Ka}(n) = \sum_{p=1}^{n} W^{n-p}e_{Ka}(p)e_{Ka}^t(p) \tag{7.66}$$

The minimization process yields

$$A_K(n + 1) = R_{KN}^{-1}(n)r_{KN}^a(n + 1) \tag{7.67}$$

The forward prediction coefficients, updated by

$$A_K(n + 1) = A_K(n) + G_K(n)e_{Ka}^t(n + 1) \tag{7.68}$$

are used to derive the a posteriori prediction error $\varepsilon_{Ka}(n + 1)$, also a K-element vector, by

$$\varepsilon_{Ka}(n + 1) = \chi(n + 1) - A_K^t(n + 1)X(n) \tag{7.69}$$

The quadratic error energy matrix can also be expressed by

$$E_{Ka}(n + 1) = \sum_{p=1}^{n+1} W^{n+1-p}\chi(p)\chi^t(p) - A_K^t(n + 1)r_{KN}^a(n + 1) \tag{7.70}$$

which, by the same approach as in Section 6.3, yields the updating recursion

$$E_{Ka}(n + 1) = WE_{Ka}(n) + e_{Ka}(n + 1)\varepsilon_{Ka}^t(n + 1) \tag{7.71}$$

The a priori adaptation gain $G_K(n)$ can be updated by reproducing the developments given in Section 6.4 and using the two partitioning equations (7.59) and (7.62) for $R_{KN1}(n + 1)$. The fast algorithm based on a priori errors is given in Figure 7.2.

If the predictor order N is sufficient, the prediction error elements, in the steady-state phase, approach white noise signals and the matrix $E_{Ka}(n)$ approaches a diagonal matrix. Its initial value can be taken as a diagonal matrix

$$E_{Ka}(0) = E_0 I_K \tag{7.72}$$

where E_0 is a positive scalar; all other initial values can be zero.

A stabilization constant, as in Section 6.8, can be introduced by modifying recursion (7.71) as follows:

$$E_{Ka}(n + 1) = WE_{Ka}(n) + e_{Ka}(n + 1)\varepsilon_{Ka}^t(n + 1) + CI_K \tag{7.71a}$$

where C is a positive scalar.

The matrix inversion in Figure 7.2 is carried out, with the help of the matrix inversion lemma (6.25b) of Chapter 6 by updating the inverse quadratic error matrix:

$$E_{Ka}^{-1}(n + 1) = W^{-1}\left[E_{Ka}^{-1}(n) - \frac{E_{Ka}^{-1}(n)e_{Ka}(n + 1)\varepsilon_{Ka}^t(n + 1)E_{Ka}^{-1}(n)}{W + \varepsilon_{Ka}^t(n + 1)E_{Ka}^{-1}(n)e_{Ka}(n + 1)}\right] \tag{7.73}$$

The computational complexity of that expression amounts to $3K^2 + 2K$ multiplications and one division or inverse calculation.

ALGORITHM F.L.S. 1-K

AVAILABLE AT TIME n:
COEFFICIENTS OF ADAPTIVE FILTER : $H(n)$
FORWARD PREDICTION MATRIX : $A_K(n)$
BACKWARD PREDICTION MATRIX : $B_K(n)$
DATA VECTOR : $X(n)$
ADAPTATION GAIN : $G_K(n)$
QUADRATIC ERROR MATRIX : $E_{Ka}(n)$
WEIGHTING FACTOR : W

NEW DATA AT TIME n:
Input signal : $x(n+1)$; Reference : $y(n+1)$

ADAPTATION GAIN UPDATING :

$$e_{Ka}(n+1) = x(n+1) - A_K^t(n) X(n)$$

$$A_K(n+1) = A_K(n) + G_K(n) e_{Ka}^t(n+1)$$

$$\epsilon_{Ka}(n+1) = x(n+1) - A_K^t(n+1) X(n)$$

$$E_{Ka}(n+1) = W E_{Ka}(n) + e_{Ka}(n+1) \epsilon_{Ka}^t(n+1)$$

$$G_{K1}(n+1) = \begin{bmatrix} 0 \\ G_K(n) \end{bmatrix} + \begin{bmatrix} I_K \\ -A_K(n+1) \end{bmatrix} E_{Ka}^{-1}(n+1) \epsilon_{Ka}(n+1) = \begin{bmatrix} M_K(n+1) \\ m_K(n+1) \end{bmatrix}$$

$$e_{KB}(n+1) = x(n+1-N) - B_K^t(n) X(n+1)$$

$$G_K(n+1) = \frac{1}{1 - e_{Kb}^t(n+1) m_K(n+1)} (M_K(n+1) + B_K(n) m_K(n+1))$$

$$B_K(n+1) = B_K(n) + G_K(n+1) e_{Kb}^t(n+1)$$

ADAPTIVE FILTER :

$$e(n+1) = y(n+1) - H^t(n) X(n+1)$$

$$H(n+1) = H(n) + G_K(n+1) e(n+1)$$

Figure 7.2 FLS algorithm for multidimensional input signals.

Note that if $N = 0$, which means that there is no convolution on the input data, then $E_{Ka}^{-1}(n)$ is just the inverse cross-correlation matrix $R_{KN}^{-1}(n)$, and it is updated directly from the input signal data as in conventional RLS techniques.

For the operations related to the filter order N, the algorithm presented in Figure 7.2 requires $7K^2N + KN$ multiplications for the adaptation gain and $2KN$ multiplications for the filter section. The FORTRAN program is given in Annex 7.1.

The ratio $\varphi(n)$ of a posteriori to a priori prediction errors is still a scalar, because

$$\varepsilon_{aK}(n + 1) = e_{aK}(n + 1)[1 - G_K^t(n)X(n)] \tag{7.74}$$

Therefore it can still serve to check the correct operation of the multidimensional algorithms. Moreover, it allows us to extend to multidimensional input signals the algorithms based on all prediction errors.

7.5 M-D ALGORITHM BASED ON ALL PREDICTION ERRORS

An alternative adaptation gain vector, which leads to exploiting a priori and a posteriori prediction errors is defined by

$$G_K'(n + 1) = R_{KN}^{-1}(n)X(n + 1) = \frac{G_K(n + 1)W}{\varphi(n + 1)} \tag{7.75}$$

The updating procedure uses the ratio of a posteriori to a priori prediction errors, under the form of the scalar $\alpha(n)$ defined by

$$\alpha(n) = W + X^t(n)R_{KN}^{-1}(n - 1)X(n) = \frac{W}{\varphi(n)} \tag{7.76}$$

The computational organization of the corresponding algorithm is shown in Figure 7.3. Indeed, it follows closely the sequence of operations already given in Figure 6.5, but scalars and vectors have been replaced by vectors and matrices when appropriate.

The operations related to the filter order N correspond to $6K^2N$ multiplications for the gain and $2KN$ multiplications for the filter section.

In the above procedure, the backward a priori prediction error vector $e_{Kb}(n + 1)$ can also be calculated directly by

$$e_{Kb}(n + 1) = E_{Kb}(n)m_K(n + 1) \tag{7.77}$$

Again that provides means to control the roundoff error accumulation, through updating the backward prediction coefficients, as in (6.123) of

ALGORITHM F.L.S.2-K

AVAILABLE AT TIME n:

$$\begin{aligned}
\text{COEFFICIENTS OF ADAPTIVE FILTER} &: H(n)\\
\text{FORWARD PREDICTION MATRIX} &: A_K(n)\\
\text{BACKWARD PREDICTION MATRIX} &: B_K(n)\\
\text{DATA VECTOR} &: X(n)\\
\text{ADAPTATION GAIN} &: G'_K(n)\\
\text{QUADRATIC ERROR MATRICES} &: E_{Ka}(n), E_{Kb}(n)\\
\text{PREDICTION ERROR RATIO} &: \alpha(n)\\
\text{WEIGHTING FACTOR} &: W
\end{aligned}$$

NEW DATA AT TIME n:

Input signal : $x(n+1)$; Reference : $y(n+1)$

ADAPTATION GAIN UPDATING:

$$e_{Ka}(n+1) = x(n+1) - A_K^t(n)X(n)$$

$$A_K(n+1) = A_K(n) + G'_K(n)e_{Ka}^t(n+1)/x(n)$$

$$E_{Ka}(n+1) = (E_{Ka}(n) + e_{Ka}(n+1)e_{Ka}^t(n+1)/\alpha(n))W$$

$$G'_{K1}(n+1) = \begin{bmatrix} 0 \\ \\ G_K(n) \end{bmatrix} + \begin{bmatrix} I_K \\ \\ -A_K(N+1) \end{bmatrix} E_{Ka}^{-1}(n+1)e_{Ka}(n+1) = \begin{bmatrix} M_K(n+1) \\ \\ m_K(n+1) \end{bmatrix}$$

$$e_{Kb}(n+1) = x(n+1-N) - B_K^t(n+1)X(n+1)$$

$$G'_K(n+1) = M_K(n+1) + B_K(n)m_K(n+1)$$

$$\alpha_1(n+1) = \alpha(n) + e_{Ka}^t(n+1)E_{Ka}^{-1}(n)e_{Ka}(n+1)$$

$$\alpha(n+1) = \alpha_1(n+1) - e_{Kb}^t(n+1)m_K(n+1)$$

$$E_{Kb}(n+1) = (E_{Kb}(n) + e_{Kb}(n+1)e_{Kb}^t(n+1)/\alpha(n+1))W$$

$$B_K(n+1) = B_K(n+1) + G'_K(n)e_{Kb}(n+1)/\alpha(n+1)$$

ADAPTIVE FILTER:

$$e(n+1) = y(n+1) - H^t(n)X(n+1)$$

$$H(n+1) = H(n) + G'_K(n+1)e(n+1)/\alpha(n+1)$$

Figure 7.3 Algorithm based on all prediction errors for M-D input signals.

Chapter 6 by

$$B_K(n + 1) = B_K(n) + G'_K(n + 1)$$
$$\times \; [e_{Kb}(n + 1) + e_{Kb}(n + 1) - E_{Kb}(n)m_K(n + 1)]/a(n + 1) \tag{7.78}$$

Up to now, the reference signal has been assumed to be a scalar sequence. The adaptation gain calculations which have been carried out only depend on the input signals, and they are valid for multidimensional reference signals as well. The case of K-dimensional (K-D) input and L-dimensional (L-D) reference signals is depicted in Figure 7.4. The only modifications with respect to the previous algorithms concern the filter section. The L-element reference vector $Y_L(n)$ is used to derive the output error vector $e_L(n)$ from the input and the $KN \times L$ coefficient matrix $H_L(n)$ as follows:

$$e_L(n + 1) = Y_L(n + 1) - H_L^t(n)X(n + 1) \tag{7.79}$$

The coefficient matrix is updated by

$$H_L(n + 1) = H_L(n) + \frac{G'_K(n + 1)e_L^t(n + 1)}{\alpha(n + 1)} \tag{7.80}$$

The associated complexity amounts to $2NKL + L$ multiplications.

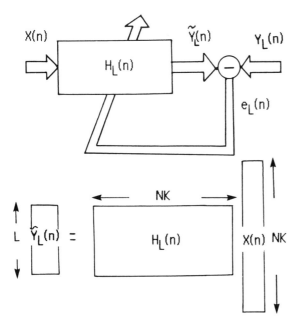

Figure 7.4   Adaptive filter with M-D input and reference signals.

The developments given in Chapter 6 and the preceding sections have illustrated the flexibility of the procedures used to derive fast algorithms. Another example is provided by filters of nonuniform length [4].

7.6 FILTERS OF NONUNIFORM LENGTH

In practice it is desirable to tailor algorithms to meet the specific needs of applications. The input sequences may be fed to filters with different lengths, and adjusting the fast algorithms accordingly can provide substantial savings.

Assume that the K filters in Figure 7.1 have lengths $N_i (1 \leqslant i \leqslant K)$. The data vector $X(n)$ can be rearranged as follows:

$$X^t(n) = [X_1^t(n), X_2^t(n), \ldots, X_K^t(n)] \tag{7.81}$$

where

$$X_i^t(n) = [x_i(n), x_i(n-1), \ldots, x_i(n+1-N_i)]$$

The number of elements ΣN is

$$\Sigma N = \sum_{i=1}^{K} N_i \tag{7.82}$$

The connecting $(\Sigma N + K)(\Sigma N + K)$ matrix $R_{\Sigma N1}(n+1)$, defined by

$$R_{\Sigma N1}(n+1) = \begin{bmatrix} x_1(n+1) \\ X_1(n) \\ \vdots \\ x_K(n+1) \\ X_K(n) \end{bmatrix} [x_1(n+1), X_1^t(n), \ldots, x_K(n+1), X_K^t(n)]$$

can again be partitioned in two different manners and provide the gain updating operations. The algorithms obtained are those shown in Figures 7.2 and 7.3. The only difference is that the prediction coefficient $\Sigma N \times K$ matrices are organized differently to accommodate the rearrangement of the data vector $X(n)$.

A typical case where filter dimensions can be different is pole-zero modeling.

7.7 FLS POLE-ZERO MODELING

Pole-zero modeling techniques are used in control for parametric system identification [5].

An adaptive filter with zeros and poles can be viewed as a filter with 2-D input data and 1-D reference signal. The filter defined by

$$\tilde{y}(n + 1) = A^t(n)X(n + 1) + B^t(n)\tilde{Y}(n) \tag{7.83}$$

is equivalent to a filter as in Figure 7.1 with input signal vector

$$\chi(n + 1) = \begin{bmatrix} x(n + 1) \\ \tilde{y}(n) \end{bmatrix} \tag{7.84}$$

For example, let us consider the pole-zero modeling of a system with output $y(n)$ when fed with $x(n)$. An approach which ensures stability is shown in Figure 4.11(b). A 2-D FLS algorithm can be used to compute the model coefficients with input signal vector

$$\chi(n + 1) = \begin{bmatrix} x(n + 1) \\ y(n) \end{bmatrix} \tag{7.85}$$

However, as pointed out in Section 4.11, that series-parallel type of approach is biased when noise is added to the reference signal. It is preferable to use the parallel approach in Figure 4.11(a). But stability can only be guaranteed if the smoothing filter with z-transfer function $C(z)$ satisfying strictly positive real (SPR) condition (4.128) in Chapter 4 is introduced on the error signal.

An efficient approach to pole-zero modeling is obtained by incorporating the smoothing filter in the LS process [6]. A 3-D FLS algorithm is employed, and the corresponding diagram is shown in Figure 7.5. The output error signal $f(n)$ used in the adaptation process is

$$f(n) = y(n) - [u_1(n) + u_2(n) + u_3(n)] \tag{7.86}$$

where $u_1(n)$, $u_2(n)$, and $u_3(n)$ are the outputs of the three filters fed by $\tilde{y}(n)$, $x(n)$, and $e(n) = y(n) - \tilde{y}(n)$, respectively. The cost function is

$$J_3(n) = \sum_{p=1}^{n} W^{n-p} f^2(p) \tag{7.87}$$

Let the unknown system output be

$$y(n) = \sum_{i=0}^{N} a_i x(n - i) + \sum_{i=1}^{N} b_i y(n - i) \tag{7.88}$$

or

$$y(n) = \sum_{i=0}^{N} a_i x(n - i) + \sum_{i=1}^{N} b_i \tilde{y}(n - i) + \sum_{i=1}^{N} b_i e(n - i) \tag{7.89}$$

From (7.86), the error signal is zero in the steady state if

$$a_i(\infty) = a_i, \quad b_i(\infty) = b_i, \quad c_i(\infty) = b_i, \qquad 1 \leqslant i \leqslant N$$

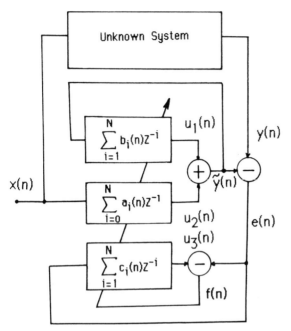

Figure 7.5 Adaptive pole-zero modeling with a 3-D FLS algorithm.

Now, assume that a white noise sequence $\eta(n)$ with power σ_η^2 is added to the system output. The cost function to be minimized becomes

$$J_{3\eta}(n) = \sum_{p=1}^{n} W^{n-p} \left[f(p) + \eta(p) - \sum_{i=1}^{N} \eta(p-i) \right]^2 \tag{7.90}$$

which, for sufficiently large n can be approximated by

$$J_{3\eta}(n) \approx \sum_{p=1}^{n} W^{n-p} \left[f^2(p) + \sigma_\eta^2 \left[1 + \sum_{i=1}^{N} c_i^2(p) \right] \right] \tag{7.91}$$

The steady-state solution is

$$a_i(\infty) = a_i, \quad b_i(\infty) = b_i, \quad c_i(\infty) = 0, \qquad 1 \leqslant i \leqslant N$$

Finally, the correct system identification is achieved, in the presence of noise or not. The smoothing filter coefficients vanish on the long run when additive noise is present. An illustration is provided by the following example.

Example [6]

Let the transfer function of the unknown system be

$$H(z) = \frac{0.05 + 0.1z^{-1} + 0.075z^{-2}}{1 - 0.96z^{-1} + 0.94z^{-2}}$$

and let the input be the first-order AR signal

$$x(n) = e_0(n) + 0.8x(n - 1)$$

where $e_0(n)$ is a white Gaussian sequence.

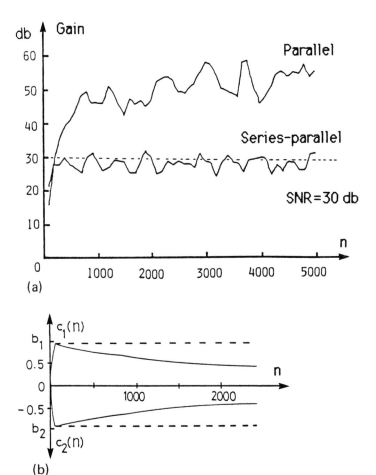

(a)

(b)

Figure 7.6 Pole-zero modeling of an unknown system: (a) System gain in FLS identification. (b) Smoothing filter coefficients.

The system gain G_S defined by

$$G_S = 10 \log \frac{E[y^2(n)]}{E[e^2(n)]}$$

is shown in Figure 7.6(a) versus time. The ratio of the system output signal to additive noise power is 30 dB. For comparison the gain obtained with the series-parallel approach is also given. In accordance with expression (4.133) in Chapter 4, it is bounded by the SNR. The smoothing filter coefficients are shown in Figure 7.6(b). They first reach the b_i values ($i = 1, 2$) and decay to zero after.

The 3-D parallel approach requires approximately twice the number of multiplications of the 2-D series-parallel approach.

7.8 MULTIRATE ADAPTIVE FILTERS

The sampling frequencies of input and reference signals can be different. In the sample rate reduction case, depicted in Figure 7.7, the input and reference sampling frequencies are f_S and $f_{S/K}$, respectively. The input signal sequence is used to form K sequences with sample rate $f_{S/K}$ which are fed to K filters with coefficient vectors $H_i(n)(0 \leqslant i \leqslant K - 1)$. The cost function to be minimized in the adaptive filter, $J_{SRR}(Kn)$, is

$$J_{SRR}(Kn) = \sum_{p=1}^{n} W^{n-p}[y(Kp) - H^t(Kp)X(Kp)]^2 \tag{7.92}$$

The data vector $X(Kn)$ is the vector of the NK most recent input values. The input may be considered as consisting of K different signals, and the

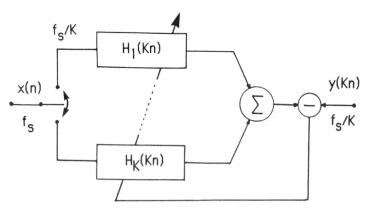

Figure 7.7 Sample rate reduction adaptive filter.

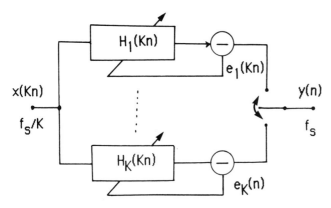

Figure 7.8 Sample rate increase adaptive filter.

algorithms presented in the preceding sections can be applied. The corresponding calculations are carried out at the frequency $f_{S/K}$.

The sample rate increase case is shown in Figure 7.8. It corresponds to 1-D input and multidimensional reference signals.

It is much more economical in terms of computational complexity than the sample rate reduction, because the adaptation gain is computed once for the K interpolating filters. All the calculations are again carried out at frequency $f_{S/K}$, the reference sequence being split into K sequences at that frequency. The system boils down to K different adaptive filters with the same input.

In signal processing, multirate aspects are often linked with DFT applications and filter banks, which correspond to frequency domain conversions [7].

7.9 FREQUENCY DOMAIN ADAPTIVE FILTERS

The power conservation principle states that the power of a signal in the time domain equals the sum of the powers of its frequency components. Thus, the LS techniques and adaptive methods worked out for time data can be transposed in the frequency domain.

The principle of a frequency domain adaptive filter (FDAF) is depicted in Figure 7.9. The N-point DFTs of the input and reference signals are computed. The complex input data obtained are multiplied by complex coefficients and subtracted from the reference to produce the output error used to adjust the coefficients.

At first glance, the approach may look complicated and farfetched. However, there are two motivations [8, 9]. First, from a theoretical point of

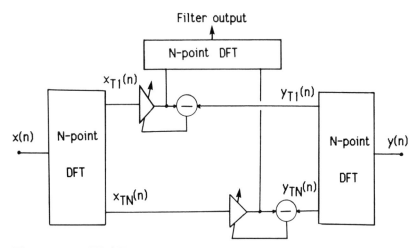

Figure 7.9 FDAF structure.

view, the DFT computer is actually a filter bank which performs some orthogonalization of the data; thus, an order N adaptive filter becomes a set of N separate order 1 filters. Second, from a practical standpoint, the efficient FFT algorithms to compute the DFT of blocks of N data, particularly for large N, can potentially produce substantial savings in computation speed, because the DFT output sampling frequency can be reduced by the factor N.

Assuming N separate complex filters and combining the results of Sections 6.1 and 7.3, we obtain the LS solution for the coefficients

$$h_i(n) = \frac{\sum_{p=1}^{n} W^{n-p} \bar{y}_{Ti}(p) x_{Ti}(p)}{\sum_{p=1}^{n} W^{n-p} x_{Ti}(p) \bar{x}_{Ti}(p)}, \qquad 0 \leqslant i \leqslant N - 1 \qquad (7.93)$$

where $x_{Ti}(n)$ and $y_{Ti}(n)$ are the transformed sequences.

For sufficiently large n, the denominator of that equation is an estimate of the input power spectrum, and the numerator is an estimate of the cross-power spectrum between input and reference signals. Overall the FDAF is an approximation of the optimal Wiener filter, itself the frequency domain counterpart of the time domain filter associated with the normal equations. Note that the optimal method along these lines, in case of stationary signals, would be to use the eigentransform of Section 3.12.

The updating equations associated with (7.93) are

$$h_i(n + 1) = h_i(n) + r_i^{-1}(n + 1) x_{Ti}(n + 1) \times [y_{Ti}(n + 1) - \bar{h}_i(n) x_{Ti}(n + 1)]$$

$$(7.94)$$

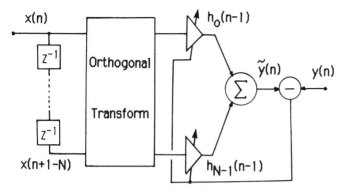

Figure 7.10 FDAF with a single orthogonal transform.

and

$$r_i(n + 1) = Wr_i(n) + x_{Ti}(n + 1)\bar{x}_{Ti}(n + 1) \qquad (7.95)$$

The FFT algorithms need about $(N/2)\log_2(N/2)$ complex multiplications each, which have to be added to the N order 1 adaptive filter operations. Altogether savings can be significant for large N, with respect to FLS algorithms.

The LMS algorithm can also be used to update the coefficients, and the results given in Chapter 4 can serve to assess complexity and performance.

It must be pointed out that the sample rate reduction by N at the DFT output can alter the adaptive filter operation, due to the circular convolution effects [8]. A scheme without sample rate reduction is shown in Figure 7.10, where a single orthogonal transform is used. If the first row of the transform matrix consists of 1's only, the inverse transformed data are obtained by just summing the transformed data [10]. Note also that complex operations are avoided if a real transform, such as the DCT [equations (3.147) in Chapter 3], is used.

A general observation about the performance of frequency domain adaptive filters is that they can yield poor results in the presence of nonstationary signals, because the subband decomposition they include can enhance the nonstationary character of the signals.

7.10 UNIFIED GENERAL VIEW AND CONCLUSION

The adaptive filters presented in Chapters 4, 6, and 7, in FIR or IIR direct form, have a strong structural resemblance, illustrated in the following coefficient updating equations:

$$\begin{vmatrix} \text{new} \\ \text{coefficient} \\ \text{vector} \end{vmatrix} = \begin{vmatrix} \text{old} \\ \text{coefficient} \\ \text{vector} \end{vmatrix} + \begin{vmatrix} \text{step} \\ \text{size} \end{vmatrix} \begin{vmatrix} \text{input} \\ \text{data} \\ \text{vector} \end{vmatrix} \begin{vmatrix} \text{innovation} \\ \text{signal} \end{vmatrix}$$

To determine the terms in that equation, the adaptive filter has only the data vector and reference signal available. All other variables, including the coefficients, are estimated. There are two categories of estimates; those which constitute predictions from the past, termed a priori, and those which incorporate the new information available, termed a posteriori. The final output of the filter is the a posteriori error signal

$$\varepsilon(n + 1) = y(n + 1) - H^t(n + 1)X(n + 1) \tag{7.96}$$

which can be interpreted as a measurement noise, a model error, or, in prediction, an excitation signal.

The innovation signal $i(n)$ represents the difference between the reference $y(n + 1)$ and a priori estimates which are functions of the past coefficients and output errors:

$$i(n + 1) = y(n + 1) - F_1[H^t(n), H^t(n - 1), \ldots]X(n + 1)$$
$$- F_2[\varepsilon(n), \varepsilon(n - 1), \ldots] \tag{7.97}$$

or, in terms of variable deviations

$$i(n + 1) = \Delta H^t(n + 1)X(n + 1) - \Delta\varepsilon(n + 1) \tag{7.98}$$

with

$$\Delta H(n + 1) = H(n + 1) - F_1[H(n), H(n - 1), \ldots]$$
$$\Delta\varepsilon(n + 1) = \varepsilon(n + 1) - F_2[\varepsilon(n), \varepsilon(n - 1), \ldots]$$

The derivation of an adaptive algorithm requires the design of predictors to generate the a priori estimates and a criterion defining how to use the innovation $i(n + 1)$ to determine the a posteriori estimates from the a priori ones.

When one takes

$$i(n + 1) = e(n + 1) = y(n + 1) - H^t(n)X(n + 1) \tag{7.99}$$

one simply assumes that the a priori estimate $H(n)$ for the coefficients is the a posteriori estimate at time n, which is valid for short-term stationary signals, and that the a priori error signal is zero, which is reasonable since the error signal is expected to be a zero mean white noise [11].

Minimizing the deviation between a posteriori and a priori estimates, with the cost function

$$J(n) = \Delta H^t(n)R(n)\Delta H + [\Delta\varepsilon(n)]^2 \tag{7.100}$$

where $R(n)$ is a symmetric positive definite weighting matrix, yields

$$H(n+1) = H(n) + \frac{i(n+1)}{1 + X^t(n+1)R(n+1)X(n+1)} R^{-1}(n+1)X(n+1)$$

$$\tag{7.101}$$

The flow graph of the general direct-form adaptive filter is given in Figure 7.11. It is valid for real, complex, or M-D data. The type of algorithm employed impacts the matrix $R(n)$, which is diagonal for the LMS algorithm and a square symmetric matrix for the LS approaches. Only the IIR filter discussed in Sections 4.11 and 7.7 uses an error prediction calculation to control the stability. The coefficient prediction filter can be used in a nonstationary environment to exploit the a priori knowledge on the nature of the nonstationarity and perform an appropriate bandlimited extrapolation.

Finally, the transversal adaptive filters form a large, diverse, and versatile family which can satisfy the requirements of applications in many technical fields. Their complexity can be tailored to the resources of the users, and their performance assessed accordingly. It is particularly remarkable to observe how flexible the FLS algorithms are, since they can provide exact solutions for different kinds of signals, observation conditions, and structures. A further illustration is given in the next chapter.

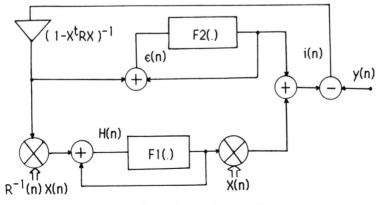

Figure 7.11 General direct-form adaptive filter.

EXERCISES

1. Use the approach in Section 7.1 to derive an algorithm based on all prediction errors as in Section 6.5, with nonzero initial input data vector. What is the additional computation load?

2. Taking

    $$e(n + 1) = y(n + 1) - H^t(n)X(n + 1)$$

    instead of (7.41) as the definition for the output error signal, give the computational organization of an alternative FLS algorithm for complex signals. Show that only forward prediction equations are modified by complex conjugation operations. Compare with the equations given in Section 7.3.

3. Give the detailed computational organization of an FLS algorithm for 2-D input signals, the coefficient vectors $H_1(n)$ and $H_2(n)$ having $N_1 = N_2 = 4$ elements. Count the memories needed. Modify the algorithm to achieve the minimum number of operations when $N_1 = 4$ and $N_2 = 2$. What reduction in number of multiplications and memories is obtained?

4. Extend the algorithm given in Section 7.4 for M-D input signals to the case of a sliding window. Estimate the additional computation load.

5. At the input of an adaptive filter with order $N = 4$, the signal is sampled at 4 kHz. The observed reference signal is available at the sampling frequency 1 kHz. Give the FLS algorithm for this multirate filter. Compare the complexities of the multirate algorithm and the standard algorithm which corresponds to a 4-kHz reference signal sampling rate. Compare also the performance of the two algorithms; what is the penalty in adaptation speed brought by undersampling the reference?

6. Use the technique described in Section 7.7 for pole-zero modeling to design an LS FIR/IIR predictor. Compare the 2-D and 3-D approaches in terms of computational complexity.

7. Consider the FDAF in Figure 7.10. The orthogonal transform of order N is the DCT which produces real outputs; describe the corresponding FLS algorithms. Compare the multiplication speed obtained with that of a direct FLS algorithm of order N. Compare also the performance of the two approaches.

ANNEX 7.1 FLS ALGORITHM WITH MULTIDIMENSIONAL INPUT SIGNAL

```
      SUBROUTINE FLS1MD(K,N,EAINV,UA,UB,VU,VU1,A,G,B,W)
C
C     COMPUTES THE ADAPTATION GAIN FOR MULTIDIMENSIONAL INPUT SIGNALS
C
C     K  = NUMBER OF INPUT SIGNALS (FILTER DIMENSION)
C     N  = NUMBER OF COEFFICIENTS IN EVERY CHANNEL
C     UA = INPUT VECTOR AT TIME (n+1)
C     UB = INPUT VECTOR AT TIME (n+1-N)
C     VU = KN ELEMENT DATA VECTOR AT TIME (n)
C     VU1= KN ELEMENT DATA VECTOR AT TIME (n+1)
C     A  = FORWARD LINEAR PREDICTION (KNxK) MATRIX
C     B  = BACKWARD LINEAR PREDICTION (KNxK) MATRIX
C     G  = ADAPTATION GAIN VECTOR
C     EAINV= PREDICTION ERROR ENERGY INVERSE (KxK) MATRIX
C     W  = WEIGHTING FACTOR
C
      DIMENSION UA(1),UB(1),VU(1),VU1(1),G(1)
      DIMENSION A(20,10),B(20,10),EAINV(10,10)
      DIMENSION SM(10),RM(20),EKA(10),EKB(10),AUX(10,10)
      DIMENSION EPKA(10),P1(10,10),P2(10),P3(10,10),P5(10,10)
      KN=K*N
C
C     FORWARD LINEAR PREDICTION ERROR :
C
      DO 1 I=1,K
      PR=0.
      P2(I)=0.
      DO 2 J=1,KN
      PR=PR+A(J,I)*VU(J)
    2 CONTINUE
      EKA(I)=UA(I)-PR
    1 CONTINUE
C
C     FORWARD PREDICTION MATRIX :
C
      DO 3 I=1,KN
      DO 4 J=1,K
      A(I,J)=A(I,J)+G(I)*EKA(J)
    4 CONTINUE
    3 CONTINUE
C
C     A POSTERIORI PREDICTION ERROR :
C
      DO 5 I=1,K
      PR=0.
      DO 6 J=1,KN
      PR=PR+A(J,I)*VU(J)
    6 CONTINUE
      EPKA(I)=UA(I)-PR
    5 CONTINUE
C
C     UPDATING OF ERROR ENERGY INVERSE MATRIX :
C
      P4=0.
      DO 7 J=1,K
      DO 8 I=1,K
```

```
          P1(J,I)=EKA(J)*EPKA(I)
          P2(J)=P2(J)+EPKA(I)*EAINV(I,J)
          P3(I,J)=0.
          P5(I,J)=0.
     8    CONTINUE
     7    CONTINUE
          DO 21 I=1,K
          DO 22 J=1,K
          DO 23 L=1,K
          P3(I,J)=P3(I,J)+EAINV(I,L)*P1(L,J)
    23    CONTINUE
    22    CONTINUE
          P4=P4+P2(I)*EKA(I)
    21    CONTINUE
          P4=P4+W
          DO 24 I=1,K
          DO 25 J=1,K
          DO 26 L=1,K
          P5(I,J)=P5(I,J)+P3(I,L)*EAINV(L,J)
    26    CONTINUE
          P5(I,J)=P5(I,J)/P4
    25    CONTINUE
    24    CONTINUE
          DO 27 I=1,K
          DO 28 J=1,K
          EAINV(I,J)=(EAINV(I,J)-P5(I,J))/W
          AUX(I,J)=EAINV(I,J)
    28    CONTINUE
    27    CONTINUE
C
C         EAINV IS IN AUX FOR SUBSEQUENT CALCULATIONS
C         KN+K ELEMENT ADAPTATION GAIN ( VECTORS RM AND SM ) :
C
          DO 9 I=1,K
          EX=0.
          DO 10 J=1,K
          EX=EX+AUX(I,J)*EPKA(J)
    10    CONTINUE
          AUX(I,1)=EX
     9    CONTINUE
          DO 11 I=K+1,KN+K
          EX=0.
          DO 12 J=1,K
          EX=EX-A(I-K,J)*AUX(J,1)
    12    CONTINUE
          AUX(I,1)=EX+G(I-K)
    11    CONTINUE
          DO 13 I=1,KN
          RM(I)=AUX(I,1)
          IF(I.LE.K)  SM(I)=AUX(KN+I,1)
    13    CONTINUE
C
C         BACKWARD PREDICTION ERROR :
C
          DO 14 I=1,K
          PR=0.
          DO 15 J=1,KN
          PR=PR+B(J,I)*VU1(J)
    15    CONTINUE
          EKB(I)=UB(I)-PR
    14    CONTINUE
```

```
C
C       KN ELEMENT ADAPTATION GAIN :
C
        EX=0.
        DO 16 I=1,K
        EX=EX+EKB(I)*SM(I)
   16   CONTINUE
        EX=1./(1.-EX)
        DO 17 I=1,KN
        PR=0.
        DO 18 J=1,K
        PR=PR+B(I,J)*SM(J)
   18   CONTINUE
        G(I)=EX*(RM(I)+PR)
   17   CONTINUE
C
C       BACKWARD PREDICTION (KNxK) MATRIX :
C
        DO 19 I=1,KN
        DO 20 J=1,K
        B(I,J)=B(I,J)+G(I)*EKB(J)
   20   CONTINUE
   19   CONTINUE
        RETURN
        END
```

REFERENCES

1. D. Lin, "On Digital Implementation of the Fast Kalman Algorithms," *IEEE Trans.* **ASSP-32**, 998–1005 (October 1984).
2. M. L. Honig and D. G. Messerschmitt, *Adaptive Filters: Structures, Algorithms and Applications*, Kluwer Academic, Boston, 1984, Chap. 6.
3. S. T. Alexander, "A Derivation of the Complex Fast Kalman Algorithm," *IEEE Trans.* **ASSP-32**, 1230–1232 (December 1984).
4. D. Falconer and L. Ljung, "Application of Fast Kalman Estimation to Adaptive Equalization," *IEEE Trans.* **COM-26**, 1439–1446 (October 1978).
5. P. Eykhoff, *System Identification: Parameter and State Estimation*, Wiley, London, 1974.
6. K. Kurosawa and S. Tsujii, "A New IIR Adaptive Algorithm of Parallel Type Structure," *Proc. IEEE/ICASSP-86* Tokyo, 1986, pp. 2091–2094.
7. R. E. Crochière and L. R. Rabiner, *Multirate Digital Signal Processing*, Prentice-Hall, Englewood Cliffs, N.J., 1983.
8. E. R. Ferrara, "Frequency Domain Adaptive Filtering," in *Adaptive Filters*, Prentice-Hall, Englewood Cliffs, N.J., 1985.
9. J. C. Ogue, T. Saito, and Y. Hoshiko, "A Fast Convergence Frequency Domain Adaptive Filter," *IEEE Trans.* **ASSP-31**, 1312–1314 (October 1983).

10. S. S. Narayan, A. M. Peterson, and M. J. Narasimha, "Transform Domain LMS Algorithm," *IEEE Trans.* **ASSP-31**, 609–615 (June 1983).
11. G. Kubin, "Direct Form Adaptive Filter Algorithms: A Unified View," in *Signal Processing III*, Elsevier, 1986, pp. 127–130.

8

Lattice Algorithms and Geometrical Approach

Although FLS algorithms for transversal adaptive structures are essentially based on time recursions, the algorithms for lattice structures make a joint use of time and order recurrence relationships. For a fixed filter order value N, they require more operations than their transversal counterparts. However, they provide adaptive filters of all the intermediate orders from 1 to N, which is an attractive feature in those applications where the order is not known beforehand and several different values have to be tried [1–3].

The order recurrence relationships introduced in Section 5.6 can be extended to real-time estimates.

8.1 ORDER RECURRENCE RELATIONS FOR PREDICTION COEFFICIENTS

Let $A_N(n)$, $B_N(n)$, $E_{aN}(n)$, $E_{bN}(n)$ and $G_N(n)$ denote the input signal prediction coefficient vectors, the error energies, and the adaptation gain at time n for filter order N. The forward linear prediction matrix equation for order $N-1$ is

$$R_N(n) \begin{bmatrix} 1 \\ -A_{N-1}(n) \end{bmatrix} = \begin{bmatrix} E_{a(N-1)}(n) \\ 0 \end{bmatrix} \tag{8.1}$$

Similarly, the backward prediction equation is

$$R_N(n) \begin{bmatrix} -B_{N-1}(n) \\ 1 \end{bmatrix} = \begin{bmatrix} 0 \\ E_{b(N-1)}(n) \end{bmatrix} \tag{8.2}$$

Now, partitioning equation (6.61) in Chapter 6 for $R_{N+1}(n)$ yields

$$\begin{bmatrix} R_N(n) & r_N^b(n) \\ [r_N^b(n)]^t & R_1(n-N) \end{bmatrix} \begin{bmatrix} 1 \\ -A_{N-1}(n) \\ 0 \end{bmatrix} = \begin{bmatrix} E_{a(N-1)}(n) \\ 0 \\ K_N(n) \end{bmatrix} \tag{8.3}$$

where the variable $K_N(n)$, corresponding to the last row, is

$$K_N(n) = \sum_{p=1}^{n} W^{n-p} x(p) x(p-N) - A_{N-1}^t(n) R_{N-1}(n-1) B_{N-1}(n-1) \tag{8.4}$$

In (8.4), forward and backward prediction coefficients appear in a balanced manner. Therefore the same variable $K_N(n)$ appears in the backward prediction matrix equation as well:

$$\begin{bmatrix} R_1(n) & [r_N^a(n)]^t \\ r_N^a(n) & R_N(n-1) \end{bmatrix} \begin{bmatrix} 0 \\ -B_{N-1}(n-1) \\ 1 \end{bmatrix} = \begin{bmatrix} K_N(n) \\ 0 \\ E_{b(N-1)}(n-1) \end{bmatrix} \tag{8.5}$$

as can be readily verified by analyzing the first row. Multiplying both sides by the scalar $K_N(n)/E_{b(N-1)}(n-1)$ gives

$$R_{N+1}(n) \begin{bmatrix} 0 \\ \begin{bmatrix} -B_{N-1}(n-1) \\ 1 \end{bmatrix} \dfrac{K_N(n)}{E_{b(N-1)}(n-1)} \end{bmatrix} = \begin{bmatrix} \dfrac{K_N^2(n)}{E_{b(N-1)}(n-1)} \\ 0 \\ K_N(n) \end{bmatrix} \tag{8.6}$$

Now, subtracting equation (8.6) from equation (8.3) and identifying with the forward prediction matrix equation (8.1) for order N, we obtain the following recursion for the forward prediction coefficient vectors:

$$A_N(n) = \begin{bmatrix} A_{N-1}(n) \\ 0 \end{bmatrix} - \dfrac{K_N(n)}{E_{b(N-1)}(n-1)} \begin{bmatrix} B_{N-1}(n-1) \\ -1 \end{bmatrix} \tag{8.7}$$

The first row yields a recursion for the forward prediction error energies:

$$E_{aN}(n) = E_{a(N-1)}(n) - \dfrac{K_N^2(n)}{E_{b(N-1)}(n-1)} \tag{8.8}$$

The same method can be applied to backward prediction equations. Matrix equation (8.3) can be rewritten as

$$R_{N+1}(n) \begin{bmatrix} \begin{bmatrix} 1 \\ -A_{N-1}(n) \\ 0 \end{bmatrix} \dfrac{K_N(n)}{E_{a(N-1)}(n)} \end{bmatrix} = \begin{bmatrix} K_N(n) \\ 0 \\ K_N^2(n) \\ E_{a(N-1)}(n) \end{bmatrix} \tag{8.9}$$

Subtracting equation (8.9) from equation (8.5) and identifying with the backward prediction matrix equation (8.2) for order N lead to recurrence relations for the backward prediction coefficients vectors

$$B_N(n) = \begin{bmatrix} 0 \\ B_{N-1}(n-1) \end{bmatrix} - \frac{K_N(n)}{E_{a(N-1)}(n)} \begin{bmatrix} -1 \\ A_{N-1}(n) \end{bmatrix} \qquad (8.10)$$

and for the backward prediction error energy

$$E_{bN}(n) = E_{b(N-1)}(n-1) - \frac{K_N^2(n)}{E_{a(N-1)}(n)} \qquad (8.11)$$

The definitions of the backward prediction a priori error

$$e_{aN}(n+1) = x(n+1) - A_N^t(n)X(n)$$

and backward prediction error

$$e_{bN}(n+1) = x(n+1-N) - B_N^t(n)X(n+1)$$

in connection with recursions (8.7) and (8.10), lead to the lattice predictor structure, which relates errors for orders N and $N-1$:

$$e_{aN}(n+1) = e_{a(N-1)}(n+1) - \frac{K_N(n)}{E_{b(N-1)}(n-1)} e_{b(N-1)}(n) \qquad (8.12)$$

and

$$e_{bN}(n+1) = e_{b(N-1)}(n) - \frac{K_N(n)}{E_{a(N-1)}(n)} e_{a(N-1)}(n+1) \qquad (8.13)$$

Similarly, for a posteriori errors,

$$\varepsilon_{aN}(n+1) = x(n+1) - A_N^t(n+1)X(n)$$

and

$$\varepsilon_{bN}(n+1) = x(n+1-N) - B_N^t(n+1)X(n+1)$$

The lattice structure operations are

$$\varepsilon_{aN}(n+1) = \varepsilon_{a(N-1)}(n+1) - k_{bN}(n+1)\varepsilon_{b(N-1)}(n) \qquad (8.14)$$

and

$$\varepsilon_{bN}(n+1) = \varepsilon_{b(N-1)}(n) - k_{aN}(n+1)\varepsilon_{a(N-1)}(n+1) \qquad (8.15)$$

where

$$k_{aN}(n+1) = \frac{K_N(n+1)}{E_{a(N-1)}(n+1)}, \qquad k_{bN}(n+1) = \frac{K_N(n+1)}{E_{b(N-1)}(n)} \qquad (8.16)$$

are the estimates of the PARCOR or reflection coefficients introduced in Section 5.5.

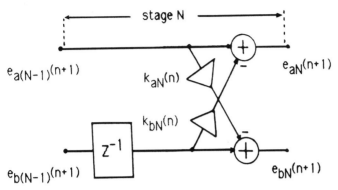

Figure 8.1   Adaptive lattice prediction error filter section.

The flow diagram of the corresponding lattice filter section is shown in Figure 8.1. The same structure can be used for a priori and a posteriori errors. A prediction error filter of order N is obtained by cascading N such sections.

Similar order recursions can be derived for the coefficients of adaptive filters, the adaptation gain, and the ratio of a posteriori to a priori errors.

8.2 ORDER RECURRENCE RELATIONS FOR THE FILTER COEFFICIENTS

An adaptive filter with N coefficients produces an output error signal $e_N(n)$:

$$e_N(n + 1) = y(n + 1) - H_N^t(n)X(n + 1) \tag{8.17}$$

The coefficient vector $H_N(n)$, which minimizes the error energy at time n, is obtained by

$$H_N(n) = R_N^{-1}(n)r_{yxN}(n) \tag{8.18}$$

with

$$r_{yxN}(n) = \sum_{p=1}^{n} W^{n-p}y(p)X_N(p)$$

For a filter with $N + 1$ coefficients, the equations are

$$e_{N+1}(n + 1) = y(n + 1) - H_{N+1}^t(n)X_{N+1}(n + 1) \tag{8.17a}$$

and

$$R_{N+1}(n)H_{N+1}(n) = \begin{bmatrix} r_{yxN}(n) \\ \sum_{p=1}^{n} W^{n-p}y(p)x(p - N) \end{bmatrix} \tag{8.18a}$$

The coefficient vector $H_{N+1}(n)$ can be obtained from $H_N(n)$ with the help of the partitioning (6.61) of Chapter 6 of the input signal AC matrix. As in the preceding section, consider the equation

$$R_{N+1}(n) \begin{bmatrix} H_N(n) \\ 0 \end{bmatrix} = \begin{bmatrix} R_N(n) & r_N^b(n) \\ [r_N^b(n)]^t & R_1(n-N) \end{bmatrix} \begin{bmatrix} H_N(n) \\ 0 \end{bmatrix}$$

$$= \begin{bmatrix} r_{yxN}(n) \\ [r_N^b(n)]^t H_N(n) \end{bmatrix} \tag{8.19}$$

The last row can also be written as

$$[r_N^b(n)]^t H_N(n) = B_N^t(n) R_N(n) H_N(n) = B_N^t(n) r_{yxN}(n) \tag{8.20}$$

Subtracting equation (8.19) from (8.18a) yields

$$R_{N+1}(n) \begin{bmatrix} H_{N+1}(n) - \begin{bmatrix} H_N(n) \\ 0 \end{bmatrix} \end{bmatrix} = \begin{bmatrix} 0 \\ K_{fN}(n) \end{bmatrix} \tag{8.21}$$

where

$$K_{fN}(n) = \sum_{p=1}^{n} W^{n-p} y(p) [x(p-N) - B_N^t(n) X(p)] \tag{8.22}$$

Now, identifying equation (8.21) with the backward linear prediction matrix equation leads to the following recurrence equation for the filter coefficients:

$$H_{N+1}(n) = \begin{bmatrix} H_N(n) \\ 0 \end{bmatrix} - \frac{K_{fN}(n)}{E_{bN}(n)} \begin{bmatrix} B_N(n) \\ -1 \end{bmatrix} \tag{8.23}$$

Substituting (8.23) into definition (8.17a) yields the relation for a priori output errors

$$e_{N+1}(n+1) = e_N(n+1) - \frac{K_{fN}(n)}{E_{bN}(n)} e_{bN}(n+1) \tag{8.24}$$

The corresponding equation for a posteriori errors is

$$\varepsilon_{N+1}(n+1) = \varepsilon_N(n+1) - \frac{K_{fN}(n+1)}{E_{bN}(n+1)} \varepsilon_{bN}(n+1) \tag{8.25}$$

Altogether, equations (8.12), (8.13), and (8.24) constitute the set of a priori equations for the lattice filter, while equations (8.14), (8.15), and (8.25) give the a posteriori version.

The error energy can also be computed recursively. According to the definition of the filter error output, we have

$$E_{N+1}(n) = \sum_{p=1}^{n} W^{n-p} y^2(p) - H_{N+1}^t(n) R_{N+1}(n) H_{N+1}(n) \tag{8.26}$$

Substituting recurrence relation (8.23) into (8.26) and using the backward prediction matrix equation, we obtain the order recursion

$$E_{N+1}(n) = E_N(n) - \frac{K_{fN}^2(n)}{E_{bN}(n)} \tag{8.27}$$

Obviously $E_{N+1}(n) \leqslant E_N(n)$, and the error power decreases as the filter order increases, which is a logical result.

Recall from Section 6.4 that the adaptation gain can be computed in a similar way. The derivation is repeated here for convenience. From the definition relation

$$R_N(n)G_N(n) = X_N(n) \tag{8.28}$$

we have

$$R_N(n)\begin{bmatrix} G_{N-1}(n) \\ 0 \end{bmatrix} = \begin{bmatrix} R_{N-1}(n) & r_{N-1}^b(n) \\ [r_{N-1}^b(n)]^t & R_1(n+1-N) \end{bmatrix}\begin{bmatrix} G_{N-1}(n) \\ 0 \end{bmatrix}$$
$$= \begin{bmatrix} X_{N-1}(n) \\ [r_{N-1}^b(n)]^t G_{N-1}(n) \end{bmatrix} \tag{8.29}$$

The last row can be expressed by

$$[r_{N-1}^b(n)]^t G_{N-1}(n) = B_{N-1}^t(n)X_{N-1}(n) = x(n+1-N) - \varepsilon_{b(N-1)}(n) \tag{8.30}$$

and equation (8.29) can be rewritten as

$$\begin{bmatrix} G_{N-1}(n) \\ 0 \end{bmatrix} = G_N(n) - R_N^{-1}(n)\begin{bmatrix} 0 \\ \varepsilon_{b(N-1)}(n) \end{bmatrix} \tag{8.31}$$

But the last row of the inverse AC matrix is proportional to the backward prediction coefficient vector; hence

$$G_N(n) = \begin{bmatrix} G_{N-1}(n) \\ 0 \end{bmatrix} + \frac{\varepsilon_{b(N-1)}(n)}{E_{b(N-1)}(n)}\begin{bmatrix} -B_{N-1}(n) \\ 1 \end{bmatrix} \tag{8.32}$$

This is equation (6.75) in Section 6.4. Recall that the other partitioning of $R_N(n)$ and the use of forward variables led to equation (6.73) in Chapter 6, which is a mixture of time and order recursions.

This expression is useful to recursively compute the ratio $\varphi_N(n)$ of a posteriori to a priori errors, defined by

$$\varphi_N(n) = \frac{\varepsilon_N(n)}{e_N(n)} = 1 - X_N^t(n)R_N^{-1}(n)X_N(n) = 1 - X_N^t(n)G_N(n)$$

Direct substitution yields

$$\varphi_N(n) = \varphi_{N-1}(n) - \frac{\varepsilon_{b(N-1)}^2(n)}{E_{b(N-1)}(n)} \tag{8.33}$$

The initial stage $N = 1$ is worth considering:

$$\varphi_1(n) = \varphi_0(n) - \frac{\varepsilon_{b0}^2(n)}{E_{b0}(n)} = 1 - \frac{x^2(n)}{\sum_{p=1}^{n} W^{n-p}x^2(p)} \tag{8.34}$$

Thus, in order to compute $\varphi_N(n)$ recursively, it is sufficient to take $\varphi_0(n) = 1$ and repeatedly use equation (8.33).

We reemphasize that $\varphi_N(n)$ is a crucial variable in FLS algorithms. It is of particular importance in lattice algorithms because it forms an essential link between order and time recursions.

8.3 TIME RECURRENCE RELATIONS

For a fixed filter order N, the lattice variable $K_N(n)$ can be computed recursively in time. According to definition (8.4), we have

$$K_{N+1}(n+1) = W \sum_{p=1}^{n} W^{n-p}x(p)x(p-N-1) + x(n+1)x(n-N)$$
$$- A_N^t(n+1)R_N(n)B_N(n) \tag{8.35}$$

Now, from the time recurrence relations (6.45), (6.26), and (6.53) in Chapter 6 for $A_N(n+1)$, $R_N(n)$, and $B_N(n)$, respectively, the following updating relation is obtained after some algebraic manipulations:

$$K_{N+1}(n+1) = WK_{N+1}(n) + e_{aN}(n+1)\varepsilon_{bN}(n) \tag{8.36}$$

Due to relations (6.49) and (6.56) of Chapter 6 between a priori and a posteriori errors, an alternative updating equation is

$$K_{N+1}(n+1) = WK_{N+1}(n) + \varepsilon_{aN}(n+1)e_{bN}(n) \tag{8.36a}$$

Clearly, the variable $K_{N+1}(n)$ represents an estimation of the cross-correlation between forward and backward order N prediction errors. Indeed, equation (8.36) is similar to the prediction error energy updating relations (6.58) and (6.59) derived and used in Chapter 6.

A similar relation can be derived for the filter output error energy $E_N(n)$. Equation (8.26) for order N and time $n + 1$ corresponds to

$$E_N(n+1) = \sum_{p=1}^{n+1} W^{n+1-p}y^2(p) - H_N^t(n+1)R_N(n+1)H_N(n+1) \tag{8.37}$$

Substituting the coefficient updating relation

$$H_N(n+1) = H_N(n) + G_N(n+1)e_N(n+1) \tag{8.38}$$

into (8.37), again yields after simplification

$$E_N(n+1) = WE_N(n) + e_N(n+1)\varepsilon_N(n+1) \tag{8.39}$$

For the filter section variable $K_{fN}(n + 1)$, definition (8.22) can be rewritten as

$$
\begin{aligned}
K_{fN}(n + 1) = \sum_{p=1}^{n+1} W^{n+1-p} y(p) x(p - N) \\
- [B_N^t(n) + G_N^t(n + 1) e_{bN}(n + 1)] \\
\times [Wr_{yxN}(n) + y(n + 1) X_N(n + 1)]
\end{aligned}
\tag{8.40}
$$

which, after simplification, leads to

$$
K_{fN}(n + 1) = WK_{fN}(n) + \varepsilon_{bN}(n + 1) e_N(n + 1)
\tag{8.41}
$$

Note that the variable $K_{fN}(n + 1)$, which according to definition (8.22) is an estimate of the cross-correlation between the reference signal and the backward prediction error, can be calculated as an estimate of the cross-correlation between the filter output error and the backward prediction error. This is due to the property of noncorrelation between the prediction errors and the data vector.

The recurrence relations derived so far can be used to build FLS algorithms for filters in lattice structures.

8.4 FLS ALGORITHMS FOR LATTICE STRUCTURES

The algorithms combine time and order recurrence relations to compute, for each set of new values of input and reference signals which become available, the lattice coefficients, the prediction and filter errors, their energies, and their cross-correlations. For a filter of order N, the operations are divided into prediction and filter operations.

To begin with, let us consider the initialization procedure. Since there are two types of recursions, two types of initializations have to be distinguished. The initializations for the order recursions are obtained in a straightforward manner: the prediction errors are initialized by the new input signal sample, the prediction error energies are set equal to the input signal power, and the variable $\varphi_0(n)$ is set to 1.

For time recursions, an approach to initialize the state variables of the order N lattice filter can be obtained as an extension of that given in Section 6.7. The input signal for $n \leq 0$ is assumed to consist of a single pulse at time $-N$, which leads to

$$
\begin{aligned}
e_{ai}(0) = e_{bi}(0) = \varepsilon_{ai}(0) = \varepsilon_{bi}(0) = 0, \qquad & 0 \leq i \leq N - 1 \\
E_{ai}(0) = W^N E_0, \qquad & 0 \leq i \leq N - 1 \\
E_{bi}(0) = W^{N-i} E_0, \qquad & 0 \leq i \leq N - 1 \\
K_i(0) = 0, \qquad & 1 \leq i \leq N
\end{aligned}
\tag{8.42}
$$

where E_0 is a real positive scalar. It can be verified that the prediction order recursions, and particularly energy relations (8.8) and (8.11), are satisfied for $n = 0$. Indeed, in these conditions, the impact of the choice of the initial error energy value E_0 on the filter performance is the same as for the transversal structure, and the relevant results given in Chapter 6 are still valid.

Many more or less different algorithms can be worked out from the basic time and order recursions, depending on the selection of internal variables and on whether the emphasis is on a priori or a posteriori error calculations and on time or order recurrence relations.

There are general rules to design efficient and robust algorithms, some of which can be stated as follows:

Minimize the number of state variables.
Give precedence to time recurrence whenever possible.
Make sure that reliable control variables are available to check the proper functioning of the adaptive filter.

Accordingly, the lattice algorithm given below avoids using the cross-correlation variable $K_i(n)$ and is based on a direct time updating of the reflection coefficients [4].

Substituting the time recursion (8.36a) and the error energy updating equation into definition (8.16) gives

$$[E_{ai}(n + 1) - e_{ai}(n + 1)\varepsilon_{ai}(n + 1)]k_{a(i+1)}(n)$$
$$= K_{i+1}(n + 1) - \varepsilon_{ai}(n + 1)e_{bi}(n) \tag{8.43}$$

Hence, using again (8.16) at time $n + 1$ gives

$$k_{a(i+1)}(n + 1) = k_{a(i+1)}(n) + \frac{\varepsilon_{ai}(n + 1)}{E_{ai}(n + 1)}[e_{bi}(n) - k_{a(i+1)}(n)e_{ai}(n + 1)] \tag{8.44}$$

Now, the time recursion (8.13) yields

$$k_{a(i+1)}(n + 1) = k_{a(i+1)}(n) + \frac{\varepsilon_{ai}(n + 1)e_{b(i+1)}(n + 1)}{E_{ai}(n + 1)} \tag{8.45}$$

which provides a time updating for the reflection coefficients involving only error variables.

The same procedure, using time recursions (8.36) and (8.12), leads to the time updating equation for the other reflection coefficients in the prediction section:

$$k_{b(i+1)}(n + 1) = k_{b(i+1)}(n) + \frac{\varepsilon_{bi}(n)e_{a(i+1)}(n + 1)}{E_{bi}(n)} \tag{8.46}$$

For the filter section, let

$$k_{fN}(n) = \frac{K_{fN}(n)}{E_{bN}(n)} \tag{8.47}$$

The same procedure again, using time recursion (8.21) and the filter error energy updating relation, yields

$$k_{fi}(n + 1) = k_{fi}(n) + \frac{\varepsilon_{bi}(n + 1)e_{i+1}(n + 1)}{E_{bi}(n + 1)} \tag{8.48}$$

The computational organization of the lattice adaptive filter based on a priori

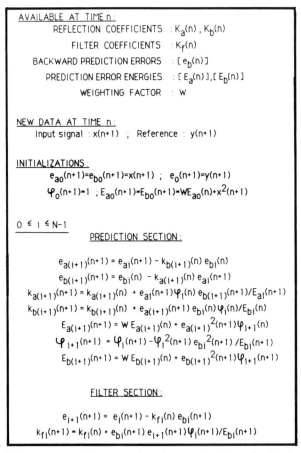

Figure 8.2 Computational organization of a lattice adaptive filter.

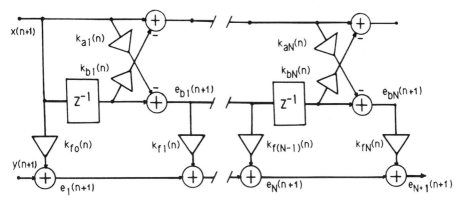

Figure 8.3 The lattice adaptive filter.

errors is given in Figure 8.2. The initial conditions are

$$e_{bi}(0) = k_{ai}(0) = k_{bi}(0) = k_{fi}(0) = 0, \qquad 0 \leqslant i \leqslant N - 1$$

$$\varphi_i(0) = 1, \qquad E_{ai}(0) = W^N E_0, \qquad E_{bi}(0) = W^{N-i} E_0, \qquad 0 \leqslant i \leqslant N - 1 \qquad (8.49)$$

and the FORTRAN program is given in Annex 8.1.

A lattice algorithm based on a posteriori errors can be derived in a similar manner.

The computational complexity of the algorithm in Figure 8.2 amounts to $16N + 2$ multiplications and $3N$ divisions in the form of inverse calculations. About $7N$ memories are required.

The block diagram of the adaptive filter is shown in Figure 8.3. The filter section is sometimes called the ladder section, and the complete system is called a lattice-ladder adaptive filter.

Since it has been shown in Section 5.3 that the backward prediction errors are uncorrelated, the filter can be viewed as a decorrelation processor followed by a set of N first-order separate adaptive filters.

In the presence of stationary signals, the two sets of lattice coefficients, like the forward and backward prediction coefficients, take on similar values in the steady state. Algorithms which use only one set of coefficients, and thus are potentially simpler, can be obtained with normalized variables [5].

8.5 NORMALIZED LATTICE ALGORITHMS

The variable $K_i(n)$ defined by equation (8.4) and updated by (8.36) corresponds to a cross-correlation calculation. A true cross-correlation coefficient, with magnitude range $[-1, 1]$, is obtained by scaling that variable with the energies of the error signals, which leads to the normalized variable, $k_i(n)$, defined by

$$k_{i+1}(n) = \frac{K_{i+1}(n)}{\sqrt{E_{ai}(n)E_{bi}(n-1)}} \tag{8.50}$$

A time recurrence relation can be derived, using (8.36) to get

$$k_{i+1}(n+1) = [E_{ai}(n+1)]^{-1/2}[WK_{i+1}(n) + e_{ai}(n+1)\varepsilon_{bi}(n)][E_{bi}(n)]^{-1/2} \tag{8.51}$$

In order to make $k_{i+1}(n)$ appear in (8.51), we have to consider the ratios of the error energies. The time updating equations can be rewritten as

$$W\frac{E_{ai}(n)}{E_{ai}(n+1)} = 1 - \frac{e_{ai}^2(n+1)}{E_{ai}(n+1)}\varphi_i(n) \tag{8.52}$$

and

$$W\frac{E_{bi}(n-1)}{E_{bi}(n)} = 1 - \frac{\varepsilon_{bi}^2(n)}{E_{bi}(n)}\frac{1}{\varphi_i(n)} \tag{8.53}$$

If the normalized forward prediction error is defined by

$$e_{nai}(n+1) = e_{ai}(n+1)\sqrt{\frac{\varphi_i(n)}{E_{ai}(n+1)}} = \varepsilon_{ai}(n+1)[\varphi_i(n)E_{ai}(n+1)]^{-1/2} \tag{8.54}$$

and the normalized backward prediction error by

$$e_{nbi}(n) = \varepsilon_{bi}(n)[\varphi_i(n)E_{bi}(n)]^{-1/2} \tag{8.55}$$

then, the recurrence equation (8.51) becomes

$$k_{i+1}(n+1) = k_{i+1}(n)[(1 - e_{nai}^2(n+1))(1 - e_{nbi}^2(n))]^{1/2} + e_{nai}(n+1)e_{nbi}(n) \tag{8.56}$$

Clearly, with the above definitions, the normalized error variables are intermediates between a priori and a posteriori errors.

To obtain an algorithm, we must derive recursions for the normalized prediction errors. The order recursion (8.14) for forward a posteriori errors can be rewritten as

$$\varphi_{i+1}(n)e_{a(i+1)}(n+1) = \varphi_i(n)e_{ai}(n+1) - \frac{K_{i+1}(n+1)}{E_{bi}(n)}\varepsilon_{bi}(n) \tag{8.57}$$

Substitution of the normalized errors in that expression leads to

$$e_{na(i+1)}(n+1) = \left[\frac{E_{ai}(n+1)}{E_{a(i+1)}(n+1)}\right]^{1/2}\left[\frac{\varphi_i(n)}{\varphi_{i+1}(n)}\right]^{1/2}e_{nai}(n+1)$$
$$- k_{i+1}(n+1)e_{nbi}(n) \tag{8.58}$$

The normalized variables can be introduced into the order recursions (8.8) and (8.33) to yield

$$E_{a(i+1)}(n+1) = E_{ai}(n+1)[1 - k_{i+1}^2(n+1)] \tag{8.59}$$

and

$$\varphi_{i+1}(n) = \varphi_i(n)[1 - e_{nbi}^2(n)] \tag{8.60}$$

Substituting into (8.58) leads to the final form of the time recurrence relation for the normalized forward prediction error:

$$e_{na(i+1)}(n+1) = [1 - k_{i+1}^2(n+1)]^{-1/2}[1 - e_{nbi}^2(n)]^{-1/2}$$
$$\times (e_{nai}(n+1) - k_{i+1}(n+1)e_{nbi}(n)) \tag{8.61}$$

The same method can be applied to backward prediction errors. Order recursion (8.15) is expressed in terms of normalized variables by

$$e_{nb(i+1)}(n+1) = \left[\frac{E_{bi}(n)}{E_{b(i+1)}(n+1)}\right]^{1/2}\left[\frac{\varphi_i(n)}{\varphi_{i+1}(n+1)}\right]^{1/2}$$
$$\times (e_{nbi}(n) - k_{i+1}(n+1)e_{nai}(n+1)) \tag{8.62}$$

Equation (8.11) for the energy can be written

$$E_{b(i+1)}(n+1) = E_{bi}(n)[1 - k_{i+1}^2(n+1)] \tag{8.63}$$

An equation relating $\varphi_{i+1}(n+1)$ and $\varphi_i(n)$ can be obtained with the help of adaptation gain recurrence relation (6.73) in Chapter 6, which yields

$$\varphi_{i+1}(n+1) = \varphi_i(n) - \frac{\varepsilon_{ai}^2(n+1)}{E_{ai}(n+1)} \tag{8.64}$$

and thus

$$\varphi_{i+1}(n+1) = \varphi_i(n)[1 - e_{nai}^2(n+1)] \tag{8.65}$$

Hence the final form of the time recurrence relation for the normalized backward prediction error is

$$e_{nb(i+1)}(n+1) = [1 - k_{i+1}^2(n+1)]^{-1/2}[1 - e_{nai}^2(n+1)]^{-1/2}$$
$$\times (e_{nbi}(n) - k_{i+1}(n+1)e_{nai}(n+1)) \tag{8.66}$$

Finally equations (8.56), (8.61), and (8.66) make an algorithm for the normalized lattice adaptive predictor.

Normalized variables can be introduced as well in the filter section. The normalized filter output errors are defined by

$$e_{ni}(n) = e_i(n)\left[\frac{\varphi_i(n)}{E_i(n)}\right]^{1/2} = \varepsilon_i(n)[\varphi_i(n)E_i(n)]^{-1/2} \tag{8.67}$$

Then order recursion (8.25) yields

$$e_{n(i+1)}(n) = \left[\frac{E_i(n)}{E_{i+1}(n)}\right]^{1/2}\left[\frac{\varphi_i(n)}{\varphi_{i+1}(n)}\right]^{1/2}\left(e_{ni}(n) - \frac{K_{fi}(n)}{\sqrt{E_i(n)\varphi_i(n)}}\frac{\varepsilon_{bi}(n)}{E_{bi}(n)}\right) \quad (8.68)$$

Defining the normalized coefficients by

$$k_{fi}(n) = \frac{K_{fi}(n)}{\sqrt{E_{bi}(n)E_i(n)}} \quad (8.69)$$

We can write the order recursion (8.27) for error energies as

$$E_{i+1}(n) = E_i(n)[1 - k_{fi}^2(n)] \quad (8.70)$$

Substituting (8.60) and (8.70) into (8.68) leads to the order recursion for filter output errors:

$$e_{n(i+1)}(n) = [1 - k_{fi}^2(n)]^{-1/2}[1 - e_{nbi}^2(n)]^{-1/2}[e_{ni}(n) - k_{fi}(n)e_{nbi}(n)] \quad (8.71)$$

Now the normalized coefficients themselves have to be calculated. Once the normalized variables are introduced in time recursion (8.41), one gets

$$k_{fi}(n+1) = \left[\frac{E_{bi}(n)}{E_{bi}(n+1)}\right]^{1/2}\left[\frac{E_i(n)}{E_i(n+1)}\right]^{1/2}Wk_{fi}(n) + e_{nbi}(n+1)e_{ni}(n+1) \quad (8.72)$$

The time recursion for filter output error energies can be rewritten as

$$W\frac{E_i(n)}{E_i(n+1)} = 1 - \frac{e_i^2(n+1)\varphi_i(n+1)}{E_i(n+1)} = 1 - e_{ni}^2(n+1) \quad (8.73)$$

Substituting (8.53) and (8.73) into (8.72), we obtain the time recursion for the normalized filter coefficients:

$$k_{fi}(n+1) = k_{fi}(n)[1 - e_{nbi}^2(n+1)]^{1/2}[1 - e_{ni}^2(n+1)]^{1/2}$$
$$+ e_{nbi}(n+1)e_{ni}(n+1) \quad (8.74)$$

which completes the normalized lattice filter algorithm. The initializations follow the definition of the normalized variables, which implies for the prediction

$$e_{na0}(n+1) = \frac{x(n+1)}{\sqrt{E_{a0}(n+1)}} = e_{nb0}(n+1) \quad (8.75)$$

and for the filter section

$$e_{n0}(n+1) = \frac{y(n+1)}{\sqrt{E_{f0}(n+1)}}, \qquad E_{f0}(n+1) = WE_{f0}(n) + y^2(n+1) \quad (8.76)$$

Other initializations are in accordance with (8.49), with the additional equation $E_{f0}(0) = E_0$.

The computational organization of the normalized lattice adaptive filter is shown in Figure 8.4, and a filter section is depicted in Figure 8.5.

In spite of its conciseness, this algorithm requires more calculations than its unnormalized counterpart. The prediction section needs $10N + 2$ multiplications, $2N + 1$ divisions, and $3N + 1$ square roots, whereas the filter

---

ALGORITHM F.L.S.L.N.

AVAILABLE AT TIME n :
  REFLECTION COEFFICIENTS : $K(n)$
  FILTER COEFFICIENTS : $K_f(n)$
BACKWARD PREDICTION ERRORS : $[e_b(n)]$
  SIGNAL ENERGIES : E_{ao}, E_{fo}
  WEIGHTING FACTOR : W

NEW DATA AT TIME n:
  input signal : $x(n+1)$; reference : $y(n+1)$

INITIALIZATIONS :

$$E_{ao}(n+1) = W E_{ao}(n) + x^2(n+1)$$
$$e_{nao}(n+1) = e_{nbo}(n+1) = x(n+1) / [E_{ao}(n+1)]^{1/2}$$
$$E_{fo}(n+1) = W E_{fo}(n) + y^2(n+1)$$
$$e_{no}(n+1) = y(n+1) / [E_{fo}(n+1)]^{1/2}$$

$0 \leq i \leq N-1$

PREDICTION SECTION :

$$k_{i+1}(n+1) = k_{i+1}(n)[(1-e_{nai}^2(n+1))(1-e_{nbi}^2(n))]^{1/2} + e_{nai}(n+1)e_{nbi}(n)$$
$$e_{na(i+1)}(n+1) = (1-k_{i+1}^2(n+1))^{-1/2}(1-e_{nbi}^2(n))^{-1/2}[e_{nai}(n+1)-k_{i+1}(n+1)e_{nbi}(n)]$$
$$e_{nb(i+1)}(n+1) = (1-k_{i+1}^2(n+1))^{-1/2}(1-e_{nai}^2(n+1))^{-1/2}[e_{nbi}(n)-k_{i+1}(n+1)e_{nai}(n+1)]$$

FILTER SECTION :

$$k_{fi}(n+1) = k_{fi}(n)(1-e_{nbi}^2(n+1))^{1/2}(1-e_{ni}^2(n+1))^{1/2} + e_{nbi}(n+1)e_{ni}(n+1)$$
$$e_{n(i+1)}(n+1) = (1-k_{fi}^2(n+1))^{-1/2}(1-e_{nbi}^2(n+1))^{-1/2}[e_{ni}(n+1)-k_{fi}(n+1)e_{nbi}(n+1)]$$

Figure 8.4 Computational organization of the normalized lattice adaptive filter.

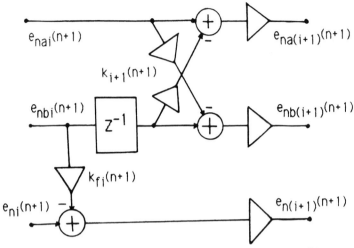

Figure 8.5 A section of normalized lattice adaptive filter.

section requires $6N + 2$ multiplications, $N + 1$ divisions, and $2N + 1$ square roots. Altogether, the algorithm complexity amounts to $16N + 4$ multiplications, $3N + 2$ divisions, and $5N + 2$ square roots. An important point is the need for square-root calculations, which are a significant burden in implementations. The number of memories needed is about $3N$.

Overall, the normalized algorithm may be attractive for handling nonstationary signals with fixed-point arithmetic because it has a built-in magnitude scaling of its variables. The resulting robustness to roundoff errors is enhanced by the fact that only one set of prediction coefficients is calculated [5–7].

The main advantage of the lattice approach is that it constitutes a set of N adaptive filters with all orders from 1 to N. Therefore it may be interesting to calculate the coefficients and adaptation gains of the corresponding transversal filters.

8.6 CALCULATION OF TRANSVERSAL FILTER COEFFICIENTS

The conversion from lattice to transversal prediction coefficients is performed with the help of the order recursions (8.7) and (8.10), which can be written as

$$A_{i+1}(n + 1) = \begin{bmatrix} A_i(n + 1) \\ 0 \end{bmatrix} - k_{b(i+1)}(n + 1)\begin{bmatrix} B_i(n) \\ -1 \end{bmatrix}$$

$$B_{i+1}(n + 1) = \begin{bmatrix} 0 \\ B_i(n) \end{bmatrix} - k_{a(i+1)}(n + 1)\begin{bmatrix} -1 \\ A_i(n + 1) \end{bmatrix} \qquad (8.77)$$

The coefficients of the transversal filters can be recursively computed from order 2 to order N. However, it may be more convenient to replace $B_i(n)$ by $B_i(n + 1)$ in order to deal with a set of variables homogeneous in time.

Substituting the time recursions of the forward and backward prediction coefficients into (8.77) and adding the order recursion (8.32) for the adaptation gain, the conversion set becomes

$$A_{i+1}(n + 1) = \begin{bmatrix} A_i(n + 1) \\ 0 \end{bmatrix} - k_{b(i+1)}(n + 1)\begin{bmatrix} B_i(n + 1) \\ -1 \end{bmatrix}$$

$$+ k_{b(i+1)}(n + 1)e_{bi}(n + 1)\begin{bmatrix} G_i(n + 1) \\ 0 \end{bmatrix}$$

$$B_{i+1}(n + 1) = \begin{bmatrix} 0 \\ B_i(n + 1) \end{bmatrix} - e_{bi}(n + 1)\begin{bmatrix} 0 \\ G_i(n + 1) \end{bmatrix}$$

$$- k_{a(i+1)}(n + 1)\begin{bmatrix} -1 \\ A_i(n + 1) \end{bmatrix}$$

$$G_{i+1}(n + 1) = \begin{bmatrix} G_i(n + 1) \\ 0 \end{bmatrix} + \frac{\varepsilon_{bi}(n + 1)}{E_{bi}(n + 1)}\begin{bmatrix} -B_i(n + 1) \\ 1 \end{bmatrix} \qquad (8.78)$$

The corresponding flow graph is shown in Figure 8.6. The implementation requires some care in handling the coefficient vectors. The operator Z^{-1} in the flow graph represents a one-element shift of an $(i + 1)$-element vector in an $(i + 2)$-element register. The input of the first section, corresponding to $i = 0$, is $(1, 1, 0)$, and the output of the last section, corresponding to $i = N - 1$, yields the prediction coefficients.

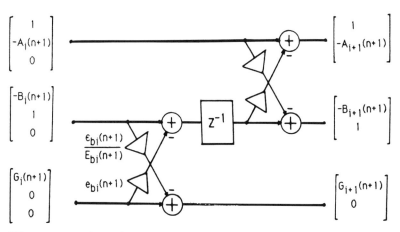

Figure 8.6 A section for calculating the transversal predictor coefficients.

The transversal coefficients $H_i(n)$ of the filter section are obtained recursively from equation (8.23).

Note that a similar computational complexity can be obtained through the direct calculation of the forward prediction transversal coefficients. Suppose we want to calculate all the coefficients from order 1 to order N: since the adaptation gain updating can use only forward variables, backward variables are no longer needed, and the algorithm obtained by simplifying the algorithms in Chapter 6 is shown in Figure 8.7. The computational complexity is about $2N(N+1)$ multiplications and N divisions per time sample.

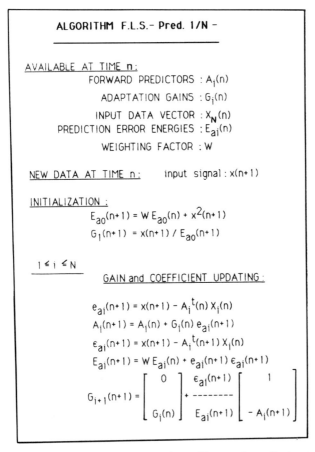

Figure 8.7 Direct calculation of forward prediction transversal coefficients for orders 1 to N.

8.7 MULTIDIMENSIONAL LATTICE ALGORITHMS

The lattice algorithms for scalar input and reference signals can be extended to vector signals. As shown in Section 7.5, for a K-element input signal the prediction errors become a K-element vector, the lattice coefficients and error energies become $K \times K$ matrices, and the prediction error ratios remain scalars. It is sufficient to change accordingly the equations in Figure 8.2 to obtain a multidimensional lattice algorithm.

As an example, let us consider the 2-D input signals $\chi^t(n) = [x_1(n), x_2(n)]$ and scalar reference $y(n)$, the notations being as in Section 7.4.

The $2i$-element filter coefficient vector $H_{2i}(n)$ which minimizes the cost function

$$J_{2i}(n) = \sum_{p=1}^{n} W^{n-p}[y(p) - H_{2i}(n)X^t_{2i}(p)]^2 \tag{8.79}$$

satisfies the relation

$$R_{2i}(n)H_{2i}(n) = r_{2i}(n)$$

The same relation at order $i + 1$ is

$$\sum_{p=1}^{n} W^{n-p} \begin{bmatrix} X_{2i}(p) \\ \chi(p-i) \end{bmatrix} [X^t_{2i}(p), \chi^t(p-i)]H_{2(i+1)}(n)$$

$$= \begin{bmatrix} r_{2i}(n) \\ \sum_{p=1}^{n} W^{n-p} y(p)\chi(p-i) \end{bmatrix} \tag{8.80}$$

The partitioning of the matrix $R_{2(i+1)}(n)$ leads to

$$\begin{bmatrix} R_{2i}(n) & r^b_{2i}(n) \\ [r^b_{2i}(n)]^t & \sum_{p=1}^{n} W^{n-p}\chi(p-i)\chi^t(p-i) \end{bmatrix} \begin{bmatrix} H_{2i}(n) \\ 0 \\ 0 \end{bmatrix} = \begin{bmatrix} r_{2i}(n) \\ [r^b_{2i}(n)]^t H_{2i}(n) \end{bmatrix} \tag{8.81}$$

Hence

$$H_{2(i+1)}(n) = \begin{bmatrix} H_{2i}(n) \\ 0 \\ 0 \end{bmatrix} + R^{-1}_{2(i+1)}(n) \begin{bmatrix} 0 \\ K_i(n) \end{bmatrix} \tag{8.82}$$

with

$$K_i(n) = \sum_{p=1}^{n} W^{n-p} y(p)[\chi(p-i) - B^t_{2i}(n)X_{2i}(p)] \tag{8.83}$$

the $2i \times 2$ backward prediction coefficient matrix being expressed by

$$B_{2i}(n) = R_{2i}^{-1}(n)r_{2i}^b(n)$$

The backward prediction matrix equation is

$$R_{2(i+1)}(n) \begin{bmatrix} -B_{2i}(n) \\ I_2 \end{bmatrix} = \begin{bmatrix} 0 \\ E_{2bi}(n) \end{bmatrix} \tag{8.84}$$

where $E_{2bi}(n)$ is the 2×2 backward error energy matrix. From the output error definition

$$e_{i+1}(n+1) = y(n+1) - H_{2(i+1)}^t(n)X_{2(i+1)}(n+1) \tag{8.85}$$

the following order recursion is obtained, from (8.82) and (8.84):

$$e_{i+1}(n+1) = e_i(n+1) - K_i^t(n)E_{2bi}^{-1}(n)e_{2bi}(n+1) \tag{8.86}$$

It is the extension of (8.24) to the 2-D input signal case.

Consequently, for each order, the filter output error is computed with the help of the backward prediction errors, which are themselves computed recursively with the forward prediction errors. The filter block diagram is in Figure 8.3.

Simplifications can be made when the lengths of the two corresponding adaptive filters, as shown in Figure 7.1, are different, say M and $N + M$. Then the overall filter appears as a combination of a 1-D section with N stages and a 2-D section with M stages. These two different sections have to be carefully interconnected. It is simpler to make the 1-D section come first [8].

At order N, the elements of the forward prediction error vector are

$$\begin{aligned}
e_{aN}^{(1)}(n+1) &= x_1(n+1) - [x_1(n), \ldots, x_1(n+1-N)]A_{11}(n) \\
e_{aN}^{(2)}(n+1) &= x_2(n+1) - [x_1(n), \ldots, x_1(n+1-N)]A_{21}(n)
\end{aligned} \tag{8.87}$$

and those of the backward prediction error vector are

$$\begin{aligned}
e_{bN}^{(1)}(n+1) &= x_1(n+1-N) - [x_1(n+1), \ldots, x_1(n+2-N)]B_{11}(n) \\
e_{bN}^{(2)}(n+1) &= x_2(n+1) - [x_1(n+1), \ldots, x_1(n+2-N)]B_{21}(n)
\end{aligned} \tag{8.88}$$

where the prediction coefficient matrices are partitioned as

$$A_{2N}(n) = \begin{bmatrix} A_{11}(n) & A_{12}(n) \\ A_{21}(n) & A_{22}(n) \end{bmatrix}, \quad B_{2N}(n) = \begin{bmatrix} B_{11}(n) & B_{12}(n) \\ B_{21}(n) & B_{22}(n) \end{bmatrix}$$

Clearly, $e_{aN}^{(1)}(n+1)$ and $e_{bN}^{(1)}(n+1)$ are the forward and backward prediction errors of the 1-D process, as expected. They are provided by the last stage of the 1-D lattice section. The two other errors $e_{aN}^{(2)}(n+1)$ and $e_{bN}^{(2)}(n+1)$ turn out to be the outputs of 1-D filters whose reference signal is $x_2(n)$.

Therefore, they can be computed recursively as shown in Section 8.2, using equations similar to (8.24) for the error signal and (8.41) for the cross-correlation estimation; the initial values are $e_{a0}^{(2)}(n + 1) = e_{b0}^{(2)}(n + 1) = x_2(n + 1)$.

Definition (8.88) and the procedure in Section 8.2 lead to

$$e_{aN}^{(2)}(n + 1) = e_{a(N-1)}^{(2)}(n + 1) - \frac{K_{a(N-1)}(n)}{E_{b(N-1)}(n - 1)} e_{b(N-1)}^{(1)}(n) \tag{8.89}$$

and for a posteriori errors

$$\varepsilon_{aN}^{(2)}(n + 1) = \varepsilon_{a(N-1)}^{(2)}(n + 1) - \frac{K_{a(N-1)}(n + 1)}{E_{b(N-1)}(n)} \varepsilon_{b(N-1)}^{(1)}(n) \tag{8.90}$$

with

$$K_{a(N-1)}(n + 1) = WK_{a(N-1)}(n) + \varepsilon_{b(N-1)}^{(1)}(n)e_{a(N-1)}^{(2)}(n + 1) \tag{8.91}$$

We can obtain $e_{bN}^{(2)}(n + 1)$ directly from the forward prediction errors, because it has the same definition as $e_{aN}^{(2)}(n + 1)$ except for the shift of the data vector. Therefore the order recursive procedure can be applied again to yield

$$\varepsilon_{bN}^{(2)}(n + 1) = \varepsilon_{a(N-1)}^{(2)}(n + 1) - \frac{K_{bN}(n + 1)}{E_{a(N-1)}(n + 1)} \varepsilon_{a(N-1)}^{(1)}(n + 1) \tag{8.92}$$

and

$$K_{bN}(n + 1) = WK_{bN}(n) + \varepsilon_{a(N-1)}^{(2)}(n + 1)e_{a(N-1)}^{(1)}(n + 1) \tag{8.93}$$

Finally, the 1-D/2-D lattice filter for nonuniform lengths is depicted in Figure 8.8.

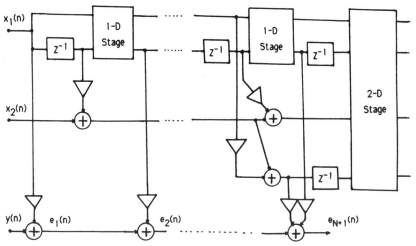

Figure 8.8 The 1-D/2-D lattice structure for nonuniform length filters.

The above technique can be extended to higher dimensions to produce cascades of lattice sections with increasing dimensions.

8.8 BLOCK PROCESSING

The algorithms considered so far assume that updating the coefficient is needed whenever new data become available. However, in a number of applications the coefficient values are used only when a set or block of n data has been received. Updating at each time index is adequate in that case too, but it may require an excessive number of arithmetic operations.

The problem is to compute the N elements of the coeficient vector $H_N(n)$ which minimizes the cost function $J_N(n)$ given by

$$J_N(n) = \sum_{p=1}^{n} [y(p) - H_N^t(n)X_N(p)]^2 \tag{8.94}$$

where the block length n is usually significantly larger than the filter order N. As seen before, the solution is

$$H_N(n) = \left[\sum_{p=1}^{n} X_N(p)X_N^t(p) \right]^{-1} \sum_{p=1}^{n} y(p)X_N(p) \tag{8.95}$$

If the initial data vector is null, $X(0) = 0$, it is recommended to carry out the calculation up to the time $n + N - 1$ while taking $X(n + N) = 0$, because the input signal AC matrix so obtained is Toeplitz. The computation of its N different elements requires nN multiplications and additions. The same amount is required by the cross-correlation vector. Once the correlation data have been calculated, the prediction coefficients are obtained through the Levinson algorithm given in Section 5.4, which requires N divisions and $N(N + 1)$ multiplications. The filter coefficients are then calculated recursively through (8.23), where the variable $k_{fi}(n)$ $(0 \leqslant i \leqslant N - 1)$ can be obtained directly from its definition (8.22), because the cross-correlation coefficients $r_{yxN}(n)$ are available; again N divisions are required as well as $N(N - 1)$ multiplications. The corresponding FORTRAN subroutine is given in Annex 5.1.

For arbitrary initial vectors or for zero initial input vector and summation stopping at n, the AC matrix estimation in (8.95) is no longer Toeplitz, and order recursive algorithms can be worked out to obtain the coefficient vector $H_N(n)$. They begin with calculating the cross-correlation variables $K_i(n)$ and $K_{fi}(n)$ from their definitions (8.3) and (8.22), and they use the recursions given in the previous sections. They are relatively complex, in terms of number of equations [9]. For example, the computational requirements are about $nN + 4.5N^2$ for prediction and $2nN + 5.5N^2$ for the filter, in the algorithm given in [10].

8.9 GEOMETRICAL DESCRIPTION

The procedure used to derive the FLS algorithms in the previous chapters consists of matrix manipulations. A vector space viewpoint is introduced below, which provides an opportunity to unify the derivations of the different algorithms [3, 11–14].

The vector space considered is defined over real numbers, and its vectors have M elements; it it denoted R^M. The vector of the N most recent input data is

$$X_M(n) = [x(n), x(n-1), \ldots, x(1), 0, \ldots, 0]^t$$

and the data matrix containing the N most recent input vectors is

$$X_{MN}(n) = [X_M(n), X_M(n-1), \ldots, X_M(n+1-N)]$$

The column vectors form a basis of the corresponding N dimensional subspace.

An essential operator is the projection matrix, which for a subspace U is defined by

$$P_U = U(U^t U)^{-1} U^t \tag{8.96}$$

It is readily verified that $P_U U = U$. If U and Y are vectors, $P_U Y$ is the projection of Y on U as shown in Figure 8.9. The following are useful relationships:

$$P_U^t = P_U, \quad (P_U Y)^t(P_U Y) = Y^t P_U Y, \quad P_U P_U = P_U \tag{8.97}$$

The orthogonal projection operator is defined by

$$P_U^o = I - U(U^t U)^{-1} U^t \tag{8.98}$$

Indeed the sum of the projections is the vector itself:

$$P_U Y + P_U^o Y = Y \tag{8.99}$$

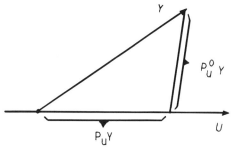

Figure 8.9   Projection operator.

Let us consider as a particular case the operator $P^o_{\{X_{MN}(n-1)\}}$ applied to the M-element vector $X_M(n)$:

$$P^o_{\{X_{MN}(n-1)\}} X_M(n) = X_M(n) - X_{MN}(n-1)$$
$$\times [X^t_{MN}(n-1)X_{MN}(n-1)]^{-1} X^t_{MN}(n-1)X_M(n)$$

The product of the last two terms is

$$X^t_{MN}(n-1)X_M(n)$$

$$= \begin{bmatrix} x(n-1) & x(n-2), & \cdots & x(2) & x(1) & 0 & \cdots & 0 \\ x(n-2) & x(n-3) & \cdots & x(1) & 0 & 0 & \cdots & 0 \\ \vdots & \vdots & & & & & & \\ x(n+1-N) & x(n-N) & \cdots & \cdots & \cdots & & \cdots & 0 \end{bmatrix} \begin{bmatrix} x(n) \\ x(n-1) \\ \vdots \\ x(1) \\ \vdots \\ 0 \end{bmatrix}$$

$$(8.100)$$

With the relations of the previous chapters, we have

$$X^t_{MN}(n-1)X_M(n) = \sum_{p=1}^{n} X_N(p-1)x(p) = r^a_N(n) \qquad (8.101)$$

Similarly

$$[X^t_{MN}(n-1)X_{MN}(n-1)] = \sum_{p=1}^{n} X_N(p)X^t_N(p) = R_N(n-1) \qquad (8.102)$$

Hence

$$[X^t_{MN}(n-1)X_{MN}(n-1)]^{-1}X^t_{MN}(n-1)X_M(n) = R_N^{-1}(n-1)r^a_N(n) = A_N(n) \qquad (8.103)$$

Thus, the M-element forward prediction error vector is obtained:

$$P^o_{\{X_{MN}(n-1)\}} X_M(n) = e_M(n) = X_M(n) - X_{MN}(n-1)A_N(n) \qquad (8.104)$$

It is such that

$$e^t_M(n)e_M(n) = \sum_{p=1}^{n} [x(p) - X^t(p-1)A_N(n)]^2 = E_{aN}(n) \qquad (8.105)$$

and the forward prediction error energy is the squared norm of the orthogonal projection of the new vector $X_M(n)$ on the subspace spanned by the N most recent input vectors.

Finally, the operator $P^o_{\{X_{MN}(n-1)\}}$, denoted in a shorter form by $P^o_x(n-1)$, is a prediction operator. Note that the first element in the error vector $e_M(n)$ is the a posteriori forward prediction error

$$\varepsilon_{aN}(n) = x(n) - X^t_N(n-1)A_N(n) \qquad (8.106)$$

It is useful to define a dual prediction operator $Q_X^o(n - 1)$ which produces the a priori forward prediction error as the first element of the error vector. It is defined by

$$Q_U^o = I - U(U^tS^tSU)^{-1}U^tS^tS \qquad (8.107)$$

where S is the $M \times M$ shifting matrix

$$S = \begin{bmatrix} 0 & 1 & 0 & \cdots & 0 & 0 \\ 0 & 0 & 1 & \cdots & 0 & 0 \\ \vdots & \vdots & \vdots & & \vdots & \vdots \\ 0 & 0 & 0 & \cdots & 0 & 1 \\ 0 & 0 & 0 & \cdots & 0 & 0 \end{bmatrix}$$

The product of S with a time-dependent $M \times 1$ vector shifts this vector one sample back. Therefore one has

$$SX_M(n) = X_M(n - 1), \qquad SX_{MN}(n) = X_{MN}(n - 1) \qquad (8.108)$$

The $M \times M$ matrix S^tS is a diagonal matrix with 0 as the first diagonal element and 1's as the other elements.

As before, the operator $Q_{\{X_{MN}(n-1)\}}^o$ is denoted by $Q_X^o(n - 1)$. Let us consider the product $Q_X^o(n - 1)X_M(n)$. Clearly,

$$X_{MN}^t(n - 1)S^tSX_M(n) = \sum_{p=1}^{n-1} X_N(p - 1)x(p) = r_N^a(n - 1) \qquad (8.109)$$

and

$$X_{MN}^t(n - 1)S^tSX_{MN}(n - 1) = \sum_{p=1}^{n-2} X_N(p)X_N^t(p) = R_N(n - 2) \qquad (8.110)$$

which leads to

$$Q_X^o(n - 1)X_M(n) = e'_M(n) = X_M(n) - X_{MN}(n - 1)A_N(n - 1) \qquad (8.111)$$

The first element of the vector $e'_M(n)$ is

$$e_{aN}(n) = x(n) - X_N^t(n - 1)A_N(n - 1) \qquad (8.112)$$

That operation itself can be expressed in terms of operators. In order to single out the first element of a vector, we use the so-called $M \times 1$ pinning vector Π:

$$\Pi = [1, 0, \ldots, 0]^t$$

Therefore the forward prediction errors are expressed by

$$\varepsilon_{aN}(n) = \Pi^t P_X^o(n - 1)X_M(n) = X_M^t(n)P_X^o(n - 1)\Pi \qquad (8.113)$$

and

$$e_{aN}(n) = \Pi^t Q_X^o(n - 1)X_M(n) = X_M^t(n)Q_X^o(n - 1)\Pi \qquad (8.114)$$

These two errors are related by the factor $\varphi_N(n-1)$, which is expressed in terms of the space operators as follows:

$$\Pi^t P_X^o(n)\Pi = 1 - X_N^t(n)R_N^{-1}(n)X_N(n) = \varphi_N(n) \tag{8.115}$$

Hence, we have the relationship between P_X^o and Q_X^o:

$$\Pi^t Q_X^o = (\Pi^t P_X^o \Pi)^{-1}\Pi^t P_X^o \tag{8.116}$$

Fast algorithms are based on order and time recursions, and it is necessary to determine the relationship between the corresponding projection operators.

8.10 ORDER AND TIME RECURSIONS

Incrementing the filter order amounts to adding a vector to the matrix $X_{MN}(n)$ and thus expanding the dimensionality of the associated subspace. A new projection operator is obtained.

Assume U is a matrix and V a vector; then for any vector Y the following equality is valid for the orthogonal projection operators:

$$P_U^o Y = P_{U,V}^o Y + P_U^o V(V^t P_U^o V)^{-1} V^t P_U^o Y \tag{8.117}$$

It is the combined projection theorem illustrated in Figure 8.10. Clearly, if U and V are orthogonal—that is, $P_U V = 0$ and $P_U^o V = V$—then equation (8.117) reduces to

$$P_U^o Y = P_{U,V}^o Y + P_V Y \tag{8.117a}$$

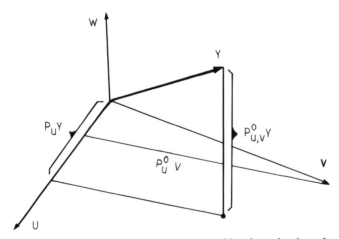

Figure 8.10 Illustration of the combined projection theorem.

For the operators one gets

$$P_{U,V} = P_U - P_U^o V(V^t P_U^o V)^{-1} V^t P_U^o \tag{8.118}$$

In Chapter 6, order recursions are involved in the adaptation gain updating process. The adaptation gain $G_N(n)$ can be viewed as the first vector of an $N \times M$ matrix

$$G_X = (X_{MN}^t X_{MN})^{-1} X_{MN}^t \tag{8.119}$$

and

$$G_N(n) = R_N^{-1}(n) X_N(n) = G_X(n)\Pi \tag{8.120}$$

In order to determine the operator associated with an expanded subspace, it is useful to notice that $X_{MN}G_X$ is the projection operator P_X. For U a matrix and V a vector, equations (8.118) and (8.99) lead to

$$[U, V)G_{U,V} = [U, V]\begin{bmatrix} G_U \\ 0 \end{bmatrix} + (V - UG_UV)(V^t P_U^o V)^{-1} V^t P_U^o$$

Hence

$$G_{U,V} = \begin{bmatrix} G_U \\ 0 \end{bmatrix} + \begin{bmatrix} -G_U V \\ 1 \end{bmatrix}(V^t P_U^o V)^{-1} V^t P_U^o \tag{8.121}$$

Similarly, if U and V are permuted, one gets

$$G_{V,U} = \begin{bmatrix} 0 \\ G_U \end{bmatrix} + \begin{bmatrix} 1 \\ -G_U V \end{bmatrix}(V^t P_U^o V)^{-1} V^t P_U^o \tag{8.122}$$

These are the basic order recursive equations exploited in the algorithms in Chapter 6.

The time recursions can be described in terms of geometrical operators as well. Instead of adding a column to the data matrix $X_{MN}(n)$, we add a row to the matrix $X_{MN}(n - 1)$ after a backward shift. Let us consider the matrices

$$S^t SX = \begin{bmatrix} 0 & 0 & \cdots & 0 \\ x(n-1) & x(n-2) & \cdots & x(n-M) \\ x(n-2) & x(n-3) & \cdots & x(n-1-M) \\ \vdots & \vdots & & \vdots \\ x(1) & 0 & & 0 \end{bmatrix}$$

$$\Pi\Pi^t X = \begin{bmatrix} x(n) & x(n-1) & \cdots & x(n+1-N) \\ 0 & 0 & \cdots & 0 \\ \vdots & \vdots & & \vdots \\ 0 & 0 & \cdots & 0 \end{bmatrix} \tag{8.123}$$

Clearly, their column vectors are orthogonal and they span orthogonal subspaces. The following equality is valid for the projectors:

$$P_X = P_{S^tSX} + P_{\Pi\Pi^t X} \tag{8.124}$$

Due to the definition of the shifting matrix, we have

$$S^tSS^t = S^t, \quad SS^tS = S, \quad S^tS + \Pi\Pi^t = I \tag{8.125}$$

Thus

$$P_{S^tSX} = S^t P_{SX} S \tag{8.126}$$

The time recursions useful in the algorithms involve the error signals, and, therefore, the orthogonal projectors are considered. Definition (8.98) yields

$$S^t P_{SX}^o S = S^tS - S^tSX(X^tS^tSX)^{-1}X^tS^tS \tag{8.127}$$

As time advances, the objective is to update the orthogonal projection operator associated with the data matrix $X_{MN}(n)$, and an equation linking P_{SX}^o and P_X^o is looked for. Definitions (8.123) lead to

$$X^tX = X^tS^tSX + X^t\Pi\Pi^t X \tag{8.128}$$

Now, using the matrix inversion lemma (6.25) of Chapter 6, one gets

$$(X^tS^tSX)^{-1} = (X^tX)^{-1} + (X^tX)^{-1}X^t\Pi(\Pi^t P_Y^o \Pi)^{-1}\Pi^t X(X^tX)^{-1} \tag{8.129}$$

Substituting into (8.127) yields, in concise form,

$$S^t P_{SX}^o S = S^tS[P_X^o - P_X^o\Pi(\Pi^t P_X^o\Pi)^{-1}\Pi^t P_X^o]S^tS$$

Using the property (8.125), we obtain the time recursion equation

$$P_X^o = S^t P_{SX}^o S + P_X^o\Pi(\Pi^t P_X^o\Pi)^{-1}\Pi^t P_X^o \tag{8.130}$$

To illustrate that result, let us postmultiply both sides by the reference signal vector $Y_M(n)$, defined by

$$Y_M(n) = [y(n), y(n-1), \ldots, y(1), 0, \ldots, 0]^t$$

Clearly

$$P_X^o Y_M(n) = \begin{bmatrix} y(n) - X_N^t(n)H_N(n) \\ y(n-1) - X_N^t(n-1)H_N(n) \\ \vdots \\ y(1) - X_N^t(1)H_N(n) \\ 0 \\ \vdots \\ 0 \end{bmatrix}; \quad H_N(n) = R_N^{-1}(n)r_{yxN}(n) \tag{8.131}$$

The same operation at time $n - 1$ leads to

$$S^t P^o_{SX} S Y_M(n) = \begin{bmatrix} 0 \\ y(n-1) - X^t_N(n-1)H_N(n-1) \\ \vdots \\ y(1) - X^t_N(1)H_N(n-1) \\ 0 \\ \vdots \\ 0 \end{bmatrix} \qquad (8.132)$$

Now

$$\Pi^t P^o_X \Pi = 1 - X^t_N(n)R_N(n)X_N(n) = \varphi_N(n) \qquad (8.133)$$

and the last term of the right side of the recursion equation (8.130) is

$$P^o_X \Pi (\Pi^t P^o_X \Pi)^{-1} \Pi^t P^o_X Y_M(n) = \begin{bmatrix} \varphi_N(n) \\ -X^t_N(n-1)G_N(n) \\ \vdots \\ -X^t_N(1)G_N(n) \\ 0 \\ \vdots \\ 0 \end{bmatrix} \dfrac{\varepsilon_N(n)}{\varphi_N(n)} \qquad (8.134)$$

The filter coefficient time updating equation

$$H_N(n) = H_N(n-1) + \frac{G_N(n)\varepsilon_N(n)}{\varphi_N(n)}$$

leads to the verifications of the result

$$P^o_X Y_M(n) = S^t P^o_{SX} S Y_M(n) + P^o_X \Pi (\Pi^t P^o_X \Pi)^{-1} \Pi^t P^o_X Y_M(n) \qquad (8.135)$$

It is important to consider the application of the time updating formula (8.130) to the gain operator G_X. Definition (8.119) and equation (8.116) lead to

$$I - XG_X = S^t(I - SXG_{SX})S + (I - XG_X)\Pi\Pi^t Q^o_X \qquad (8.136)$$

Then, the properties of the shifting matrix S and pinning vector Π yield, after simplification, the following time updating formula for the gain operator:

$$G_X = G_{SX}S + G_X\Pi\Pi^t Q^o_X \qquad (8.137)$$

With the geometrical operators presented so far, all sorts of algorithms can be derived.

8.11 UNIFIED DERIVATION OF FLS ALGORITHMS

The FLS algorithms are obtained by applying the basic order and time recursions with different choices of signal matrices and vectors.

In order to derive the transversal algorithm based on a priori errors and presented in Section 6.4, one takes $U = X_{MN}(n-1)$ and $V = X_M(n)$. The following equalities are readily verified:

$$V^t P_U^o \Pi = \varepsilon_{aN}(n), \qquad V^t Q_U^o \Pi = e_{aN}(n)$$
$$G_U V = A_N(n), \qquad V t P_U^o V = E_{aN}(n)$$

(8.138)

Therefore, the time updating of the forward prediction coefficients is obtained by postmultiplying (8.137) by $X_M(n)$. The time and order updating equation for the adaptation gain is obtained by postmultiplying (8.122) by Π. The recursion for the error energy $E_{aN}(n)$ corresponds to premultiplying the time updating formula (8.130) by $X_M^t(n)$ and postmultiplying by $X_M(n)$. The backward variables are obtained in the same manner as the forward variables, $X_M(n-N)$ replacing $X_M(n)$.

The algorithm based on all prediction errors and given in Section 6.5 uses the error ratio $\varphi_N(n) = \Pi^t P_X^o(n)\Pi$, which is calculated through a time and order updating equation.

Postmultiplying (8.118) by Π and premultiplying by Π^t yields after simplification

$$\varphi_{N+1}(n) = \varphi_N(n-1) - \frac{\varepsilon_{aN}^2(n)}{E_{aN}(n)}$$

(8.139)

Now, substituting (6.49) of Chapter 6 and the time recursion for the error energy into (8.139) gives

$$\varphi_{N+1}(n) = \varphi_N(n-1) \frac{E_{aN}(n-1)}{E_{aN}(n)}$$

(8.140)

A similar relation can be derived for the backward prediction error energies, taking $U = X_{MN}(n)$ and $V = X_M(n-N)$. It is

$$\varphi_N(n) = \frac{\varphi_{N+1}(n)E_{bN}(n)}{E_{bN}(n-1)}$$

(8.141)

In order to get a sequential algorithm, we must calculate the updated energy $E_{bN}(n)$. Postmultiplying (8.121) by $U = X_{MN}(n)$ and $V = X_M(n-N)$ yields the adaptation gain recursion (6.75) of Chapter 6, which shows that the last element of $G_{N+1}(n)$ is

$$m(n) = \frac{\varepsilon_{bN}(n)}{E_{bN}(n)}$$

Hence

$$\varphi_N(n) = \frac{\varphi_{N+1}(n)}{1 - e_{bN}(n)m(n)} \tag{8.142}$$

Finally, the error ratio $\varphi_N(n)$ can be updated by equations (8.140) and (8.142). The algorithm is completed by taking into account the backward coefficient time updating equation and rewriting (6.75) of Chapter 6 as

$$G_{N+1}(n) = \begin{bmatrix} G_N(n)[1 - e_{bN}(n)m(n)] \\ 0 \end{bmatrix} + \begin{bmatrix} -B_N(n-1) \\ 1 \end{bmatrix} m(n) \tag{8.143}$$

Dividing both sides by $\varphi_{N+1}(n)$ and substituting (8.142) lead to

$$\frac{G_{N+1}(n)}{\varphi_{N+1}(n)} = \begin{bmatrix} \dfrac{G_N(n)}{\varphi_N(n)} \\ 0 \end{bmatrix} + \begin{bmatrix} -B_N(n-1) \\ 1 \end{bmatrix} \frac{m(n)}{\varphi_{N+1}(n)} \tag{8.144}$$

Therefore the a priori adaptation gain $G'_N(n) = G_N(n)/\varphi_N(n)$ can be used instead of $G_N(n)$, and the algorithm of Section 6.5 is obtained. In Figure 6.5 $\varphi_N^{-1}(n)$ is updated.

The geometrical approach can also be employed to derive the lattice structure equations. The lattice approach consists of computing the forward and backward prediction errors recursively in order. The forward a posteriori prediction error for order i is

$$\varepsilon_{ai}(n) = X_M^t(n)P_{U,V}^o\Pi \tag{8.145}$$

where

$$U = X_{M(i-1)}(n-1), \qquad V = X_M(n-i)$$

Substituting projection equation (8.118) into (8.145) yields

$$\varepsilon_{ai}(n) = \varepsilon_{a(i-1)}(n) - X_M^t(n)P_U^o V(V^t P_U^o V)^{-1}V^t P_U^o\Pi \tag{8.146}$$

The factors in the second term on the right side are

$$V^t P_U^o\Pi = \varepsilon_{b(i-1)}(n-1), \qquad V^t P_U^o V = E_{b(i-1)}(n-1)$$

$$X_M^t(n)P_U^o V = \sum_{p=1}^{n} x(p)x(p-i) - A_{i-1}^t(n)R_{i-1}(n-1)B_{i-1}(n-1) = K_i(n) \tag{8.147}$$

Hence

$$\varepsilon_{ai}(n) = \varepsilon_{a(i-1)}(n) - \frac{K_i(n)}{E_{b(i-1)}(n-1)}\varepsilon_{b(i-1)}(n-1)$$

which is equation (8.14). The corresponding backward equation (8.15) is

$$\varepsilon_{bi}(n) = X_M^t(n - i)P_{U,V}^o\Pi \tag{8.148}$$

with $U = X_{M(i-1)}(n - 1)$, $V = X_M(n)$.

The a priori equations are obtained by using the operator $Q_{U,V}^o$ instead of $P_{U,V}^o$.

Algorithms with nonzero initial conditions in either transversal or lattice structures are obtained in the same manner; block processing algorithms are also obtained similarly.

8.12 SUMMARY AND CONCLUSION

The flexibility of LS techniques has been further illustrated by the derivation of order recurrence relationships for prediction and filter coefficients and their combination with time recurrence relationships to make fast algorithms. The lattice structures obtained are based on reflection coefficients which represent a real-time estimation of the cross-correlation between forward and backward prediction errors. A great many different algorithms can be worked out by varying the types and arrangements of the recursive equations. However, if the general rules for designing efficient and robust algorithms are enforced, the actual choice reduces to a few options, and an algorithm based on direct time updating of the reflection coefficients has been presented.

The LS variables can be normalized in such a way that time and order recursions be kept. For the lattice structure, a concise and robust algorithm can be obtained, which uses a single set of reflection coefficients. However, the computational complexity is significantly increased by the square-root operations involved.

The lattice approach can be extended to M-D signals with uniform and nonuniform filter lengths. The 1-D/2-D case has been investigated.

Overall, the lattice approach requires more computations than the transversal method. However, besides its academic interest, it provides all the filters with orders from 1 to N and can be attractive in those applications where the filter order is not known beforehand and when the user can be satisfied with reflection coefficients.

A vector space viewpoint provides an elegant description of the fast algorithms and their computational mechanisms. The calculation of errors corresponds to a projection operation in a signal vector space. Order and time updating formulae can be worked out for the projection operators. By choosing properly the matrices and vectors for these projection operators, one can derive all sorts of algorithms in a simple and concise way. The method applies to transversal or lattice structures, with or without initial conditions, with exponential or sliding time windows. Overall, the geometric description offers a unified derivation of the FLS algorithms.

EXERCISES

1. The signal

   $$x(n) = \sin(n\pi/3) + \sin(n\pi/4)$$

   is fed to an order 4 adaptive FIR lattice predictor. Give the values of the four optimal reflection coefficients. The weighting factor in the adaptive algorithm is $W = 0.98$; give upper bounds for the magnitudes of the variables $K_i(n)$. What are their steady-state values?

2. Give the computational organization of an FLS lattice algorithm in which the cross-correlation estimation variables $K_i(n)$ are updated in time and the a priori and a posteriori forward and backward prediction errors are calculated. Count the multiplications, divisions, and memories needed.

3. Consider the filter section in the block diagram in Figure 8.3. Calculate the coefficient $h_i(n)$ of an order 1 LS adaptive filter whose input sequence is $e_{bi}(n + 1)$ and whose reference signal is $e_i(n + 1)$. Compare with the expression of $k_{fi}(n)$ and comment on the difference.

4. Derive the lattice algorithm with direct time updating of the coefficients as in Figure 8.2, but with a posteriori errors. *Hint:* Use the error ratios $\varphi_i(n)$ to get the a posteriori errors and then find the updating equations for the reflection coefficients.

5. Let X_N be an N-element vector such that $0 < X_N^t X_N < 1$. Prove the identities

   $$(I_N - X_N X_N^t)^{1/2} = I_N - \frac{1 - (1 - X_N^t X_N)^{1/2}}{X_N^t X_N} X_N X_N^t$$

   $$(I_N - X_N X_N^t)^{-1/2} = I_N + \frac{(1 - X_N^t X_N)^{-1/2} - 1}{X_N^t X_N} X_N X_N^t$$

   $$(I_N - X_N X_N^t)^{1/2} X_N = X_N (1 - X_N^t X_N)^{1/2}$$

   Show that the square roots of these matrices can be obtained with $\frac{N^2}{2} + \frac{N}{2}$ multiplications and one square-root calculation.

6. In order to derive normalized versions of the transversal FLS algorithms, we define the normalized variables

   $$e_{an}(n) = \frac{e_a(n)}{E_a(n-1)}, \qquad \varepsilon_{an}(n) = \frac{\varepsilon_a(n)}{E_a(n)}$$

   Define normalized versions of the prediction coefficients and the adaptation gain. Give the corresponding time updating relationships. Give the updating equations for the error energies. Give the computational organization of a normalized transversal FLS algorithm and compare the complexity with that of the standard algorithm.

7. In order to visualize the vector space approach, consider the case where $M = 3$, $N = 2$ and the signal input sequence is

$$x(n) = 0, \qquad n \leqslant 0$$
$$x(1) = 4, \quad x(2) = 2, \quad x(3) = 4$$

In the 3-D space $(0_x, 0_y, 0_z)$, draw the vectors $X_M(1)$, $X_M(2)$, $X_M(3)$, and the vector Π. Calculate and show the vector $P_X^o(2)X_M(3)$. Show the forward and backward prediction errors at time $n = 3$. Show how the adaptation gains $G_2(2)$ and $G_2(3)$ are formed.

8. Find the order updating equation for the prediction operator Q. Use it to geometrically derive the lattice equations for a priori prediction errors.

ANNEX 8.1 FLS ALGORITHM FOR A PREDICTOR IN LATTICE STRUCTURE

```
      SUBROUTINE FLSL(N,X,EAB,EA,EB,KA,KB,W,IND)
C
C     COMPUTES THE PARAMETERS OF A LATTICE PREDICTOR
C     N  = FILTER ORDER
C     X  = INPUT SIGNAL
C     EAB= VECTOR OF BACKWARD PREDICTION ERRORS (A PRIORI)
C     EA = VECTOR OF FORWARD PREDICTION ERROR ENERGIES
C     EB = VECTOR OF BACKWARD PREDICTION ERROR ENERGIES
C     KA,KB= LATTICE COEFFICIENTS
C     W  = WEIGHTING FACTOR
C     IND= TIME INDEX
C
      REAL KA,KB
      DIMENSION EAB(1),EA(1),EB(1),KA(1),KB(1),EAV(15),PHI(15)
C
C     INITIALIZATION
C
      IF(IND.GT.1)GOTO30
      X1=0
      E0=1.
      DO20I=1,N
      EAB(I)=0.
      EA(I)=E0*W**N
      EB(I)=E0*W**(N-I)
      KA(I)=0.
      KB(I)=0.
      PHI(I)=1.
   20 CONTINUE
   30 CONTINUE
C
C     ORDER : 1
C
      E01=E0
      E0=W*E0+X*X
      EAV(1)=X-KB(1)*X1
      EAB1=EAB(1)
      EAB(1)=X1-KA(1)*X
      KA(1)=KA(1)+X*EAB(1)/E0
      KB(1)=KB(1)+EAV(1)*X1/E01
      EA(1)=W*EA(1)+EAV(1)*EAV(1)*PHI(1)
```

```
        PHI1=PHI(1)
        PHI(1)=1-X*X/E0
        EB1=EB(1)
        EB(1)=W*EB(1)+EAB(1)*EAB(1)*PHI(1)
        X1=X
C
C       ORDERS > 1
C
        N1=N-1
        DO50I=1,N1
        EAV(I+1)=EAV(I)-KB(I+1)*EAB1
        EAB2=EAB(I+1)
        EAB(I+1)=EAB1-KA(I+1)*EAV(I)
        KA(I+1)=KA(I+1)+EAV(I)*PHI1*EAB(I+1)/EA(I)
        KB(I+1)=KB(I+1)+EAV(I+1)*EAB1*PHI1/EB1
        EA(I+1)=W*EA(I+1)+EAV(I+1)*EAV(I+1)*PHI(I+1)
        PHI1=PHI(I+1)
        PHI(I+1)=PHI(I)*(1-PHI(I)*EAB(I)*EAB(I)/EB(I))
        EB1=EB(I+1)
        EB(I+1)=W*EB(I+1)+EAB(I+1)*EAB(I+1)*PHI(I+1)
        EAB1=EAB2
   50   CONTINUE
        RETURN
        END
```

REFERENCES

1. B. Friedlander, "Lattice Filters for Adaptive Processing," *Proc. IEEE* **70**, 829–867 (August 1982).

2. J. M. Turner, "Recursive Least Squares Estimation and Lattice Filters," in *Adaptive Filters*, Prentice-Hall, Englewood Cliffs, N.J., 1985, Chap. 5.

3. M. Honig and D. Messerschmitt, "Recursive Least Squares," in *Adaptive Filters: Structures, Algorithms and Applications*, Kluwer Academic, Boston, 1984, Chap. 6.

4. F. Ling, D. Manolakis, and J. Proakis, "Numerically Robust L.S. Lattice-Ladder Algorithms with Direct Updating of the Coefficients," *IEEE Trans.* **ASSP-34**, 837–845 (August 1986).

5. D. Lee, B. Friedlander, and M. Morf, "Recursive Square Root Ladder Estimation Algorithms," *IEEE Trans.* **ASSP-29**, 627–641 (June 1981).

6. C. Samson and V. U. Reddy, "Fixed Point Error Analysis of the Normalized Ladder Algorithm," *IEEE Trans.* **ASSP-31**, 1177–1191 (October 1983).

7. P. Fabre and C. Gueguen, "Improvement of the Fast Recursive Least Squares Algorithms via Normalization: A Comparative Study," *IEEE Trans.* **ASSP-34**, 296–308 (April 1986).

8. F. Ling and J. Proakis, "A Generalized Multichannel Least Squares Lattice Algorithm Based on Sequential Processing Stages," *IEEE Trans.* **ASSP-32**, 381–389 (April 1984).

9. N. Kaloupsidis, G. Carayannis, and D. Manolakis, "Efficient Recursive in Order L.S. FIR Filtering and Prediction," *IEEE Trans.* **ASSP-33**, 1175–1187 (October 1985).

10. J. Cioffi, "The Block-Processing FTF Adaptive Algorithm," *IEEE Trans.* **ASSP-34** 77–90 (February 1986).

11. C. Samson, "A Unified Treatment of Fast Algorithms for Identification," *Int. J. Control* **35**, 909–934 (1982).

12. H. Lev-Ari, T. Kailath, and J. Cioffi, "Least Squares Adaptive Lattice and Transversal Filters: A Unified Geometric Theory," *IEEE Trans.* **IT-30**, 222–236 (March 1984).

13. Jin-Der Wang and H. J. Trussel, "A Unified Derivation of the Fast RLS Algorithms," *Proc. ICASSP-86, Tokyo,* 1986, pp. 261–264.

14. S. T. Alexander, *Adaptive Signal Processing: Theory and Applications,* Springer-Verlag, New York, 1986.

9
Spectral Analysis

The estimation of prediction coefficients which is performed in the adaptation gain updating section of FLS algorithms corresponds to a real-time analysis of the input signal spectrum. Therefore, in order to make correct decisions when choosing the algorithm parameter values, we need a good knowledge of the signal characteristics, particularly its spectral parameters.

Independently of FLS algorithms, adaptive filters in general are often used to perform signal analysis. Thus, it is clear that the fields of adaptive filtering and spectral analysis are tightly interconnected.

In this chapter, the major spectrum estimation techniques are reviewed, with emphasis on the links with adaptive filtering. To begin with, the objectives are stated [1].

9.1 DEFINITION AND OBJECTIVES

In theory the spectral analysis of a stationary signal $x(n)$ consists of computing the Fourier transform $X(f)$ defined by

$$X(f) = \sum_{n=-\infty}^{\infty} x(n)e^{-j2\pi nf} \tag{9.1}$$

The function $X(f)$ consists of a set of pulses, or spectral lines, if the signal is periodic or predictable. It has a continuous component if it is random. These aspects are discussed in Chapter 2.

In practical situations, for many different reasons, only a finite set, or record, of input data is available, and it is an estimate of the true spectrum which is obtained. The set of N data, $x(n)(0 \leqslant n \leqslant N-1)$, is considered as a realization of a random process whose power spectral density, or spectrum, $S(f)$ is defined, as stated in Section 2.4, by

$$S(f) = \sum_{p=-\infty}^{\infty} r(p)e^{-j2\pi pf} \tag{9.2}$$

where the ACF values for complex signals are defined by

$$r(p) = E[x(n)\bar{x}(n-p)]$$

The spectral analysis techniques aim at providing estimates of that true spectrum $S(f)$. To judge the performance, we envisage three criteria: resolution, fidelity, and variance.

The limitation of the data record length produces a smoothing effect in the frequency domain which distorts and obscures details. If the estimated spectrum is smoothed to the degree that two spectral lines of interest cannot be distinguished, the estimator is said to have inadequate or low resolution. The resolution is often judged subjectively [2]. Here, it is taken as the minimum frequency interval necessary to separate two lines.

The fidelity can be measured by the distance of the estimated spectrum from the true spectrum. It takes into account the error or bias, when estimating the frequency of a line, as well as its amplitude.

The variance, as usual, measures the confidence one can have in the estimator.

An ideal spectrum estimator would equally well, with respect to specified criteria as above, represent the true spectrum, irrespective of its characteristics. Unfortunately, it is not possible, and the different methods are in general linked to particular signals and emphasize a specific criterion. The presentation given below corresponds, to a certain degree, to an order of increasing resolution. Therefore emphasis is put on line spectra, which represent a significant share of the applications and permit simple and clear comparisons.

9.2 THE PERIODOGRAM METHOD

From the available set of N_0 data, an estimate $S_{pd}(f)$ of the spectrum is obtained from

$$S_{pd}(f) = \frac{1}{N_0} \left| \sum_{n=0}^{N_0-1} x(n)e^{-j2\pi nf} \right|^2 \tag{9.3}$$

In order to relate $S_{pd}(f)$ to the true spectrum $S(f)$, let us expand the right side and rearrange the summation:

$$S_{pd}(f) = \sum_{p=-(N_0-1)}^{N_0-1} \left[\frac{1}{N_0} \sum_{n=p}^{N_0-1} x(n)\bar{x}(n-p) \right] e^{-j2\pi pf} \tag{9.4}$$

The expression in brackets is the estimate $r_1(p)$ of the ACF studied in Section 3.2. Taking the expectation of both sides of (9.4) yields

$$E[S_{pd}(f)] = \sum_{p=-(N_0-1)}^{N_0-1} r(p) \frac{N_0-|p|}{N_0} e^{-j2\pi pf} \tag{9.5}$$

which, due to the properties of the Fourier transform, leads to

$$E[S_{pd}(f)] = S(f) * \frac{\sin^2 \pi f N_0}{N_0 \sin^2 \pi f} \tag{9.6}$$

where $*$ denotes the convolution operation, which corresponds to a filtering operation in the frequency domain. The filtering function is shown in Figure 9.1.

When $N_0 \to \infty$, one gets

$$\lim_{N_0 \to \infty} E[S_{pd}(f)] = S(f) \tag{9.7}$$

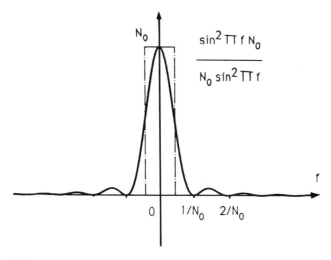

Figure 9.1 The frequency domain filtering function in the periodogram method.

Thus, the estimate is unbiased. If the spectrum consists of just a line, it is correctly found by that approach, as the peak of the estimate $S_{pd}(f)$. However, two lines associated with two sinusoids in the signal can only be distinguished if they are at least separated by the interval $\Delta f = 1/N_0$, which, therefore, is the frequency resolution of the analysis.

The variance can be calculated. The following simple expression is obtained for Gaussian signals:

$$\text{var}\{S_{pd}(f)\} = S^2(f)\left[1 + \frac{\sin^2 2\pi f N_0}{N_0 \sin^2 2\pi f}\right] \tag{9.8}$$

Equation (9.8) shows that for large N_0 the standard deviation of the estimator is equal to its expectation. The estimator is not consistent.

In order to reduce the variance, we can divide the set of N_0 data into K subsets, with or without overlap. The periodogram is computed on each subset, and the spectrum estimate is taken as the average of the values obtained. By so doing, the variance is approximately $S^2(f)/K$, and if K is made proportional to the record length N_0, then the estimator is consistent. However, the counterpart is a decrease in resolution by the factor K.

Data windowing can be incorporated in the method to counter the sidelobe leakage effect. Let $w(n)$ denote the weighting function, the estimate is

$$S_{pdw}(f) = \frac{1}{K}\sum_{i=1}^{K}\frac{1}{\sum_{n=0}^{M-1}w^2(n)}\left|\sum_{n=0}^{M-1}x(iM+n)w(n)e^{-j2\pi nf}\right|^2 \tag{9.9}$$

where $M = N_0/K$ is the number of data per section. In practice, some degree of overlap is generally taken in the sectioning process to gain on the estimator variance. For example, with $M = 2N_0/K$ and a square cosine window, $w(n) = \cos^2(\pi/M)(n - M/2)$, the variance is reduced by almost 25% with respect to the case $M = N_0/K$ and $w(n) = 1$.

The above technique is also called the weighted periodogram or Welch method [3]. It is made computationally efficient by using the FFT algorithm to compute the periodograms of the data subsets. In that case, the spectrum is estimated at discrete frequencies, which are integer multiples of f_s/M, f_s being the input signal sampling frequency.

9.3 THE CORRELOGRAM METHOD

A critical point in the previous approach is the choice of the sectioning parameter K value. It has to be a trade-off between resolution and variance, but the information for making the decision is not readily available. Consider equation (9.6); the effect of the convolution operation is negligible if the following approximation holds:

$$S(f) * \frac{\sin^2 \pi f M}{M \sin^2 \pi f} \approx S(f) \tag{9.10}$$

or, in the time domain,

$$r(p) \left[1 - \frac{|p|}{M} \right] \approx r(p) \tag{9.11}$$

Consequently, the length M of each section of the data record should be significantly greater than the range of index values P_0, over which the correlation function is not negligible.

The correlogram method is in a better position in that respect. It consists of the direct computation of the spectrum according to its definition (9.2). If P ACF values are available, the estimate is

$$S_{CR}(f) = \sum_{p=-(P-1)}^{P-1} r(p) e^{-j2\pi pf} \tag{9.12}$$

or, as a function of the true spectrum,

$$S_{CR}(f) = S(f) * \frac{\sin \pi f (2P-1)}{\sin \pi f} \tag{9.13}$$

If the correlation values are computed using N_0 data, as shown in Section 3.2, an estimate of the true correlation function is obtained, which in turn is reflected in the spectrum estimation $S_{CR}(f)$. It can be shown that the variance is approximately

$$\text{var}\{S_{CR}(f)\} \approx \frac{2P-1}{N_0} S^2(f) \tag{9.14}$$

Therefore the number of correlation values must be taken as small as possible. The optimal conditions are obtained if P is chosen as P_0, assuming the AC function can be neglected for $P > P_0$, as shown in Figure 9.2. The estimation, according to (9.13), can become negative. In real applications, the ACF values have to be estimated; another window $w(p)$ is used instead of the rectangular window, leading to the estimate

$$S_{CRw}(f) = \sum_{p=-(P-1)}^{P-1} w(p) r(p) e^{-j2\pi pf} \tag{9.15}$$

With the triangular window, $w(p) = 1 - |p|/P$, the estimate is positive, as in the previous section, and the information is available for choosing the

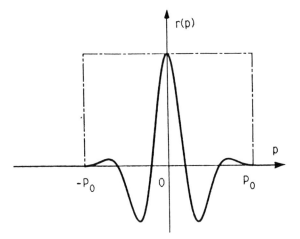

Figure 9.2 Optimal number of AC values in the correlogram method.

summation range P. The variance is

$$\operatorname{var}\{S_{CRw}(f)\} \approx S^2(f)\frac{1}{N_0}\sum_{p=-(P-1)}^{P-1} w^2(p) \tag{9.16}$$

The correlogram is also called the Blackman–Tukey method [4]. Concerning the computational complexity, the calculation of the ACF values has a significant impact [5]. However, if the simplified techniques described in Section 3.2 can be used, the need for multiplications is avoided and the approach is made efficient.

In the methods described above, and particularly the periodogram, the Fourier transform operates as a bank of filters whose coefficients are the same for all the frequency values. Instead, the filter coefficients can be adjusted for each frequency to minimize the estimation variance.

9.4 THE MINIMUM VARIANCE (MV) METHOD

The principle of that approach, also called the maximum likelihood or Capon method [6], is to calculate the coefficients of a filter matched to the frequency under consideration and take the filter output signal power as the value of the power spectrum. Consequently, a sinusoid at that frequency is undistorted, and the variance of the output is minimized.

The filter output is

$$y(n) = \sum_{i=0}^{N} h_i x(n-i) = H^t X(n) \tag{9.17}$$

The $N + 1$ coefficients are subject to the constraint

$$\sum_{i=0}^{N} \bar{h}_i e^{-j2\pi i f} = 1 \tag{9.18}$$

to preserve a cisoid at frequency f. Let

$$F = [1, e^{-j2\pi f}, \ldots, e^{-j2\pi Nf}]^t$$

be the vector with complex elements. The filter coefficients which minimize the output power minimize the expression

$$E[\bar{H}^t[X(n)X^t(n)]H] + \alpha(1 - \bar{H}^t F) \tag{9.19}$$

where α is a Lagrange multiplier.
The optimum coefficients are

$$H_{\text{opt}} = \frac{\alpha}{2} R_{N+1}^{-1} F \tag{9.20}$$

Using equation (9.16) to get the value of α and substituting into the above expression yields

$$H_{\text{opt}} = \frac{1}{\bar{F}^t R_{N+1}^{-1} F} R_{N+1}^{-1} F \tag{9.21}$$

The output signal power $S_{\text{MV}}(f)$ is

$$S_{\text{MV}}(f) = \bar{H}_{\text{opt}}^t R_{N+1} H_{\text{opt}} = \frac{1}{\bar{F}^t R_{N+1}^{-1} F} \tag{9.22}$$

If such a filter is calculated for every frequency value, an estimate of the power spectrum is

$$S_{\text{MV}}(f) = \frac{1}{\sum_{k=0}^{N} \sum_{l=0}^{N} \rho_{kl} e^{-j2\pi(k-l)f}} \tag{9.23}$$

where the values ρ_{kl} are the elements of the inverse input signal AC matrix, which have to be estimated from the input data set.
The function $S_{\text{MV}}(f)$ can be related to the prediction filter frequency responses, denoted by $A_i(f)$ and defined by

$$A_i(f) = 1 - \sum_{k=1}^{i} a_{ki} e^{-j2\pi kf}, \qquad 0 \leqslant i \leqslant N \tag{9.24}$$

The triangular decomposition of the inverse AC matrix [equation (5.69) of Chapter 5] yields

$$\frac{1}{S_{MV}(f)} = \sum_{i=0}^{N} |A_i(f)|^2 \tag{9.25}$$

or

$$\frac{1}{S_{MV}(f)} = \sum_{i=0}^{N} \frac{1}{S_{ARi}(f)} \tag{9.26}$$

where

$$S_{ARi}(f) = \frac{1}{\left|1 - \sum_{k=1}^{i} a_{ki} e^{-j2\pi kf}\right|^2} \tag{9.27}$$

is the AR spectrum estimate, taken as the inverse of the squared prediction error filter response.

Therefore, $S_{MV}(f)$ turns out to be the harmonic average of the AR estimations for all orders from 0 to N; consequently it exhibits less resolution than the AR estimate with the highest order N. The emphasis with that approach is on minimizing the variance.

The resolution of the MV method can be significantly improved. According to the definition, it provides an estimate of the signal power at each frequency. A better resolution is obtained by techniques which estimate the power spectral density instead. As seen in Chapter 3 with eigenfilters, the power spectral density is kept if the filter is unit norm, which leads to the minimum variance with normalization estimate [7]

$$S_{MVN}(f) = \frac{\bar{H}_{opt}^t R_{N+1} H_{opt}}{\bar{H}_{opt}^t H_{opt}} \tag{9.28}$$

Using (9.21), we have

$$S_{MVN}(f) = \frac{\bar{F}^t R_{N+1}^{-1} F}{\bar{F}^t R_{N+1}^{-2} F} \tag{9.29}$$

As an illustration, the functions $S_{MV}(f)$ and $S_{MVN}(f)$ are shown in Figure 9.3 for a signal consisting of two sinusoids in white noise. The data record length is 64; the filters have $N + 1 = 10$ coefficients; the power of each sinusoid is 10 dB above the noise power. Clearly, the resolution of the normalized method is significantly improved, since it can distinguish the two sine waves, whereas the standard method cannot.

The MV method with normalization comes closer to the AR method, as far as resolution is concerned. The price to be paid is a significant increase in computations.

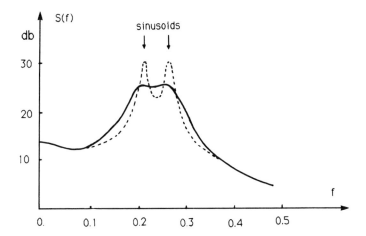

Figure 9.3 Minimum variance spectral estimation of two sinusoids in white noise: (a) standard method, (b) normalized method.

The methods presented so far are based on filtering the data with a filter matched to a single sinusoid. Consequently, they are not optimal to resolve several sinusoids. Methods for that specific case are presented next.

9.5 HARMONIC RETRIEVAL TECHNIQUES

The principle of harmonic decomposition has been introduced in Section 2.11 as an illustration of the fundamental decomposition of signals. It is based on the assumption that the signal consists of real sinusoids with uncorrelated random phases in white noise, and the spectrum is calculated as

$$S_{HR}(f) = \sigma_e^2 + \sum_{i=1}^{M} |S_i|^2 [\delta(f-f_i) + \delta(f+f_i)] \tag{9.30}$$

The corresponding ACF is

$$r(p) = 2 \sum_{i=1}^{M} |S_i|^2 \cos(2\pi p f_i) + \sigma_e^2 \delta(p) \tag{9.31}$$

where σ_e^2 is the white noise power, f_i are the sinusoid frequencies, and S_i are their amplitudes.

According to the results of Chapters 2 and 3, the noise power corresponds to the smallest eigenvalue of the $(2M + 1) \times (2M + 1)$ AC matrix R_{2M+1}:

$$\sigma_e^2 = \lambda_{\min, 2M+1} \tag{9.32}$$

and the sinusoid frequencies are the zeros of the associated eigenvector polynomial

$$H_{2M+1}(z) = 1 + \sum_{i=1}^{2M} v_{\min,i} z^{-i} \tag{9.33}$$

Once the M sinusoid frequencies have been obtained, the powers are calculated by solving the system of equations (9.31) for $p = 0, 1, \ldots, M - 1$, corresponding to the matrix equation (2.128) of Chapter 2.

If it is not known a priori, the order M is determined by calculating the minimum eigenvalues $\lambda_{\min, 2M+1}$ for increasing orders until they become virtually constant.

This method, called the *Pisarenko* method, provides, when the ACF values are known, unbiased spectral estimates of signals consisting of the sum of M sinusoids with uncorrelated random phases in white noise [8]. It is very elegant and appealing from a theoretical point of view. However, with real data, when the ACF has to be estimated, it exhibits a number of limitations.

First of all, there is a fundamental practical difficulty related to the hypothesis on the noise. Estimation bounds and experiments show that it takes numerous samples for the characteristics of a white noise to appear. It has been shown in Section 3.11 that errors on the AC matrix elements affect the eigendirections; therefore, a bias is introduced on the sinusoid frequencies which becomes significant for low SNRs and small data records. That bias can be reduced by taking the order in the procedure greater than the number of sinusoids M. In these conditions, the occurrence of spurious lines can be prevented by choosing the minimum eigenvector in the noise space in such a way that the zeros of the eigenpolynomial in excess be not on the unit circle, as discussed in Section 9.8.

Concerning computational complexity, the procedure contains two demanding operations: the eigenparameter calculation, and the extraction of the roots of the eigenpolynomial. The determination of the frequencies f_i becomes inaccurate for large order M. It can be avoided if only the shape of the spectrum is of interest by calculating

$$S'_{\mathrm{HR}}(f) = \frac{|H_m|^2}{\left| 1 + \sum_{i=1}^{2M} v_{i,\min} e^{-j2\pi i f} \right|^2} \tag{9.34}$$

where H_m is the peak of the modulus of $H_{2M+1}(e^{j2\pi f})$.

For the eigenparameter calculation, the conjugate gradient technique given in Annex 3.2 is an efficient approach. Iterative gradient techniques, mentioned in Section 3.10, can also be used; they permit an adaptive realization.

The hypothesis on the noise is avoided in the damped sinusoid decomposition method, called the *Prony* method [9]. The principle consists of fitting the set of $N_0 = 2P$ data to P damped cisoids:

$$x(n) = \sum_{i=1}^{P} S_i z_i^n \qquad (9.35)$$

with

$$z_i = e^{-(\alpha_i + j2\pi f_i)}$$

As pointed out in Chapter 2, this is equivalent to assuming that the data satisfy the recurrence relationship

$$x(n) = \sum_{i=1}^{P} a_i x(n - i) \qquad (9.36)$$

The coefficients can be obtained by solving the system

$$\begin{bmatrix} x(p) & x(p-1) & \cdots & x(1) \\ x(p+1) & x(p) & \cdots & x(2) \\ \vdots & \vdots & & \vdots \\ x(2p-1) & x(2p-2) & \cdots & x(p) \end{bmatrix} \begin{bmatrix} a_1 \\ a_2 \\ \vdots \\ a_p \end{bmatrix} = \begin{bmatrix} x(p+1) \\ x(p+2) \\ \vdots \\ x(N_0) \end{bmatrix} \qquad (9.37)$$

The values z_i are computed as the roots of the equation

$$1 - \sum_{i=1}^{P} a_i z^{-i} = 0 = \prod_{i=1}^{P} (1 - z_i z^{-1}) \qquad (9.38)$$

Finally the amplitudes S_i are obtained by solving the system

$$\begin{bmatrix} 1 & 1 & \cdots & 1 \\ z_1 & z_2 & \cdots & 2P \\ \vdots & \vdots & & \vdots \\ z_1^{P-1} & z_2^{P-1} & \cdots & z_P^{P-1} \end{bmatrix} \begin{bmatrix} S_1 \\ S_2 \\ \vdots \\ S_P \end{bmatrix} = \begin{bmatrix} x(1) \\ x(2) \\ \vdots \\ x(P) \end{bmatrix} \qquad (9.39)$$

The spectral estimate corresponds to the Fourier transform of equation (9.35); if we assume symmetry, $x(n) = x(-n)$, the result, for continuous signals, is

$$S_{DS}(f) = \left| \sum_{i=1}^{P} \frac{2\alpha_i S_i}{\alpha_i^2 + 4\pi^2 (f - f_i)^2} \right|^2 \qquad (9.40)$$

The method can be extended to real undamped sinusoids. It is well suited to finding out the modes in a vibration transient.

System (9.37) may be under- or overdetermined and solved as indicated in Section 3.5. The overdetermination case corresponds to the AR approach.

9.6 AUTOREGRESSIVE MODELING

The method is associated with the calculation of the linear prediction coefficients $a_i (1 \leqslant i \leqslant N)$ through the normal equations

$$R_{N+1} \begin{bmatrix} 1 \\ -a_1 \\ \vdots \\ -a_N \end{bmatrix} = \begin{bmatrix} E_N \\ 0 \\ \vdots \\ 0 \end{bmatrix} \tag{9.41}$$

where E_N is the prediction error power and R_{N+1} is an estimate of the signal AC matrix. The spectrum is derived from the coefficients by

$$S_{AR}(f) = \frac{E_N}{\left| 1 - \sum\limits_{i=1}^{N} a_i e^{-j2\pi i f} \right|^2} \tag{9.42}$$

The resolution capability of that approach is illustrated in Figure 9.4, which shows the spectrum estimated from $N = 64$ samples of a signal consisting of two sinusoids separated by $\frac{1}{5}(\frac{1}{N_0})$, the SNR being 50 dB. Clearly, the AR method provides a good analysis, but the Fourier transform approach cannot distinguish the two components; it is a high-resolution technique.

The matrix R_{N+1} used in (9.41) can be calculated from the set of N_0 data in various ways [10]. Let us consider the $(N_0 + N) \times (N + 1)$ input signal matrix

$$X_{N_0(N+1)} = \begin{bmatrix} x(N_0) & \cdots & 0 \\ x(N_0-1) & \cdots & 0 \\ \vdots & & \vdots \\ x(N_0-N+1) & \cdots & 0 \\ x(N_0-N) & \cdots & x(N_0) \\ \vdots & \cdots & \vdots \\ x(1) & \cdots & x(N+1) \\ 0 & \cdots & x(N) \\ \vdots & & \vdots \\ 0 & \cdots & x(1) \end{bmatrix} \begin{array}{l} \left.\rule{0pt}{40pt}\right\} U \\ \left.\rule{0pt}{40pt}\right\} X_a \\ \left.\rule{0pt}{40pt}\right\} L \end{array} \tag{9.43}$$

and denote by U, X_a, and L the upper, center, and lower sections, respectively, as indicated. The choice

$$R_{N+1} = \bar{X}^t X = \bar{U}^t U + \bar{X}_a^t X_a + \bar{L}^t L \tag{9.44}$$

corresponds to the so-called AC equations, because the matrix obtained is Hermitian and Toeplitz, like the theoretical AC matrix.

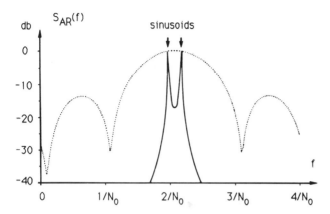

Figure 9.4   AR spectral estimation of two sinusoids.

Another choice is

$$R_{N+1} = \bar{X}_a^t X_a \qquad (9.45)$$

which leads to the so-called covariance equations. The matrix is near Toeplitz. To complete the picture, the prewindowed equations correspond to

$$R_{N+1} = \bar{X}_a^t X_a + \bar{L}^t L = \bar{X}^t X - \bar{U}^t U \qquad (9.46)$$

and the postwindowed equations to

$$R_{N+1} = \bar{X}_a^t X_a + \bar{U}^t U = \bar{X}^t X - \bar{L}^t L \qquad (9.47)$$

Of these four types of equations, the covariance type is, in general, the most efficient for resolution with short data records because it does not imply any assumptions on the data outside the observation interval. The method is well suited to adaptive implementation, as emphasized in the previous chapters.

The values taken above for the matrix R_{N+1} are based on forward linear prediction. If the signal is stationary, the coefficients of the backward prediction filter are identical to the coefficients of the forward prediction filter, but reversed in time and conjugated, in the complex case. Therefore, the most complete estimation procedure is based on minimizing the sum of the forward and backward prediction error powers, and the corresponding expression for the matrix R_{N+1} is

$$R_{N+1} = \bar{X}_a^t X_a + \bar{X}_b^t X_b \qquad (9.48)$$

where X_b is the backward $(N_0 - N) \times (N + 1)$ matrix

$$X_b = \begin{bmatrix} \bar{x}(N_0) & \cdots & \bar{x}(N_0 - N) \\ \vdots & \cdots & \vdots \\ \bar{x}(N + 1) & \cdots & \bar{x}(1) \end{bmatrix}$$

An efficient approach to solve the normal equations consists of calculating the reflection coefficients. The LS lattice structure analysis given in Section 8.1 leads to the calculation of two sets of reflection coefficients $k_{ai}(n)$ and $k_{bi}(n)$ for forward and backward prediction, respectively, defined by equation (8.16). For stationary signals, the PARCOR coefficient k_i is unique and given by expression (5.51) in Chapter 5. Estimates for k_i from $k_{ai}(n)$ and $k_{bi}(n)$ can be obtained in different ways. A first estimate is the geometric mean

$$|k_i| = \sqrt{|k_{ai}(n)| \, |k_{bi}(n-1)|} \tag{9.49}$$

which, according to (8.50), corresponds to the normalized lattice structure.

Another estimate is the harmonic mean, corresponding to the so-called *Burg* method [1]

$$\frac{2}{k_i} = \frac{1}{k_{ai}(n)} + \frac{1}{k_{bi}(n)} \tag{9.50}$$

Accordingly, the coefficients are calculated for complex data by

$$k_i = \frac{2 \sum\limits_{j=i+1}^{N_0} e_{a,(i-1)}(j) \bar{e}_{b,(i-1)}(j-1)}{\sum\limits_{j=i+1}^{N_0} |e_{a,(i-1)}(j)|^2 + |e_{b,(i-1)}(j-1)|^2} \tag{9.51}$$

Their absolute values are bounded by unity, which corresponds to a stable prediction error filter and a finite spectrum estimate at all frequencies.

The Burg procedure is summarized as follows:

1. Calculate k_1 by

$$k_1 = \frac{2 \sum\limits_{j=2}^{N_0} x(j)x(j-1)}{\sum\limits_{j=2}^{N_0} |x(j)|^2 + |x(j-1)|^2} \tag{9.52}$$

2. For $1 \leqslant i \leqslant N$:

a. Calculate the prediction errors by

$$e_{ai}(j) = e_{a(i-1)}(j) - k_i e_{b(i-1)}(j-1)$$
$$e_{bi}(j) = e_{b(i-1)}(j-1) - k_i e_{a(i-1)}(j) \tag{9.53}$$

b. Calculate the reflection coefficients k_{i+1} by equation (9.51)

c. Calculate the prediction coefficients by

$$a_{ii} = k_i$$
$$a_{ji} = a_{j(i-1)} - k_i a_{(i-j)(i-1)} e; \qquad 1 \leqslant j \leqslant i-1 \tag{9.54}$$

Independently of adaptive LS lattice filters, the procedure itself can be made adaptive by the introduction of a weighting factor W in the summations of the numerator and denominator of equation (9.51). The updating equations are

$$D_i(n + 1) = WD_i(n) + |e_{a(i-1)}(n + 1)|^2 + |e_{b(i-1)}(n)|^2$$

$$k_i(n + 1) = k_i(n) + \frac{1}{D_i(n + 1)} [e_{ai}(n + 1)\bar{e}_{b(i-1)}(n)$$

$$+ e_{a(i-1)}(n + 1)\bar{e}_{bi}(n + 1)] \tag{9.55}$$

The above updating technique can be simplified by making constant the variable $D_i(n + 1)$, which leads to the gradient approach defined by expression (5.104) of Chapter 5.

The adaptive technique associated with the geometric mean approach (9.49) is based on the adaptive normalized lattice algorithms described in Section 8.5.

An interesting aspect of the lattice approach is that it provides the linear prediction coefficients for all orders from 1 to N, and, consequently, the corresponding spectral estimations.

Given a data record of length N_0, the selection of the optimal predictor order N is not straightforward. If it is too small, a smoothed spectrum is obtained, which produces a poor resolution. On the contrary, if it is to large, spurious peaks may appear in the spectrum.

The results given in Section 5.7 indicate that the predictor order N can be chosen as the value which corresponds to the maximum of the reflection coefficients $|k_i|$. Another choice is based on the prediction error power; it minimizes the final prediction error (FPE) [11]

$$\text{FPE}(N) = \frac{N_0 + N + 1}{N_0 - N - 1} E_N \tag{9.56}$$

Experience has shown that a reasonable upper bound for N is [12]

$$N \leqslant \frac{N_0}{3} \tag{9.57}$$

The resolution of the AR method strongly depends on the noise level. Extending the results in Section 5.2, we can state that it is not possible to distinguish between two real sinusoids separated by Δf if the noise power exceeds the limit σ_L^2 given approximately by

$$\sigma_L^2 \approx 40(\Delta f)^3 \tag{9.58}$$

when the predictor order N is twice the number M of real sinusoids in the signal. When N increases, the results of Section 5.8 show that the limit σ_L^2 increases as a function of $N - 2M$.

The variance of the AR spectrum estimate is shown to be proportional to $\frac{1}{N^2 N_0}$ for an AR signal and to $\frac{1}{N^2 N_0^2}$ for a signal composed of sinusoids in noise.

A noisy signal expressed by

$$x(n) = x_p(n) + e(n) \tag{9.59}$$

where $x_p(n)$ is a predictable signal satisfying the recursion

$$x_p(n) = \sum_{i=1}^{M} a_i x_p(n - i) \tag{9.60}$$

and $e(n)$ a noise, can be viewed as an ARMA signal because

$$x(n) = \sum_{i=1}^{M} a_i x(n - i) + e(n) - \sum_{i=1}^{M} a_i e(n - i) \tag{9.61}$$

The gain obtained by increasing the predictor order corresponds to an approximation of the MA section of the model [13]. However, a direct ARMA modeling approach can be more efficient.

9.7 ARMA MODELING

The spectrum estimation is

$$S_{ARMA}(f) = E_N \frac{\left| 1 + \sum_{i=1}^{N} b_i e^{-j2\pi i f} \right|^2}{\left| 1 + \sum_{i=1}^{N} a_i e^{-j2\pi i f} \right|^2} \tag{9.62}$$

where E_N is an error power and the signal $x(n)$ is assumed to follow the model

$$x(n) = \sum_{i=1}^{N} a_i x(n - i) + e(n) + \sum_{i=1}^{N} b_i e(n - i) \tag{9.63}$$

A detailed analysis of ARMA signals is provided in Section 2.6.

The results can be used to calculate the model coefficients from an estimate of the $2N + 1$ first values of the signal ACF. The spectrum is then calculated from (9.62). Recall, that the spectrum can be calculated without explicitly determining the MA coefficients. The AR coefficients are found from the extended normal equations (2.68), the ACF of the auxiliary signal is derived from (2.78), and the spectrum is obtained from (2.81). All equations are from Chapter 2.

Adaptive aspects of ARMA modeling are dealt with in Section 4.11, where the application of the LMS algorithm is discussed, and in Section 7.7, which covers FLS techniques. A particular simplified case worth pointing out is the notch filter presented in Section 5.7, whose transfer function is

$$H_N(z) = \frac{1 + \sum_{i=1}^{N} a_i z^{-i}}{1 + \sum_{i=1}^{N} a_i (1 - \varepsilon)^i z^{-i}} \tag{9.64}$$

where the notch parameter ε is a small positive scalar ($0 \leqslant \varepsilon \ll 1$). When predictable signals are analyzed, because of the respective locations of its zeros and poles in the z-plane, it can be a useful intermediate between the prediction error filter, whose zeros are prevented from reaching the unit circle by the noise, and the minimum eigenfilter, whose zeros are on the unit circle.

The coefficients can be derived from a set of N_0 data through iterative techniques. The filter can also be made adaptive using gradient-type algorithms [14]. With LS the nonlinear character of the problem can be overcome in a way which illustrates the flexibility of that technique [15]. Consider the diagram in Figure 9.5. The input signal is fed to the recursive section of the filter first. Then the output sequence obtained is fed to a prediction filter, whose coefficients are updated using an FLS algorithm.

If the same coefficients at each time are also used in the recursive section, an FLS adaptive notch filter is achieved. The value of the fixed-notch parameter ε reflects the a priori information available about the signal: for sinusoids in noise it can be close to zero, whereas noiselike signals lead to choosing ε close to unity. The results obtained for two sinusoids in noise are shown in Figure 9.6. The SNR is 3 dB and $\varepsilon = 0.1$. The coefficient learning curves and the locations of the corresponding filter zeros in the complex plane demonstrate that the two sinusoids are clearly identified by an order 4 filter.

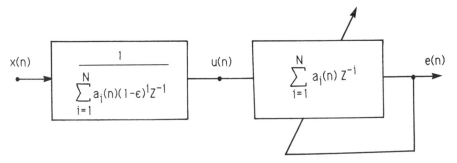

Figure 9.5 FLS adaptive notch filter.

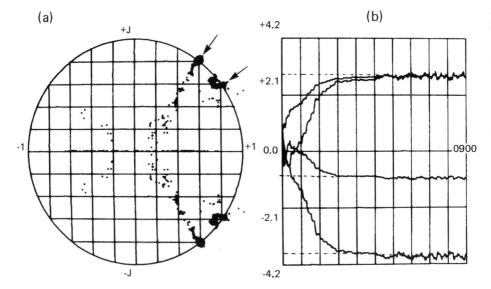

Figure 9.6 Identification of two sinusoids in noise by an adaptive notch filter: (a) filter zeros in the complex plane, (b) coefficient learning curves.

In that approach, which can be used for efficient real-time analysis, the recursive section placed in front of the predictor enhances the lines in the spectrum and helps the predictor operation. A similar effect can be obtained with signal and noise space methods [16].

9.8 SIGNAL AND NOISE SPACE METHODS

The signal AC matrix estimates used in AR methods are degraded by noise. Improvements can be expected from attempts to remove that degradation [17].

Assume that M real sinusoids are searched in a signal. From the set of N_0 data an estimate R_N of the $N \times N$ AC matrix is computed. Its eigendecomposition, as seen in Chapter 3, is

$$R_N = \sum_{i=1}^{N} \lambda_i U_i U_i^t; \qquad \lambda_1 \geqslant \lambda_2 \geqslant \cdots \geqslant \lambda_N \tag{9.65}$$

and the prediction coefficient vector is

$$A_N = R_N^{-1} r_N^a = \sum_{i=1}^{N} \frac{1}{\lambda_i} U_i U_i^t r_N^a \tag{9.66}$$

For $N > 2M$, if R_N were the true AC matrix, the $N - 2M$ last eigenvalues would be zero. Therefore the $N - 2M$ smallest eigenvalues of R_N can be assumed to be associated with the noise space, and the optimum approximation of the true AC matrix is R'_N obtained by

$$R'_N = \sum_{i=1}^{2M} \lambda_i U_i U_i^t \tag{9.67}$$

Thus, an improved estimate of the prediction coefficients is

$$A'_N = \sum_{i=1}^{2M} \frac{1}{\lambda_i} U_i (U_i^t r_N^a) \tag{9.68}$$

If not known, the number of sinusoids can be determined from the observation of the eigenvalues λ_i. Concerning the computational complexity, the eigenvalue decomposition is a significant load.

Another estimate of the prediction coefficient vector can be found by noticing that, for predictable signals in the absence of noise, the prediction error is zero and the filter coefficient vector is orthogonal to all signal vectors [18]. Therefore the estimate A''_N is a vector satisfying

$$U_i^t \begin{bmatrix} 1 \\ -A''_N \end{bmatrix} = 0, \qquad 1 \leqslant i \leqslant 2M \tag{9.69}$$

where U_i now denotes an $(N + 1)$-element eigenvector of the estimated $(N + 1) \times (N + 1)$ AC matrix R_{N+1}. The eigenvectors $U_i (1 \leqslant i \leqslant 2M)$ span the signal space.

In matrix form the system (9.69) is

$$\begin{bmatrix} u_{12} & \cdots & u_{1(N+1)} \\ u_{22} & \cdots & u_{2(N+1)} \\ \vdots & \cdots & F \\ u_{2M2} & \cdots & u_{2M(N+1)} \end{bmatrix} \begin{bmatrix} a''_1 \\ a''_2 \\ \vdots \\ a''_N \end{bmatrix} \begin{bmatrix} u_{11} \\ u_{22} \\ \vdots \\ u_{2M1} \end{bmatrix} \tag{9.70}$$

or, in concise form,

$$U_{2M,N} A''_N = U_{2M,1}$$

The system is underdetermined, since there are $2M$ equations in N unknowns and $N > 2M$. The minimum norm solution is given by expression (3.66) in Chapter 3, which here yields

$$A''_N = U^t_{2M,N} [U_{2M,N} U^t_{2M,N}]^{-1} U_{2M,1} \tag{9.71}$$

Because the eigenvectors are orthogonal and have unit norm, the above equation simplifies to

$$A''_N = (1 - U^t_{2M,1} U_{2M,1})^{-1} U^t_{2M,N} U_{2M,1} \tag{9.72}$$

Once the estimation of the prediction coefficients has been calculated, the spectrum is obtained by (9.42).

A further, efficient, approach to exploit the orthogonality of signal and noise spaces consists of searching the set of M cisoids which are the most closely orthogonal to the noise space, spanned by the $N - M$ eigenvectors $U_i (M + 1 \leqslant i \leqslant N)$ [19]. The M frequencies are taken as the peaks of the function

$$S(\omega) = \frac{1}{\sum\limits_{i=M+1}^{N} |\bar{F}^t(\omega) U_i|^2} \tag{9.73}$$

where $F(\omega)$ is the cisoid vector

$$\bar{F}^t(\omega) = [1, e^{j\omega}, \ldots, e^{j(N-1)\omega}]$$

Weighting factors can also be introduced in the above summation.

The signal and noise space methods are essentially suited to batch processing. Nevertheless, they provide a deep insight into the structures of real signals and therefore are useful for adaptive filter design.

9.9 ESTIMATION BOUNDS

Adaptive techniques can be used for estimation purposes, and their performance can be tested against estimation bounds in these cases [12–20, 22].

Let us consider an indirect estimation procedure in which a set of N parameters θ_i is derived from a set of N measurements γ_i. These measurements are related to the parameters by

$$\gamma_i = f_i(\theta_1, \ldots, \theta_N), \qquad 1 \leqslant i \leqslant N \tag{9.74}$$

The absence of perturbations corresponds to optimal values, and a Taylor series expansion in the vicinity of the optimum yields

$$\gamma_i - \gamma_{i,\text{opt}} = \frac{\partial f_i}{\partial \theta_1}(\theta_{1,\text{opt}})[\theta_1 - \theta_{1,\text{opt}}] + \cdots + \frac{\partial f_i}{\partial \theta_N}(\theta_{N,\text{opt}})[\theta_N - \theta_{N,\text{opt}}] \tag{9.75}$$

or, in matrix form,

$$(\Gamma - \Gamma_{\text{opt}}) = M_G^t(\theta - \theta_{\text{opt}}) \tag{9.76}$$

where M_G is the parameter measurement function gradient matrix.

The measurement covariance matrix is

$$E[(\Gamma - \Gamma_{\text{opt}})(\Gamma - \Gamma_{\text{opt}})^t] = M_G^t E[(\theta - \theta_{\text{opt}})(\theta - \theta_{\text{opt}})^t] M_G \tag{9.77}$$

Assuming unbiased estimation, the parameter covariance matrix, denoted var$\{\theta\}$, is

$$\text{var}\{\theta\} = (M_G^{-1})^t \, \text{var}\{\Gamma\} M_G^{-1} \tag{9.78}$$

This provides a lower bound for the variance of the parameter estimation θ, obtained from the measurement vector Γ in the presence of perturbations. If complex functions are involved, the transpose operation is replaced by conjugate transpose.

If the numbers of parameters and measurements are different, pseudoinverses can be used. For example, with more measurements than parameters, equation (9.78) can be written as

$$\text{var}\{\theta\} = [M_G M_G^t]^{-1} M_G \, \text{var}\{\Gamma\} M_G^t [M_G M_G^t]^{-1} \tag{9.79}$$

An important simplification occurs when the perturbation in the measurements is just caused by a white noise with power σ_b^2, because equation (9.79) becomes

$$\text{var}\{\theta\} = [M_G M_G^t]^{-1} \sigma_b^2 \tag{9.80}$$

For example, let us consider the cisoid in noise

$$x(n) = ae^{jn\omega} + b(n) \tag{9.81}$$

and assume the amplitude a and the angular frequency ω have to be estimated from N samples ($1 \leqslant n \leqslant N$). The $2 \times N$ matrix M_G is

$$M_G = \begin{bmatrix} \dfrac{\partial f_i}{\partial a} & \cdots & \dfrac{\partial f_N}{\partial a} \\[2mm] \dfrac{\partial f_i}{\partial \omega} & \cdots & \dfrac{\partial f_N}{\partial \omega} \end{bmatrix} = \begin{bmatrix} e^{j\omega} & \cdots & e^{jN\omega} \\[1mm] jae^{j\omega} & \cdots & jNae^{jN\omega} \end{bmatrix}$$

and therefore

$$[M_G \bar{M}_G^t] = \begin{bmatrix} N & -ja\dfrac{N(N+1)}{2} \\[3mm] ja\dfrac{N(N+1)}{2} & a^2\dfrac{N(N+1)(2N+1)}{6} \end{bmatrix} \tag{9.82}$$

The variances of the estimates are

$$\text{var}\{a\} = 2\sigma_b^2 \frac{2N+1}{N(N-1)}, \qquad \text{var}\{\omega\} = 2\sigma_b^2 \frac{6}{a^2 N(N^2-1)} \tag{9.83}$$

The bound in equation (9.78) can also be expressed as

$$\text{var}\{\theta\} = [M_G \, \text{var}\{\Gamma\}^{-1} M_G^t]^{-1}$$

This expression is reminiscent of the definition of joint probability densities of Gaussian variables. In fact, the above procedure can be generalized in estimation theory, using the log-likelihood function [20].

Let $\text{Pr}(\Gamma|\theta)$ denote the conditional joint probability density function of the random vector represented by the measurement vector Γ. The log-likelihood function $L(\theta)$ is defined by

$$L(\theta) = \ln[\text{Pr}(\Gamma|\theta)] \tag{9.84}$$

The matrix I_{nf} of the derivatives of the log-likelihood function is called Fisher's information matrix:

$$I_{\text{nf}} = -E \begin{bmatrix} \left(\dfrac{\partial L}{\partial \theta_1}\right)^2 & \cdots & \dfrac{\partial L}{\partial \theta_1}\dfrac{\partial L}{\partial \theta_N} \\ \vdots & & \vdots \\ \dfrac{\partial L}{\partial \theta_N}\dfrac{\partial L}{\partial \theta_1} & \cdots & \left(\dfrac{\partial L}{\partial \theta_N}\right)^2 \end{bmatrix} \tag{9.85}$$

For example, let us consider the case of M-D Gaussian signals with probability density $p(x)$ expressed by [equation (2.135) of Chapter 2]

$$P(X) = \frac{1}{(2\pi)^{N/2}} \frac{1}{(\det R)^{1/2}} \exp[-\tfrac{1}{2}(X - m)^t R^{-1}(X - m)]$$

where the AC matrix R and the mean vector m are functions of a set of variables θ. The information matrix elements are

$$\text{inf}(k, l) = \frac{1}{2}\text{trace}\left[R^{-1}\frac{\partial R}{\partial \theta_k} R^{-1}\frac{\partial R}{\partial \theta_l}\right] + \frac{\partial m^t}{\partial \theta_k} R^{-1}\frac{\partial m}{\partial \theta_l} \tag{9.86}$$

The lower bound of the variance of the parameter vector estimation is called the Cramer–Rao bound, and it is defined by

$$\text{CRB}(\theta) = \text{diag}[I_{\text{nf}}^{-1}] \tag{9.87}$$

When the functional form of the log-likelihood function is known, for unbiased estimates, a lower bound of the parameter estimates can be calculated, and the following set of inequalities hold:

$$\text{var}\{\theta_i\} \geqslant \text{CRB}(\theta_i), \qquad 1 \leqslant i \leqslant N \tag{9.88}$$

An unbiased estimator is said to be efficient if its variance equals the bound.

9.10 CONCLUSION

The analysis techniques discussed in this chapter provide a set of varied and useful tools to investigate the characteristics of signals. These characteristics are helpful in studying, designing, and implementing adaptive filters. In particular, they can provide pertinent information on how to select filter parameter values or to assess the dynamic range of the variables, which is necessary for word-length determination.

As an illustration, consider the initial value E_0 of the prediction error energy in FLS algorithms, used for example in prediction applications. As pointed out in Chapter 6, it controls the initial adaptation speed of the filter. If the SNR is poor, it does not help to take a small value, and E_0 can be chosen close to the signal power; on the contrary, with high SNRs, small values of the initial error energy make the filter fast and can lead to quick detection of sinusoids for example.

Several analysis techniques, particularly the AR method, are well suited to adaptive approaches, which lead to real-time signal analysis.

EXERCISES

1. Calculate the order $N = 16$ DFT of the sequences

$$x_1(n) = \sin\left(2\pi\frac{2.5}{16}n\right), \qquad 0 \leqslant n \leqslant 15$$

$$x_2(n) = \sin\left(2\pi\frac{3.5}{16}n\right), \qquad 0 \leqslant n \leqslant 15$$

$$x_3(n) = x_1(n) + x_2(n)$$

Discuss the possibility of recognizing the sinusoids in the spectrum.

2. The real signal $x(n)$ is analyzed with an N-point DFT operator. Show that the signal power spectrum can be estimated by

$$S(k) = X(k)X(N - k)$$

where $X(k)$ is the DFT output with index $k\,(0 \leqslant k \leqslant N - 1)$. If $x(n) = b(n)$, where $b(n)$ is a white noise with power σ_b^2, calculate the mean and variance of the estimator $S(k)$.

Now assume that the signal is

$$x(n) = \sin\left(2\pi\frac{k_0}{N}n\right) + b(n), \qquad 1 < k_0 < \frac{N}{2}$$

with k_0 integer. Calculate the mean and variance of the estimator and comment on the results. Is the analysis technique efficient?

3. A signal has ACF $r(0) = 1.0$, $r(1) = 0.866$, $r(2) = 0.5$. Perform the eigenvalue decomposition of the 3×3 AC matrix and give the harmonic decomposition of the signal. How is it modified if
   (a) a white noise with power σ^2 is added to the signal:
   (b) The ACF $r(p)$ is replaced by $0.9^P r(p)$?
   Give the shape of the spectrum using expression (9.34).

4. Consider the signal sequence

   $$n = 1 \quad 2 \quad 3 \quad 4 \quad 5 \quad 6$$
   $$x(n) = 1.05 \quad 0.72 \quad 0.45 \quad -0.32 \quad -0.61 \quad -0.95$$

   Perform the damped sinusoid decomposition following the procedure in Section 9.5 and calculate the spectrum $S_{DS}(f)$.

5. For the signal

   $$n = 1 \quad 2 \quad 3 \quad 4 \quad 5 \quad 6 \quad 7 \quad 8 \quad 9 \quad 10$$
   $$x(n) = 1.41 \quad 1.43 \quad 1.36 \quad 1.22 \quad 1.14 \quad 0.91 \quad 0.84 \quad 0.67 \quad 0.51 \quad 0.31$$

   calculate the matrix R_4 according to expression (9.45) and use it to derive three forward prediction coefficients. Calculate the AR spectrum and draw the curve $S_{AR}(f)$ versus frequency.
   Repeat the above operations with R_4 calculated according to the forward-backward technique (9.48). Compare the spectra obtained with both approaches.

6. Consider the cisoids in noise

   $$x(n) = e^{jn\omega_1} + e^{jn\omega_2} + b(n)$$

   and assume the angular frequencies have to be estimated from N samples. Calculate the variance estimation bounds and show the importance of the quantity $\omega_2 - \omega_1$.
   Perform the same calculations when a phase parameter φ_1 is introduced

   $$x(n) = e^{j(n\omega_1 + \varphi_1)} + e^{jn\omega_2} + b(n)$$

   Comment on the impact of the phase parameter.

REFERENCES

1. S. L. Marple, *Digital Spectral Analysis with Applications*, Prentice-Hall, Englewood Cliffs, N.J., 1986.
2. S. M. Kay and C. Demeure, "The High Resolution Spectrum Estimator–A Subjective Entity," *Proc. IEEE* No. 72, 1815–1816 (December 1984).

3. P. Welch, "The Use of Fast Fourier Transform for the Estimation of Power Spectra: A Method Based on Time Averaging of Short, Modifed Periodograms," *IEEE Trans.* **AU-15**, 70–73 (June 1967).

4. R. B. B. Blackman and J. W. Tukey, *The Measurement of Power Spectra from the Point of View of Communications Engineering.* Dover, New York, 1959.

5. H. Clergeot, "Choice between Quadratic Estimators of Power Spectra," *Ann. Telecomm.* **39**, 113–128 (1984).

6. J. Capon, "High Resolution Frequency Wave Number Spectrum Analysis," *Proc. IEEE* **57**, 1408–1418 (August 1969).

7. M. A. Lagunas-Hernandez and A. Gazull-Llampallas, "An Improved Maximum Likelihood Method for Power Spectral Density Estimation," *IEEE Trans.* **ASSP-32**, 170–173 (February 1984).

8. V. F. Pisarenko, "The Retrieval of Harmonics from a Covariance Function," *Geophys. J. Roy. Astronom. Soc.* **33**, 347–366 (1967).

9. F. B. Hildebrand, *Introduction to Numerical Analysis*, McGraw-Hill, New York, 1956, Chap. 9.

10. S. Haykin, ed., *Nonlinear Methods of Spectral Analysis*, Springer-Verlag, New York, 1983.

11. H. Akaike, "A New Look at the Statistical Model Identification," *IEEE Trans.* **AC-19**, 716–723 (1974).

12. S. W. Lang and J. H. McClellan, "Frequency Estimation with Maximum Entropy Spectral Estimators," *IEEE Trans.* **ASSP-28**, 716–724 (December 1980).

13. J. A. Cadzow, "Spectral Estimation: An Overdetermined Rational Model Equation Approach," *Proc. IEEE* **70**, 907–939 (September 1982).

14. D. V. Bhaskar Rao and S. Y. Kung, "Adaptive Notch Filtering for the Retrieval of Sinusoids in Noise," *IEEE Trans.* **ASSP-32**, 791–802 (August 1984).

15. J. M. Travassos-Romano and M. Bellanger, "FLS Adaptive Notch Filtering," *Proc. ECCTD Conf.*, Paris, 1987.

16. G. Bienvenu and L. Kopp, "Optimality of High Resolution Array Processing Using the Eigensystem Approach," *IEEE Trans.* **ASSP-31**, 1235–1243 (October 1983).

17. D. W. Tufts and R. Kumaresan, "Estimation of Frequencies of Multiple Sinusoids: Making Linear Prediction Perform Like Maximum Likelihood," *Proc. IEEE* **70**, 975–989 (1983).

18. R. Kumaresan and D. W. Tufts, "Estimating the Angles of Arrival of Multiple Plane Waves," *IEEE Trans.* **AES-19**, 134–139 (1983).

19. M. Kaveh and A. Barabell, "The Statistical Performance of the Music and Minimum Norm Algorithms in Resolving Plane Waves in Noise," *IEEE Trans.* **ASSP-34**, 331–341 (April 1986).

20. H. L. Van Trees, *Detection, Estimation and Modulation Theory*, Wiley, New York, 1971.

21. C. Delhote, "The High Resolution Concept: Its Reality and Limits," *Traitement du signal* **2**, 111–120 (1985).

22. S. Kay and J. Makhoul, "On the Statistics of the Estimated Reflection Coefficients of an Autoregressive Process," *IEEE Trans.* **ASSP-31**, 1447–1455 (December 1983).

10

Circuits and Applications

The integrated circuits, programmable microprocessors, and array processors designed for general signal processing can be used for the implementation of adaptive filters. However, several specific aspects can make the realization of dedicated architectures worthwhile. A major feature of adaptive techniques is the real-time updating of a set of internal variables, with the related checking operations. The optimization of some of the functions included in that process may be justified; a typical function is the multiplication of the elements of a vector by a scalar. An important point also is the introduction of arithmetic operations which are not widely used in other areas of signal processing, namely division and, to a lesser extent, square root.

10.1 DIVISION AND SQUARE ROOT

Division can be implemented in arithmetic logic units (ALUs) as a sequence of shifts, subtractions, and tests. The procedure is time consuming, and a more efficient approach is obtained by dedicated circuitry.

Let n and d be two positive numbers satisfying $0 < n < d$. To calculate n/d, use the following algorithm [1]:

$$t_0 = 2n - d, \quad \begin{array}{l} t_0 < 0, q_1 = 0 \\ t_0 \geqslant 0, q_1 = 1 \end{array}$$

$$t_1 = 2t_0 - (2q_1 - 1)d; \quad \begin{array}{l} t_1 < 0, q_2 = 0 \\ t_1 \geqslant 0, q_2 = 1 \end{array} \qquad (10.1)$$

$$t_i = 2t_{i-1} - (2q_i - 1)d; \quad \begin{array}{l} t_i < 0, q_{i+1} = 0 \\ t_i \geqslant 0, q_{i+1} = 1 \end{array}$$

It is readily verified that the following equations hold:

$$n = d \left(\sum_{j=1}^{i} q_j 2^{-j} \right) + 2^{-i} \frac{t_i + d}{2}, \qquad |t_i| < d \qquad (10.2)$$

The bits q_j are the most significant bits of the quotient $q = n/d$. The algorithm can be implemented in a sequential manner as shown in Figure 10.1, assuming the same word length for n and d.

The number of full adders equals the number of bits of the factors plus one. A parallel realization leads to an array of adders as with multipliers.

As an illustration, consider the following example: $n = 9 = 001001$; $d = 21 = 010101$.

$$
\begin{array}{lll}
& 001001 & (n) \\
& 010010 & (2n) \\
+ & 101011 & (-d;\ 2\text{'s complement}) \\
\hline
& 111101 & (t_0 = 2n - d) \qquad\qquad (10.3) \\
q_1 = 0 \quad (1)\ & 111010 & (2t_0) \\
+ & 010101 & \\
\hline
& 001111 & (t_1) \\
q_2 = 1 \quad (0)\ & 011110 & (2t_1) \\
+ & 101011 & (-d) \\
\hline
& 001001 & (t_2) \\
q_3 = 1 \quad (0)\ & 010010 & \\
\end{array}
$$

The result is $q = \frac{3}{7} = 0.011 \ldots$.

When the word length of the divider d is smaller than the word length of the dividend n, it may be advantageous to perform the operation as $q = (\frac{1}{d})$, which corresponds to an inverse calculation followed by a multiplication.

The square-root extraction can be viewed as a division by an unknown divider. The two operations have many similarities, and in both cases the most significant bits of the result are obtained first.

In order to show how the square-root extraction can be performed recursively, let us assume that the square root of a given number X has P bits and that i bits $(s_0, \ldots s_{i-1})$ are available after stage i of the extraction procedure. The remainder R_i is expressed by

$$R_i = X - (S_i 2^{P-i})^2 \qquad (10.4)$$

where

$$S_i = \sum_{j=0}^{i-1} s_{i-1-j} 2^j$$

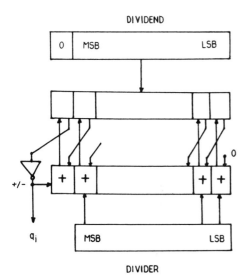

DIVIDEND

| 0 | MSB | | LSB |

q_i

MSB LSB

DIVIDER

Figure 10.1 A sequential divider.

At the next stage, the remainder R_{i+1} is

$$R_{i+1} = X - [(2S_i + s_i)2^{P-(i+1)}]^2 \tag{10.5}$$

The unknown s_i is a binary number, and thus $s_i^2 = s_i$. Now, expanding the product on the right side of (10.5) yields

$$R_{i+1} = R_i - (4S_i + 1)s_i 2^{2(P-i-1)} \tag{10.6}$$

Consequently, to obtain s_i it suffices to calculate the quantity

$$Q_{i+1} = R_i - (4s_i + 1)2^{2(P-i-1)} \tag{10.7}$$

and take its sign bit.

Hence the procedure to perform the square-root extraction for a number X having N bits ($N \leqslant 2P$) is as follows:

Initialization: $R_0 = X, S_0 = 0$

For $i = 0, \ldots, P - 1$,

$$Q_{i+1} = R_i - (4S_i + 1)2^{2(P-i-1)}$$

$$Q_{i+1} \geqslant 0, \quad s_i = 1, \quad R_{i+1} = Q_{i+1}, \quad S_{i+1} = 2S_i + 1$$

$$Q_{i+1} < 0, \quad s_i = 0, \quad R_{i+1} = R_i, \quad S_{i+1} = 2S_i \tag{10.8}$$

The desired square root is S_p.

Example

$$X = 25 = 011001, \quad N = 6 = 2P$$

$$Q_1 = 001001, \quad s_0 = 1, \quad R_1 = 001001, \quad S_1 = 000001$$

$$Q_2 = 001001 - 010100, \quad s_1 = 0, \quad R_2 = R_1, \quad S_2 = 000010$$

$$Q_3 = 0, \quad s_2 = 1, \quad R_3 = 0, \quad S_3 = 000101 = 5$$

The procedure can be implemented on a standard ALU as a sequence of shifts, additions, subtractions, and tests. Dedicated circuits can also be worked out for sequential or parallel realization.

Overall, considering the algorithms (10.1) and (10.8) it appears that, using a standard ALU, the division is more complex than the multiplication because it requires a test to decide between addition and subtraction at each stage; the square root is more complex than the division because it requires an addition, a subtraction, and two shifts at each stage. However, if a dedicated circuit is built, the test needed in the division can be included in the logic circuitry, and the division becomes equivalent to the multiplication. The square-root extraction is still more complex than the division.

10.2 A MULTIBUS ARCHITECTURE

Signal processing machines are characterized by the separation of the data and control paths. For adaptive processing, additional flexibility is desirable, due to the real-time updating of the internal variables. Three data paths can be distinguished: two for the factors of the arithmetic operations, and one for the results. Therefore, a high level of efficiency and speed is obtained with the four-bus programmable architecture sketched in Figure 10.2.

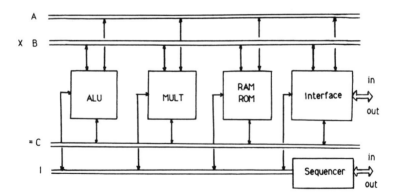

Figure 10.2　A four-bus architecture for adaptive processing.

The data buses A and B are used for the arithmetic factors, bus C for the results. The various system units, ALU, multiplier, and memories interact with these buses in an adequate manner. An interface unit handles the data exchanges with the external world. The control unit is connected through the instruction bus I to the system units and external control signal sources.

The multibus approach brings a certain level of complexity in hardware and software as well. However, the parallelism introduced that way offers a pipelining capacity which leads to fast and efficient realizations [2, 3].

A wide range of programmable integrated microsignal processors is now available, as well as specific parts to build multichip architectures. Machines can now be designed for any kind of application. A selection of applications from various fields is given below.

10.3 LINE CANCELING AND ENHANCEMENT

A frequently encountered problem is the canceling of a line while preserving the rest of the spectrum. As mentioned in Section 5.7, the notch filter is the appropriate structure. If the frequency of the line is not known or changing with time, an adaptive version can be used, with gradient or FLS algorithms as pointed out in Section 9.7. Good performance can be obtained under a wide range of conditions [4].

The recursive section of the notch filter actually performs a line enhancement. The general approach is based on linear prediction, as shown in Figure 10.3.

Let us assume that the signal $x(n)$ consists of M sinusoids in noise. The output $\tilde{x}(n)$ of the adaptive prediction filter $A(z)$ contains the same spectral lines, with virtually no deviations in amplitudes and phases, provided the filter order N exceeds $2M$ with a sufficient margin. However, as seen in Section 5.2, the noise component power is reduced in $\tilde{x}(n)$ since the power of the output $e(n)$ is minimized. The delay Δ in front of the prediction is chosen

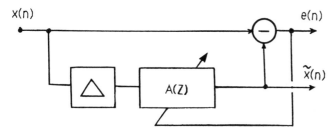

Figure 10.3 Adaptive line enhancement.

as a function of the correlation radius p_0 of the noise ($\Delta \geq p_0$); in case of white noise, a one-step predictor is adequate [5].

The improvement in SNR for the enhancer output $\tilde{x}(n)$ is the enhancement gain G_e, which can be calculated using the results in Chapter 5; it is proportional to the prediction filter order.

10.4 ADAPTIVE DIFFERENTIAL CODING

Besides signal analysis, linear prediction techniques can be used to condense the representation of signals. The information in a signal is essentially contained in the unpredictable components. Therefore, if the predictable components are attenuated, the amplitude range of the samples is reduced, fewer bits are needed to encode them, and a denser representation is obtained. In practice, for the sake of simplicity and ease of manipulation, it is generally desirable that the original signal be retrievable from the prediction error sequence only. Therefore, in an adaptive approach the filter has to be implemented in a loop configuration as shown in Figure 10.4, in order to take into account the effects of output sequence quantization.

The prediction error filter is of the FIR/IIR type. The coefficient update section can use LMS algorithms as in Section 4.11, or the FLS algorithms as pointed out in Section 7.7 on pole-zero modeling. Typical uses of the above scheme are for signal storage or efficient transmission [6]. For example, in communications the technique is known as ADPCM (adaptive differential

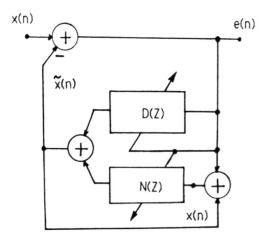

Figure 10.4   Adaptive differential encoding to condense information.

pulse code modulation), and it permits a telephone conversation to be transmitted with unnoticeable degradations through a digital link with 32 kbit/s capacity.

10.5 ECHO CANCELLATION

Echo cancellation is an illustrative application of system identification techniques, the system under study being the echo path. In communications, unwanted echoes have to be removed in full-duplex two-wire data transmission and in satellite voice transmission; they are due to electric signal reflections. Acoustic signal reflections can also be disturbing, and they can be digitally canceled in hands-free telephone terminals for video and audio conference rooms. The latter case is illustrated in Figure 10.5.

The adaptive filter is implemented according to Figure 6.8, for example, $x(n)$ being the original signal and $y(n)$ the echo.

In general, very long filters are necessary, with several hundred and even thousands of coefficients [7]. Adaptation speed and accuracy are critical. For example, acoustic couplings in a room are characterized by a rapidly evolving spectrum with numerous and deep zeros. Moreover, signals like speech and sounds are nonstationary. It is a challenging application.

The same approach can be envisaged to remove disturbing noise signals, such as engine noise, for example, in mobile radio telephones.

10.6 CHANNEL EQUALIZATION AND MEASUREMENT

Imperfect transmission channels distort the signals. If a replica of the original signal can be derived at the receiving end, the channel can be adaptively equalized. The case of data transmission is illustrated in Figure 10.6.

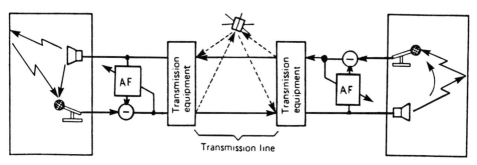

Figure 10.5 Audio conference system.

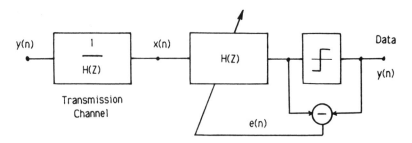

Figure 10.6 Channel equalization in data transmission.

The error signal is generated as the residue of a decision process, which supplies the data. It is then used to adjust the equalizer coefficients. The scheme can be used to carry out real-time measurements of evolving transmission paths.

The case of multipath propagation in radiocommunications leads to a specific filter structure. Consider a received signal $x(n)$ which is the sum of weighted and delayed versions of an original signal $y(n)$:

$$x(n) = y(n) + \sum_{i=1}^{N} h_i(n)\,y(n - i)$$

The original signal can be retrieved by

$$y(n) = x(n) - \sum_{i=1}^{N} h_i(n)y(n - i)$$

which corresponds to an IIR adaptive filter [8].

In the applications considered so far, the input signal has been assumed to be known, and the processing system to be unknown. Now, if the system is known, the input signal can be retrieved from the system output.

10.7 ADAPTIVE DECONVOLUTION

Deconvolution is applied to experimental data in order to remove distortions caused by a measurement system to a desired inaccessible signal. Let us assume that the experimental sequence $y(p)(1 \leqslant p \leqslant n)$ is generated by filtering the desired signal $x(p)$ as follows:

$$y(p) = \sum_{i=0}^{N-1} h_i x(p - i) \tag{10.11}$$

The operation is described in matrix notation by

$$
\begin{bmatrix} y(n) \\ y(n-1) \\ \vdots \\ y(1) \end{bmatrix} = \begin{bmatrix} h_0 & h_1 & h_2 & \cdots & 0 & 0 \\ 0 & h_0 & h_1 & \cdots & 0 & 0 \\ \vdots & \vdots & \vdots & & \vdots & \vdots \\ 0 & 0 & 0 & \cdots & h_{N-2} & h_{N-1} \end{bmatrix} \begin{bmatrix} x(n) \\ x(n-1) \\ \vdots \\ x(2-N) \end{bmatrix} \tag{10.12}
$$

or

$$
y(n) = H^t X(n) \tag{10.13}
$$

According to the results in Section 3.5, an LS solution is obtained by

$$
X(n) = H(H^t H)^{-1} Y(n) \tag{10.14}
$$

The desired sequence $x(n)$ can be retrieved in an adaptive manner through the technique depicted in Figure 10.7.

The estimated data $\tilde{x}(n)$ are fed to the distorting FIR filter, whose coefficients are assumed to be known, and the output $\hat{y}(n)$ is subtracted from the experimental data to produce an error $e(n)$ used to update the estimate at time $n + 1$.

The simplest approach one can think of consists of calculating $\tilde{x}(n + 1)$ by

$$
\tilde{x}(n + 1) = \frac{1}{h_0} \left[y(n + 1) - \sum_{i=1}^{N-1} h_i \tilde{x}(n + 1 - i) \right] \tag{10.15}
$$

However, it is unrealistic, due to initial conditions and the presence of noise added in the measurement process.

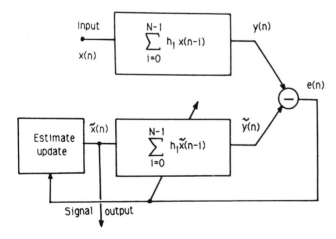

Figure 10.7 Signal restoration by adaptive deconvolution.

The gradient method corresponds to the updating equation

$$\begin{bmatrix} x_1(n+1) \\ x_2(n+1) \\ \vdots \\ x_N(n+1) \end{bmatrix} = \begin{bmatrix} 0 \\ x_1(n) \\ \vdots \\ x_{N-1}(n) \end{bmatrix} + \delta \begin{bmatrix} h_0 \\ h_1 \\ \vdots \\ h_{N-1} \end{bmatrix} \left[y(n+1) - \sum_{i=1}^{N-1} h_i x_i(n) \right] \qquad (10.16)$$

where δ is the adaptation step, the $x_i(n)$ ($1 \leqslant i \leqslant N$) are state variables, and the restored signal at the system output is

$$\tilde{x}(n+1-N) = x_N(n) \qquad (10.17)$$

The technique can be refined by using a more sophisticated adaptation gain. If matrix manipulations can be afforded, LS techniques based on equation (10.14) can be worked out, associated with recursive procedures to efficiently perform the computations [9, 10].

10.8 ADAPTIVE PROCESSING IN RADAR

Signal processing techniques are employed in radar for target detection by whitening filters, separation of targets by high-resolution spectral analysis methods, and target or terrain recognition by comparison with models or by inverse filtering [11, 12].

The target detection method is depicted in Figure 10.8.

When the signal $s(t)$ is emitted, the signal received can be expressed by

$$y(t) = Gs(t - t_0) + P(t) \qquad (10.18)$$

where G is a complex parameter representing the propagation conditions, t_0 is the delay of the signal reflected on the target, and $P(t)$ is a perturbation

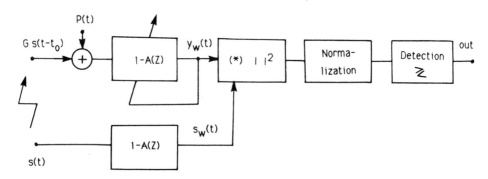

Figure 10.8 Target detection in the presence of colored evolving noise.

representing the multiple undesired reflections on various obstacles, or clutter. The useful signal $s(t - t_0)$ can be much smaller than the perturbation, which can be modeled by a colored and evolving noise or, in other words, by an AR signal with variable parameters.

The perturbation can be significantly attenuated by an adaptive prediction error filter, which performs a whitening operation and delivers the signal $y_w(t)$. The signal $s(t)$ is fed to a filter with the same coefficients, and the output is $s_w(t)$. Now the detection process consists of comparing to a threshold the quantity

$$c(\tau) = \frac{|\int y_w(t)\bar{s}_w(t - \tau)\,dt|^2}{\int |s_w(t)|^2\,dt} \tag{10.19}$$

The corresponding operations are the correlation, squared modulus calculations, normalization, and decision.

The order N of the adaptive whitening filter is typically close to 10, and the sampling frequency is several megahertz.

The applications reviewed so far deal with one-dimensional signals. In antenna arrays and image processing, M-D signals are involved.

10.9 ADAPTIVE ANTENNAS

The outputs from the elements of an antenna array can be combined to produce a far field beam pattern which optimizes the reception of a desired signal [13]. The beam can be directed towards the bearing of the signal, and it can be configured to have sidelobes which attenuate jamming signals. Moreover, an equalization of the transmission channel can be achieved. In the block diagram in Figure 10.9, the N elements collect delayed versions of the useful signal $x(t)$. For a linear array whose elements are separated by a distance D, the basic delay is

$$\Delta t = \frac{d \sin \theta}{v} \tag{10.20}$$

where θ is the incidence angle and v is the signal velocity. The delays are compensated through N interpolators whose outputs are summed. However, the filters connected to the antenna elements can do more than just compensate the delays. If a replica of the signal is available, as in digital transmission, and used as a reference, an error can be derived and the system can become an M-D adaptive filter employing the algorithms presented in Chapter 7.

An interesting simplification occurs if the antenna elements are narrowband, because the received signal can be considered as a sine wave. The

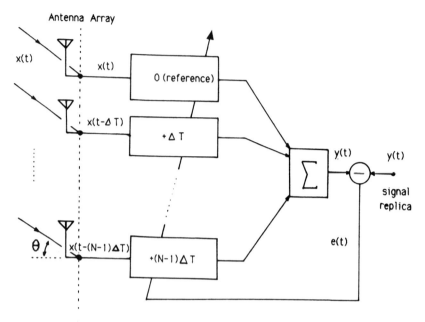

Figure 10.9   Adaptive antenna array.

equation of a propagating wave is

$$s(x, t) = S \exp 2\pi \left(ft - \frac{x}{\lambda} \right) \qquad (10.21)$$

where $\lambda = v/f$ is the wavelength associated to the frequency f. From equation
(10.21) it appears that adequate beamforming can take place only if

$$\frac{D \sin \theta}{\lambda} < \frac{1}{2} \qquad \text{or} \qquad D < \frac{\lambda}{2} \qquad (10.22)$$

Therefore $\lambda/2$ is an upper bound for the spatial sampling interval. The
filtering paths reduce to multiplications by weights $w_i = \exp(j\omega\Delta Ti)$ with
$0 \leqslant i \leqslant N - 1$ and $\Delta T = (D \sin \theta)/v$. The coefficients w_i can be calculated to
minimize the cross-correlation between the output $y(n)$ and the inputs in the
absence of the desired signal. The corresponding equation is

$$R_N \begin{bmatrix} 1 \\ -W_{N-1} \end{bmatrix} = \begin{bmatrix} E \\ 0 \end{bmatrix} \qquad (10.23)$$

where R_N is the input covariance matrix, W_{N-1} an $(N - 1)$-element coefficient vector, and E the output error power. The coefficients can be found and
updated through gradient techniques.

Another approach consists of maximizing the output SNR, which leads to the N-coefficient vector

$$W_N = \frac{1}{\bar{F}^t R^{-1} \bar{F}} R^{-1} \bar{F} \qquad (10.24)$$

with

$$\bar{F} = [1, e^{j\omega \Delta T}, \ldots, e^{(N-1)j\omega \Delta T}]^t$$

The similarity with linear prediction of time sequences is worth pointing out.

10.10 IMAGE SIGNAL PREDICTION

Linear models are useful in image processing for region classification and segmentation and also for the detection of small regions which differ from their surroundings [14].

A picture element (pixel) of a 2-D image can be represented by a white-noise-driven linear model as

$$x(n, m) = \sum_{l,k \in M} a(l, k)x(n - l, m - k) + e(n, m) \qquad (10.25)$$

where $e(n, m)$ is a white noise with power σ_e^2 and M represents the support region for the filter coefficients, called the mask. The mask M can take on several forms. In prediction, it is associated with the past of the point (n, m). The past of a point (n_0, m_0) is related to causality and defined as the set of points

$$\{(n, m)|n = n_0, m < m_0; n < n_0, -\infty < m < \infty\}$$

The model equation (10.25) can also be considered as a prediction operation in which the signal $x(n, m)$ is predicted by the summation $\tilde{x}(n, m)$ and $e(n, m)$ is the prediction error. The prediction coefficients $a(l, k)$ can be calculated from the 2-D normal equations

$$r(l, k) - \sum_{p,q \in M} a(p, q)r(l - p, k - q) = \sigma_e^2 \delta(l, k) \qquad (10.26)$$

where $r(l, k)$ is the correlation function

$$r(l, k) = E[x(n, m)x(n - l, m - k)] \qquad (10.27)$$

The image predictor can be made adaptive by using either gradient or LS algorithms.

A third dimension can be introduced with television signals, considering the sequence of frames. The principle of a 3-D predictor is shown in Figure 10.10.

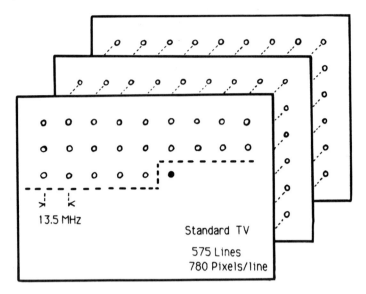

Figure 10.10   Principle of 3-D prediction of TV signals.

A pixel is predicted from its past in the same frame and from elements of the previous frames. Applications are for reduced rate encoding, for analysis such as edge detection, and for noise reduction [15]. The complexity issue is crucial in that case, since the sampling rate is 13.5 MHz. Filters with only a small number of coefficients can actually be implemented in real-time hardware.

10.11 ARTIFICIAL INTELLIGENCE

Artificial intelligence (AI) techniques attempt to reproduce and automatize simple learning and reasoning processes in order to give machines the ability to reason at a level approaching human performance in limited domains. Another of their goals is to extend and structure human-machine interaction [16].

Successful achievements in that field are the so-called knowledge-based or expert systems. A description of such a system is given in Figure 10.11. It is essentially a structured software technique of performing logical inference on symbolic data.

The expert system efficiency rests on the quality of its inference rules, exploited by the inference engine, and of the information stored in the data base.

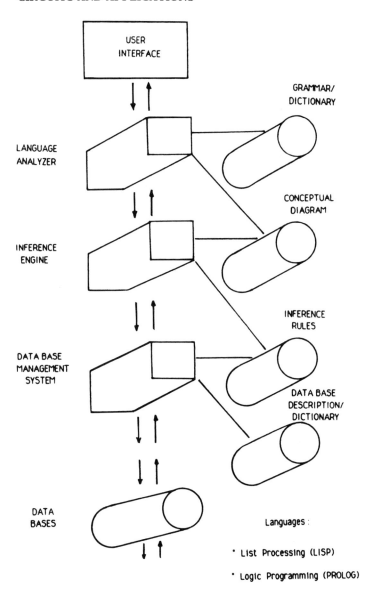

Figure 10.11 A knowledge-based system.

In some areas, signal processing is involved in the process of constituting the data bases on which AI works. For example, it can be needed to convert the real data, which carry the relevant information into symbolic data. Adaptive techniques can be useful in that process because of the improvements in accuracy and speed achieved in analyzing signals and extracting the parameters or features to condense the information.

In automation, it is increasingly important to equip machines or robots with the capability to communicate with their environment in real time, through acoustic and visual signals, like humans. To that purpose, signal generation and recognition are fundamental operations.

Recognition can be defined as the automatic assignment of a signal to a predetermined category in order to stimulate a predetermined subsequent action. Clearly, adaptive processing methods, in one-dimensional and M-D forms are instrumental in accurately and efficiently performing that task [17].

Overall, adaptive methods can contribute to the advance of AI techniques. In return, AI techniques can contribute to the diffusion of adaptive methods. For example, expert systems can be dedicated to adaptive filtering and signal analysis and exploited by practitioners as a valuable help in their efforts to optimize their realizations.

10.12 CONCLUSION

The wide range of applications which have been briefly introduced illustrate the versatility of adaptive signal processing techniques and the universality of their principles. In practice, it turns out that, for particular cases, variations and adequate refinements are often introduced in the methods to tailor them to the applications and enhance their efficiency. Therefore, these applications may look as many different fields. To a certain extent, it is true, however, there is a common ground, and it is the corresponding common knowledge which is presented in the previous chapters.

The diffusion of adaptive techniques and the extension of their application fields are highly dependent on the advances in technology. Considering the successive generations of large and very large-scale integrated circuits, considering the growing family of integrated signal microprocessor chips, some of which are even specially designed to suit the needs of adaptive filtering, it appears that more and more technological resources are available.

Finally, it can be stated that adaptive methods are bound to have an increasing impact in science and industry.

REFERENCES

1. M. David, J. P. Deschamps, and A. Thayse, *Digital Systems with Algorithm Implementation*, Wiley, London, 1983.
2. B. Barazech and L. Mary, "A Cascadable Multimode VLSI Signal Processor," *Proc. IEEE ICASSP-86*, Tokyo, 1986.
3. F. J. Van Wijk et al., "On the IC Architecture and Design of a $2\,\mu$m CMOS 8 MIPS Digital Signal Processor with Parallel Capability: The PCB 5010/5011," *Proc. IEEE ICASSP-86*, Tokyo, 1986.
4. A. Nehorai, "A Minimal Parameter Adaptive Notch Filter with Constrained Poles and Zeros," *IEEE Trans.* **ASSP-3**, 983–996 (August 1985).
5. J. R. Zeidler, E. H. Satorius, and D. M. Chabries, H. T. Wexler, "Adaptive Enhancement of Multiple Sinusoids in Uncorrelated Noise," *IEEE Trans.* **ASSP-26**, 240–254 (June 1978).
6. T. Claasen and W. Mecklenbrauker, "Adaptive techniques for Signal Processing in Communications," *IEEE Communications Mag.* **23**, 8–19 (November 1985).
7. N. Furuya, Y. Itoh, Y. Maruyama, and T. Araseki, "Audio Conference Equipment with Acoustic Echo Canceller," *NEC Res. Dev. J.* No. 76. 18–23 (January 1985).
8. A. Leclert and P. Vandamme, "Decision Feedback Equalization of Dispersive Radio Channels," *IEEE Trans.* **COM-33**, 676–684 (July 1985).
9. D. Commenges, "The Deconvolution Problem: Fast Algorithms Including the Preconditioned Conjugate-Gradient to Compute MAP Estimator," *IEEE Trans.* **AC-29**, 229–243 (1984).
10. G. Demoment and R. Reynaud, "Fast Minimum Deconvolution," *IEEE Trans.* **ASSP-33**, 1324–1326 (October 1985).
11. B. L. Lewis, F. K. Kretschmer and W. Shelton, *Aspects of Radar Signal Processing*, Artech House, London, 1986.
12. C. Delhote, "Parametric Modelling and Radar Signal Processing," La Recherche Aérospatiale, No. 2, 1985, 73–80.
13. N. G. Taylor, ed., "Adaptive Antennas," *Proc. IEEE* **130**, 1–151 (January 1983).
14. D. Dudgeon and R. Mersereau, *Multidimensional Digital Signal Processing*, Prentice-Hall, Englewood Cliffs, N.J., 1984.
15. C. Richard, A. Benveniste, and F. Kretz, "Recursive Estimation of Local Characteristics of Edges in TV Pictures as Applied to ADPCM Coding," *IEEE Trans.* **COM-32**, 718–728 (June 1984).
16. Y. Shirai and J. Tsujii, *Artificial Intelligence Concepts, Techniques and Applications*, Wiley, New York, 1986.
17. J. G. Dunn, "Signal Processing Technology and Prospects," *Electrical Communication-ITT* **59**, 252–259 (1985).

Index

Accumulation, roundoff error, 193
AC matrix, 65
  estimation, 83
  inverse, 148
Adaptation gain, 10, 175
  updating, 179
Adaptive beam former, 318
Adaptive equalizer, 314
Adaptive estimation, 120
ADPCM, 312
Algorithm
  fast least squares, 179
  gradient, 97
  recursive, 9
Analysis
  signal, 2
  spectral, 281
Antenna array, 318
AR signal, 28
Architecture, 310
ARMA signal, 30
Artificial intelligence, 320
Autocorrelation function (ACF),
  19, 23, 55

Backward linear prediction, 140
Bias
  coefficient, 112
  estimation, 58
Blackman–Tukey method, 286
Block processing, 266
Bounds, estimation, 300
Burg methods, 294

Cancellation
  echo, 313
  line, 311
Capon method, 286
Cascade form, 123
Channel equalization, 314
Coefficient updating, 9
Complex signals, 220
Composite sinusoidal modeling
  (CSM), 161
Computational complexity
  FLS, 183, 186
  gradient, 98
Condition number, 78

Consistent (estimator), 59, 284
Control variable, 176
Conventional estimation, 58
Correction system, 12
Correlation
    coefficient, 56
    function, 23
    matrix, 65
Correlogram method, 284
Cost function, 8
Covariance, 22
Covariance algorithm, 213
Cramer–Rao bound, 302
Criterion
    optimization, 6
    order, 295
Cross-correlation, 56
Cross spectrum, 58

Damped sinusoids, 15
Decomposition
    eigenvalue, 69
    harmonic, 44
    Wold, 42
Deconvolution, 317
Discrete Fourier transform (DFT),
    70
Division (circuit), 307

Echo cancellation, 313
Efficient estimation, 302
Eigenfilters, 73
Eigentransform, 86
Eigenvalue decomposition, 69
Energy, error signal, 179
Enhancement, line, 311
Ergodicity, 23
Error
    a posteriori, 99, 178
    a priori, 9, 99, 178

Error signal, 9
    energy, 179
Estimation
    AC function, 58
    AC matrix, 83
    Adaptive, 120
    recursive, 63
    spectral, 281
Extremal eigenvalues, 77

Fast Kalman algorithm, 183
Fast least squares (FLS), 169
Filter
    adaptive, 7
    eigen, 73
    lattice, 146
    prediction, 135
Final prediction error (FPE), 295
Finite impulse response (FIR)
    filter, 7
Fisher information matrix, 302
Forward-backward linear
    prediction (FBLP), 141
Forward prediction, 140
Frequency domain adaptive filter
    (FDAF), 235

Gain
    adaptation, 10
    prediction, 39
    system, 100
Gaussian signals, 24
Geometrical description, 267
Gradient algorithm, 97

Harmonic decomposition, 44, 119
Harmonic retrieval technique, 119
Hermitian matrix, 65
Hybrid sign estimation, 61

Identification, 11
Image processing, 319
Implementation, 203, 311
Independent identically
    distributed (IID) variables, 25
Infinite impulse response (IIR)
    filter, 7, 125
Information matrix, 302
Initial values, 189
Interpolation, 40
Inverse AC matrix, 148

Joint process (lattice-ladder), 255

Kalman (adaptation) gain, 10
Karhunen–Loève transform, 86

Lattice-ladder adaptive filter, 255
Lattice-prediction filter, 146
Leakage factor, 112, 203
Learning curve
    gradient, 104
    FLS, 199
Least absolute value (LAV)
    criterion, 114
Least mean absolute value
    (LMAV) criterion, 114
Least mean squares (LMS),
    criterion, 10, 97
Least squares (LS) criterion, 8
Leroux–Gueguen algorithm, 145
Levinson–Durbin algorithm, 144
Likelihood variable, 176
Limitation, word-length, 106, 203
Line enhancement/cancellation,
    311
Line spectrum pair (LSP), 162
Linear equation system, 67
Linear prediction, 37
    error filter, 135
Log-likelihood function, 302

MA signals, 27
Markov signals, 35
Matrix inversion lemma, 174
Mean square error (MSE), 101
Millions of instructions per
    second (MIPS), 11
Millions of operations per second
    (MOPS), 11
Minimum variance method, 286
Misadjustment, 103
Modeling, 5, 27
Multidimensional signals, 46, 222
    lattice, 263
Multirate filters, 234

Noise, subspace, 73, 298
Nonlinear filtering, 128
Nonstationary signals, 48
Nonuniform length filter, 230
Normal equations, 7
Normalization, 255
Normalized algorithm, 259
Notch filter, 150
    Adaptive, 297

Optimization criterion, 6
Order recursion, 142, 245
Orthogonalization, 37, 142

Parallel type IIR filter, 126
PARCOR coefficients, 147
Performance, LS filter, 195
Periodic signals, 18
Periodogram method, 282
Pinning vector, 269
Pisarenko method, 46, 290
Polarity coincidence estimation,
    60
Poles of IIR filters, 155

Pole-zero modeling, 231
Positive matrix, 66
Predictable signals, 41
Prediction error ratio, 178
Prediction filter, 135
Projection matrix/operator, 267

Quadratic coefficients, 128
Quantization, 106

Radar, 316
Radius
  blocking, 110
  spectral, 77
Random signals, 22
Ratio
  Prediction error, 178
  signal-to-noise, 73
Rayleigh density, 26
Recursive algorithm, 9
Recursive estimation, 63
Recursive least squares (RLS), 9,
  174
Reflection coefficients, 147
Regular signals, 42
Residual error, 100, 198
Resolution, spectral, 282
Roundoff error accumulation, 193

Sample rate
  reduction, 234
  increase, 235
Second order filter section, 19, 124
Second order Volterra filter (SVF),
  128
Series parallel IIR filter, 127, 233
Sign algorithm, 116
Signals
  AR, 28
  error, 7
  MA, 27
  natural, 49

Signals
  periodic, 18
  random, 22
Signal-to-noise ratio (SNR), 73
Sliding window, 216
Space (signal, noise), 73, 288
Specifications, 99
Spectral
  analysis, 281
  radius, 77
Square root operation, 308
Stability condition, 99, 187
Stabilization constant, 191
Stationary signal, 22
State vector, 35
Step response, 172
Step size, 98

Time constant, 100
Time recursion, 251
Toeplitz matrix, 66
Transversal filter, 99

Unified view of adaptive filters,
  237
Uniform distribution, 26
Unit norm constraint, 81
Updating, coefficient, 9

Variance, 24
Volterra filter, 128

Welch method, 284
White noise, 25
Whitening filter, 38
Wide sense stationary, 22
Wold decomposition, 42
Wordlength limitation, 106, 203

Zeros of linear prediction filters,
  151

DATE DUE